FAIRFAX COUNTY [VIRGINIA] ROAD ORDERS

1749-1800

Virginia Genealogical Society
Richmond, Virginia

Published With Permission from the

Virginia Transportation Research Council
(A Cooperative Organization Sponsored Jointly by the Virginia
Department of Transportation and
the University of Virginia)

HERITAGE BOOKS
2008

HERITAGE BOOKS
AN IMPRINT OF HERITAGE BOOKS, INC.

Books, CDs, and more—Worldwide

For our listing of thousands of titles see our website
at
www.HeritageBooks.com

Published 2008 by
HERITAGE BOOKS, INC.
Publishing Division
100 Railroad Avenue #104
Westminster, Maryland 21157

Copyright © 2003 Virginia Genealogical Society

All rights reserved. No part of this book may be reproduced or transmitted in any form or by any means, electronic or mechanical, including photocopying, recording or by any information storage and retrieval system without written permission from the author, except for the inclusion of brief quotations in a review.

International Standard Book Number: 978-0-7884-3369-6

HISTORIC ROADS OF VIRGINIA

Louisa County Road Orders, 1742-1748, by Nathaniel Mason Pawlett. 57 pages, indexed, map.

Goochland County Road Orders, 1728-1744, by Nathaniel Mason Pawlett. 120 pages, indexed, map.

Albemarle County Road Orders, 1744-1748, by Nathaniel Mason Pawlett. 57 pages, indexed, map.

The Route of the Three Notch'd Road, by Nathaniel Mason Pawlett and Howard Newlon. 26 pages, illustrated, 2 maps.

An Index to Roads in the Albemarle County Surveyor's Books, 1744-1853, by Nathaniel Mason Pawlett. 10 pages, map.

A Brief History of the Staunton and James River Turnpike, by Douglas Young. 22 pages, illustrated, map.

Albemarle County Road Orders, 1783-1816, by Nathaniel Mason Pawlett. 421 pages, indexed.

A Brief History of Roads in Virginia, 1607-1840, by Nathaniel Mason Pawlett. 41 pages.

A Guide to the Preparation of County Road Histories, by Nathaniel Mason Pawlett. 26 pages, 2 maps.

Early Road Location: Key to Discovering Historic Resources? by Nathaniel Mason Pawlett and K. Edward Lay. 47 pages, illustrated, 3 maps.

Albemarle County Roads, 1725-1816, by Nathaniel Mason Pawlett. 98 pages, illustrated, 8 maps.

"Backsights," A Bibliography, by Nathaniel Mason Pawlett. 29 pages, revised edition.

Orange County Road Orders, 1734-1749, by Ann Brush Miller. 323 pages, indexed, map.

Spotsylvania County Road Orders, 1722-1734, by Nathaniel Mason Pawlett. 159 pages, indexed.

Brunswick County Road Orders, 1732-1749, by Nathaniel Mason Pawlett. 81 pages, indexed.

Orange County Road Orders, 1750-1800, by Ann Brush Miller. 394 pages, indexed, map.

Lunenburg County Road Orders, 1746-1764, by Nathaniel Mason Pawlett and Tyler Jefferson Boyd. 394 pages, indexed.

Culpeper County Road Orders, 1763-1764, by Ann Brush Miller. 22 pages, indexed, map.

Augusta County Road Orders 1745-1769, by Nathaniel Mason Pawlett, Ann Brush Miller, Kenneth Madison Clark and Thomas Llewellyn Samuel, Jr. 270 pages, indexed, map.

Amelia County Road Orders 1735-1753, by Nathaniel Mason Pawlett, Ann Brush Miller, and Kenneth Madison Clark. 159 pages, indexed, map.

Requests for information as to availability and
a current price list should be directed to:

Historic Roads of Virginia
Virginia Transportation Research Council
530 Edgemont Road
Charlottesville, VA 22903

www.virginiadot.org/vtrc/history/roadordr.html

FOREWORD

by

Ann Brush Miller
Virginia Transportation Research Council

Fairfax County Road Orders 1749-1800 is a cooperative project of the Fairfax County History Commission and the Virginia Transportation Research Council. This volume covers the surviving northern Virginia transportation records for the last half of the 18th century and includes most of the earliest surviving transportation-related records for that area. It is the fourth volume of road orders produced cooperatively by a private group and the Virginia Transportation Research Council, following similar projects with the Orange County Historical Society (which sponsored the production of *Orange County Road Orders 1734-1749* and *Orange County Road Orders 1750-1800*) and the Culpeper County Historical Society (which sponsored *Culpeper County Road Orders 1763-1764*).

A copy of Orange's first volume came by chance into the hands of Donie Rieger (Mrs. Charles Rieger), then serving as a member of the Fairfax County History Commission. After examining it, reading the introductory sketch, and discussing it with other members of the commission, she decided to contact the Research Council about a similar effort for Fairfax County's early road orders. Accordingly, a meeting was arranged with members of the Fairfax County History Commission to discuss what would be necessary to undertake such an effort in that area. An agreement was concluded similar to that with the Orange County Historical Society by which the Research Council would provide secretarial, editing, and publishing services and the Fairfax County History Commission would sponsor Beth Mitchell, a researcher with extensive knowledge of Northern Virginia history, to do the actual research, transcription, and checking of the final draft. This publication is the result.

The road orders contained within this volume constitute a large portion of the surviving 18th century transportation record for Northern Virginia. At its creation from Prince William County in 1742, Fairfax County included within its territory the present-day counties of Arlington, Fairfax, and Loudoun. Loudoun County was created from the western portion of Fairfax in 1757. Initially the boundary line between Fairfax and Loudoun counties ran along Difficult Run; the present boundary line is the result of a 1798 boundary adjustment between the two counties. Fairfax reached its present boundaries with the transfer of the northeastern section of the county to the District of Columbia in 1801 (this area was later ceded back to Virginia and became Alexandria County in 1847; it was renamed Arlington County in 1920). This publication marks the twenty-first entry in the *Historic Roads of Virginia* series, first initiated by the Virginia Transportation Research Council (then the Virginia Highway & Transportation Research Council) in 1973.

A NOTE ON THE METHODS, EDITING, AND DATING SYSTEM

by

Nathaniel Mason Pawlett
(Faculty Research Historian, Virginia Transportation Research Council, 1973-1995)

The road and bridge orders contained in the order books of an early Virginia county are the primary source of information for the study of its roads. When extracted, indexed, and published by the Virginia Transportation Research Council, they greatly facilitate this. All of the early county court order books are in manuscripts, sometimes so damaged and faded as to be almost indecipherable. Usually rendered in the rather ornate script of the time, the phonetic spellings of this period often serve to complicate matters further for the researcher and recorder.

With these road orders available in an indexed and cross-indexed published form, it will be possible to produce chronological chains of road orders illustrating the development of many of the early roads of a vast area from the threshold of settlement through much of the eighteenth century. Immediate corroboration for these chains of road orders will usually be provided by other evidence such as deeds, plats, and the Confederate Engineers maps. Often, in fact, the principal roads will be found to survive in place under their early names.

With regard to the general editorial principles of the project, it has been our perception over the years as the road orders of Louisa, Hanover, Goochland, Albemarle, and other counties have been examined and recorded that road orders themselves are really a variety of "notes," often cryptic, incomplete, or based on assumptions concerning the level of knowledge of the reader. As such, any further abstracting or compression of them would tend to produce "notes" taken from "notes," making them even less comprehensible. The tendency, therefore, has been in the direction of restraint in editing, leaving any conclusions with regard to meaning up to the reader or researcher using these publications. In pursuing this course, we have attempted to present the reader with a typescript text that is as near a type facsimile of the manuscript itself as we can come.

Our objective is to produce a text that conveys as near the precise form of the original as we can, reproducing all the peculiarities of the eighteenth-century orthography. Although some compromises have had to be made because of the modern keyboard, this was really not that difficult a task. Most of their symbols can be accommodated by modern typography, and most abbreviations are fairly clear as to meaning.

Punctuations may appear misleading at times, with unnecessary commas or commas placed where periods should be located; appropriate terminal punctuation is often missing or else takes the form of a symbol such as a long dash, etc. The original capitalization has been retained insofar as it was possible to determine from the original manuscript whether capitals were intended. No capitals have been inserted in place of those originally omitted. The original spelling and syntax have been retained throughout, even including the obvious errors in various places, such as repetitions of words and simple clerical errors. Ampersands have been retained throughout to include such forms as "&c" for "etc." Superscript letters have also been retained where used in y^e, y^t, s^d. The thorn symbol (y), pronounced as "th," has been retained in the aforesaid "y^e," pronounced "the," and "y^t" (that). The tailed "p" (resembling a capital "p" with the tail extended into a loop) has also been retained. This symbol has no counterpart in modern typography; given the limitations of the modern keyboard, we have rendered it as a capital "p" (P). This should be taken to mean either "per" (by), "pre," or "pro" (and sometimes "par" as in "Pish" for parish) as the context of the order may demand. For damaged and missing portions of the manuscripts, we have used square brackets to denote the [blank], [torn], or [illegible] portions. Because of the large number of ancient forms of spelling, grammar, and syntax, it was deemed impracticable to insert the form *[sic]* after each one to indicate a literal rendering. Therefore, the reader must assume that apparent errors are merely the result of our literal transcription of the road orders, barring the introduction of typographical errors, of course. If, in any case, this appears to present insuperable problems, resort should be made to the original records.

As to dating, most historians and genealogists who have worked with early Virginian records will be aware of the English dating system in use down to 1752. Although there was an eleven-day difference from our calendar in the day of the month, the principal difference lay in the fact that the beginning of the year was dated from March 25 rather than January 1, as was the case from 1752 onward to the present. Thus January, February, and March (to the 25th) were the last three months in a given year, and the new year came in only on March 25.

Early Virginian records usually follow this practice, though in some cases, dates during these three months will be shown in the form 1732/3, showing both the English date and that in use on the Continent, where the year began January 1. For researchers using material with dates in the English style, it is important to remember that under this system (for instance) a man might die in January 1734 yet convey property or serve in public office in June 1734 since, under this system, June came *before* January in a given year.

INTRODUCTION

by

Beth Mitchell

for the

Fairfax County History Commission

The roads are under the government of the county courts, subject to be controuled by the general court. They order new roads to be opened whenever they think them necessary. The inhabitants of the county are by them laid off into precincts, to each of which they allot a convenient portion of the public roads to be kept in repair. Such bridges as may be built without the assistance of artificers, they are to be built. If the stream be such as to require a bridge of regular workmanship, the court employs workmen to build it, at the expense of the whole county. If it be too great for the county, application is made to the general assembly, who authorize individuals to build it, and to take a fixed toll from all passengers, or give sanction to such other proposition as to them appears reasonable.

Thomas Jefferson
Notes on the State of Virginia, 1781

The search for the most convenient way to travel from one place to another continues to be of great importance today, just as it was in the colonial period. A road order from the justices of the county court always began with the instruction that those who were to lay out the route should find "the most convenient way" for the new road to be established. The "most convenient way" might, or might not, be the shortest route, but it certainly would be the most level route and would avoid crossing as many streams or marshy areas as possible without unduly lengthening the distance between the specified two points.

Court-appointed "viewers" were instructed either to find the most convenient way and report to the next court or to view and mark the most convenient way. The court then ordered the road to be cleared and opened and either allotted the tithables to do the work or appointed one or two men to make the allocation of labor. Occasionally, the court simply ordered that a road be opened under supervision of an appointed surveyor, or overseer, of the road with the laboring tithables, or hands, from another road.

Alterations in the road were given serious consideration; proposed changes were viewed by a court-appointed committee who decided whether the new route was as good and convenient to the public as the former one. Often the proposed alteration of the route was rejected.

When a road was found in need of repairs, presentments, or charges, were made against the overseers of the road. Three or four presentments were normally made each time the grand jury was called into session. The citizens' frustration with the miserable condition of the roads was evident on May 20, 1772, when the grand jury brought a record twenty-three presentments against the overseers of the road. The list of presentments gives a fair representation of the roads existing in the county at that time.

Included with these road orders are entries concerning ferries, bridges, streets, and a wharf—the rudimentary components of the colonial transportation system of Fairfax County. Two other categories, ordinaries and mills, are also included since they concern businesses that of necessity would either be located on roads or require roads to be built to the premises. Taverns and inns, called ordinaries, were licensed by the court; those who "retailed spirituous Liquors" without a license were fined by the court. The majority of the ordinaries were in the two towns, Alexandria and Colchester; but many were located on major roads, often at a crossroads. Many roads were developed so that mills could serve the surrounding neighborhoods. It seems that most mills lasted a relatively short period of time, but some endured and were rebuilt many times in nearly the same area, with the adjoining road still carrying the name of the mill.

For a short period of time, from its formation in 1742 until 1757, Fairfax County included the present-day counties of Arlington, Fairfax, and Loudoun. Cut off from Prince William County in 1742, Fairfax County stretched along the Potomac River from the mouth of the Occoquan River to the Blue Ridge Mountains and was bounded on the south by the Occoquan River and Bull Run and from the heads of the main branch of Bull Run by a straight course to the Thoroughfare (a synonym for "pass" or "gap") of the Blue Ridge Mountains known by the name of Ashby's Gap, or Bent. This was the same area that had been placed in a new parish, named Truro, in 1732.

Prince William, formed in 1731, was previously a part of Stafford and King George counties. The part of Prince William that became Fairfax was part of Stafford County from 1664 to 1731. Before that, from 1653 to 1664, it was a part of Westmoreland County, which, in turn, had been formed from Northumberland in 1653. Northumberland, the "parent" county, was formed and organized in 1645.

The first division in Fairfax, as created in 1742, was in 1748 when Truro Parish was divided and Cameron Parish was formed from the area north and west of Difficult Run and Popeshead Run. This division was the basis for the creation of Loudoun County in 1757 and followed the established pattern of creating first a parish and then a county. A minor adjustment was subsequently made in the division line for the new county; instead of Popeshead Run, the division line was to be a straight line drawn from the head spring of Popeshead Run to the mouth of Rocky Run. This left a portion of Fairfax County in Cameron Parish, but this was resolved in 1762 when the parish line was changed to the same line as the county line. In 1798, the boundary line between Fairfax and Loudoun counties was redrawn and the portion of Loudoun County west of Difficult Run was returned to Fairfax County. It should be remembered that road orders included in this book from 1757 through 1798 pertained only to this smaller Fairfax County.

The court orders for roads covered in this volume begin in 1749, more than one hundred years after the formation of the parent county. The court records of Fairfax County between December 1742 and May 1749 are missing, as are the court records for Prince William County between 1731 and 1742, leaving a void during the crucial period of growth when many roads were established to serve the upward surge of settlement. For this earlier period, roads will have to be identified from surveys made for Northern Neck Grants, from surveys for court suits, from references in patents and early deeds, and from references to old roads in later deeds.

The preparation of "Fairfax County Road Orders 1749-1800" was simplified through the use of the indexing project sponsored by the Fairfax County History Commission. Under the direction of Edith Moore Sprouse, the court order books of Fairfax County from 1749-1872 have been indexed and abstracted. This extremely valuable resource tool is available on microfiche at major libraries and on interfiled cards at the Archives of the Circuit Court, Fairfax County Judicial Center.

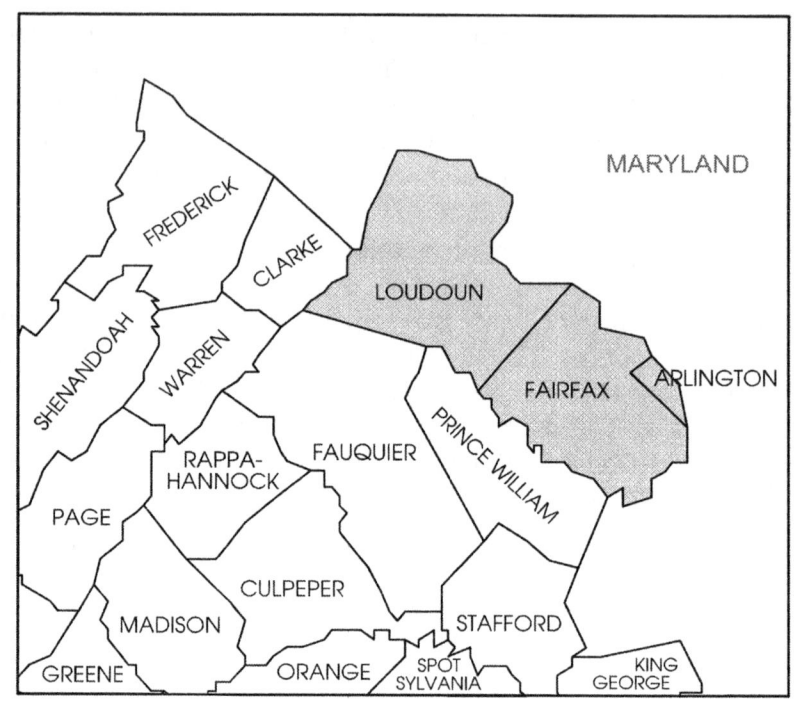

Fairfax County 1742

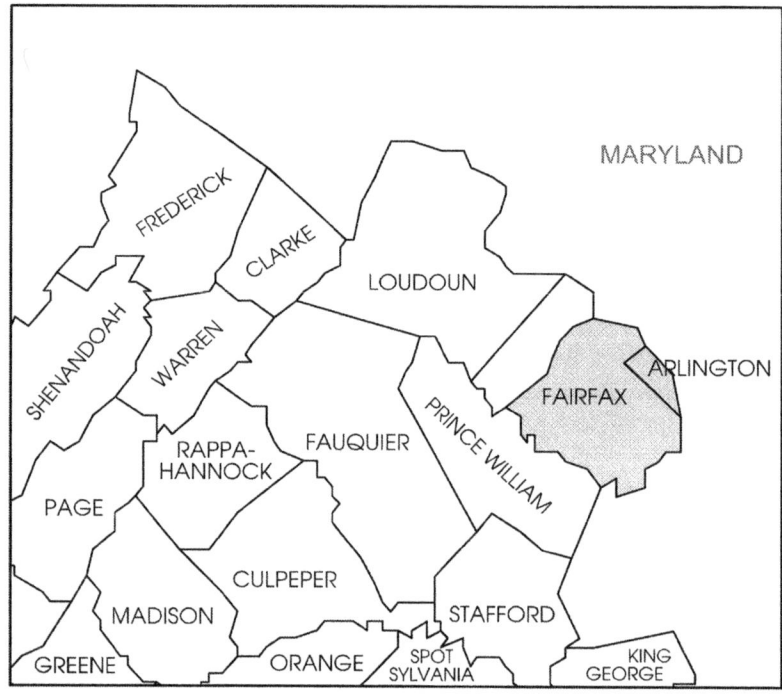

Fairfax County 1757

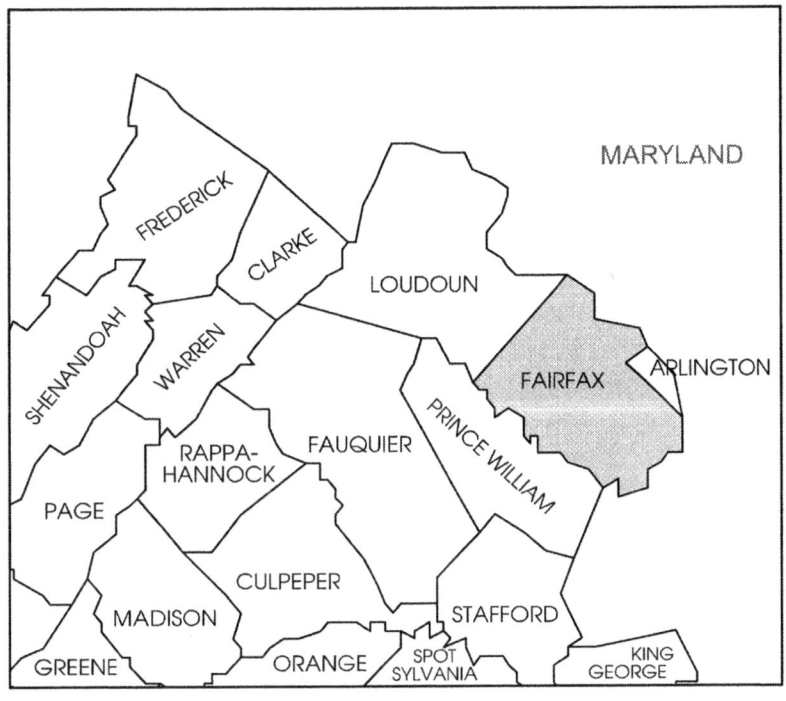

Fairfax County 1798

FAIRFAX COUNTY COURT ORDER BOOK 1749-1754

16 May 1749 Old Style, page 3
We present Thomas Lewis for not keeping the Road in repair between Ravensworth and the Ox Road by the knowledge of two of the Jury

16 May 1749 Old Style, page 3
William Hall Junr. is appointed Surveyor of the Road from his house to Major Catesbys Cockes Road where it intersects. Ordered that he keep the same in repair according to Law -

16 May 1749 Old Style, page 3
Fielding Turner is appointed Surveyor of the Road from Major Cockes Road to Cub Run. Ordered that he keep the same in repair according to Law.

17 May 1749 O. S., page 8
Ordered that John Turley Robert Lindsey and William Simpson view the most convenient way for a Road from the Prince William Road to the Ox Road. and report to the next court whether it will be convenient to the Publick to clear and open the same

17 May 1749 O. S., page 11
On the Motion of Stephen Lewis Gen^t. to rebuild a water Grist Mill on Difficult Run It is Ordered that Townshend Dade and Charles Broadwater Gen^t. View and Value an Acre of Land whereon the old Mill was Erected and that the Surveyor lay the same off and report to the next Court.

22 June 1749 O. S., page 23
On the Motion of Elisha Hall Licence is granted him to keep an Ordinary at his house the Ensueing year he having Entered into and Executed a Bond according to Law.

27 September 1749 O. S., page 35
On the motion of James Hardage Surveyor of the Prince William Road between the Ox Road and where it comes into the Road to this court house it is Ordered that Daniel Sanders Francis Summers Lewis Sanders Junior Thomas Felangdigam and William Kitchen do work on the said Road and assist in keeping the same in repair.

27 September 1749 O. S., page 35
Ordered that the Road from the Honourable Thomas Lees Quarter to the Church be Viewed by Charles Broadwater William Harle and Thomas Wren and make report to the next court

27 September 1749 O. S., page 35
Licence is granted John Lucas to keep an Ordinary at his house near the Falls Church the Ensueing Year he having Entered into and Executed a Bond accordingly.

COB 1749-1754

27 September 1749 O. S., page 35
Licence is granted Nathaniel Smith to keep an Ordinary in the Town of Alexandria the Ensueing Year he having Entered into and Executed a Bond accordingly.

28 September 1749 O. S., page 39
Jeremiah Bronaugh and John Graham Gen^t. are appointed and desired to View the Road and mark out the most convenient way from Pohick Run to Occoquan Ferry and make their report to the next Court

28 September 1749 O. S., page 39
George Fairfax Gen^t. is appointed Surveyor of the Road whereof Thomas Lewis in his lifetime was Surveyor. Ordered that he keep the same in repair and Erect posts or Stones according to the directions of the Act of Assembly in that case made and provided.

28 September 1749 O. S., page 39
Daniel M^c.Carty Gen^t. is appointed Surveyor of the Road from Pohick to Accotink and the Road to the Warehouse from the cross Road Ordered that he keep the same in repair and Erect Stones or posts according to the directions of the Act of Assembly in that case made and provided.

26 December 1749 O. S., page 46
On the petition of John Ashford to have a Road turned Ordered that William Trammell Robert Sandford and Thomas Colter or any two of them View the Road Prayed for and that they make report thereof to the next Court.

26 December 1749 O. S., page 46
On the Motion of Hugh West Ordered that John Jenkins James Jenkins and Henry Gunnell View the Road from the Wolf Trap Run to the main Road. and return their Report thereof on Oath to the next Court

26 December 1749 O. S., page 46
Lycence is granted John Templeman to keep an Ordinary at his house at the Falls the Ensueing he having Entered into and Executed a Bond according to Law.

26 December 1749 O. S., page 48
To George Mason Gen^t. for Occoquon Ferry 1600

26 December 1749 O. S., page 48
To Thomas Evans for Ferry over Goose Creek 1000

26 December 1749 O. S., page 50
On the Motion of Hugh West Lycence is granted him to keep an Ordinary at his house at the Ferry at Alexandria the Ensueing Year he having Entered into and Executed Bond according to Law.

COB 1749-1754

27 March 1750 O. S., page 54
On the Motion of William Reardon Ordered that Daniel M^c.Carty Abraham Barnes and William Peake View the Road by him cleared and report to the next Court whether the same be as good and convenient to the Publick as the former one.

27 March 1750 O. S., page 54
We Present John Anderson at Goose Creek for keeping a Tipling house and breaking the Sabbath by the Information of John Sweet.

27 March 1750 O. S., page 54
We Present William West for not keeping the Road according to Law. from Vincent Lewis Road to his house by the Information of Mr. George Johnston -

27 March 1750 O. S., page 54
We Present Vincent Lewis for not keeping the Road according to Law. from his Road to Col^o. Tayloes Quarter by Information of M^r. George Johnston -

27 March 1750 O. S., page 55
We Present Alexander Jennings for not keeping the Road in repair from Goose Creek. to Limestone Run to the knowledge of two of us.

28 March 1750 O. S., page 57
On the Petition of Edward Garret to build a Water Grist Mill on the beaver dam branch of Goose Creek. the same is granted and Ordered there issue an ad quod damnum according to Law.

28 March 1750 O. S., page 58
John Jackson is appointed Surveyor of the Road whereof John Higgerson was lately Overseer. Ordered that he do keep the same in repair according to Law -

28 March 1750 O. S., page 58
Ordered that George Mason Gen^t. be paid for keeping Occoquon Ferry in Proportion for what he was paid last year until November next, and then an agreement to be made with the Court of Prince William County concerning the said Ferry.

28 March 1750 O. S., page 58
Ordered that the Surveyors of the Roads from Sugarland Run to Goose Creek dig the Bank down at Goose Creek and make the same fit and convenient for taking in Tobaccos into the Skew.

28 March 1750 O. S., page 59
On the Motion of Richard Brown Lycence is granted him to keep an Ordinary at his house the Ensueing Year he having Executed Bond according to Law.

COB 1749-1754

28 March 1750 O. S., page 59
On the Motion of Charles Tyler Lycence is granted him to keep an Ordinary at his house the Ensueing Year he having Executed Bond according to Law.

29 March 1750 O. S., page 62
The referrees appointed to View the Road petitioned for by Christopher Strother report to the Court here. that the same is a good and Convenient way with a Small Amendment. Ordered that the same be Established and that the said Strother make the amendment

30 March 1750 O. S., page 64
William Ramsay Gen^t. is appointed Overseer of the Road. from the Ferry Road at the Town of Alexandria. to Hunting Creek Ford. and from thence to Four Mile Run Ford. Ordered that he keep the same in repair and Erect Stones or posts according to Law.

30 March 1750 O. S., page 67
Burgess Berkley is appointed Overseer of the Road whereof William West was late Surveyor and Ordered that he do keep the same in repair according to Law. and Erect posts or Stones &c^a.

30 March 1750 O. S., page 67
William West being Presented by the Grand jury for not keeping the Road in repair came into court and the matter being heard. Ordered that the same be dismist

30 March 1750 O. S., page 67
Ordered that Benjamin Grayson Gen^t. be Summoned to appear at the next Court to answer such things as shall be objected against him, or the Court shall ask him concerning a Road by his Mill in this County.

30 March 1750 O. S., page 67
William Adams is appointed Overseer of the Road Opposite to his house. to Hunting Creek and that the hands from Awbreys Road to Cameron work on the said Road which he is Ordered to keep in repair and to Erect Stones or posts according to Law.

31 March 1750 O. S., page 68
Robert Boggess on his Motion hath a Lycence granted him to keep an Ordinary at his house the Ensueing year he having Executed Bond according to Law.

26 June 1750 O. S., page 78
Alexander Jennings by his Petition Setting forth that he was Presented by the Grand jury for neglecting to keep the Road in repair whereof he was Surveyor, and praying releif in the Premisses being read and considered. Ordered the same be rejected -

COB 1749-1754

26 June 1750 O. S., page 80
It is Ordered that the former Orders of this Court for clearing a Road from the head of Dogue Creek into the back lick Road is Ordered to be renewed. and that Edward Violet be appointed Surveyor of the Middle Road and that Joining his Company to Lewis and Stephens clear the said Road and keep the same in repair according to Law -

26 June 1750 O. S., page 80
On the Motion of John Herryford Licence is granted him to keep an Ordinary at his house at the Ferry for the Ensueing Year he having Entered into and Executed Bond according to Law

26 June 1750 O. S., page 83
Thomas Lewis is appointed Surveyor of the Road whereof Henry Gunnell was late Surveyor. Ordered that he do keep the same in repair and that he Erect posts or Stones according to the directions of the Act of Assembly.

28 June 1750 O. S., page 91
On the Petition of Edward Garret to Erect and Build a Water Grist Mill on the Beaver dam Branch of Goose Creek it is Ordered that an ad quod damnum do issue according to Law.

28 June 1750 O. S., page 91
On the Motion of John Dalton Lycence is granted him to keep an Ordinary at the Town of Alexandria the Ensueing Year he having Entered into and Executed a Bond accordingly.

30 June 1750 O. S., page 98
The Motion of Daniel M^c.Carty Gen^t. to restrain Robert Boggess and take away his Licence, to prevent him from keeping a Publick house for the future being considered is Ordered to be rejected.

3 July 1750 O. S., page 99
Moses Linton Gen^t. is appointed Surveyor of the Road whereof Paul Turley was late Surveyor Ordered that he do keep the same in repair and Erect Stones or posts according to the directions of the Act of Assembly in that case made and Provided.

3 July 1750 O. S., page 99
Licence is granted John Higgerson to keep an Ordinary at his House the Ensueing Year he having Entered into and Executed a Bond according to Law.

COB 1749-1754

3 July 1750 O. S., page 102
John Graham Gen^t. is appointed Overseer of the Road from the Bridge at Graysons Mill to Occoquon Ferry. Ordered that he do keep the same in repair according to the directions of the Act of Assembly and that he have liberty to open and clear another Road if it shall appear to be a more convenient one for the public than the former Way.

26 September 1750 O. S., page 111
The Petition of Stephen Lewis Gen^t. to Erect a Water Grist Mill on Difficult Run is Ordered to be continued until the next Court.

26 September 1750 O. S., page 111
John Anderson is appointed Surveyor of the Road from William's Gap to the upper fork of Goose Creek. Ordered that he do keep the same in repair and Erect Stones or posts. according to the directions of the Act of Assembly in such cases made and Provided.

26 September 1750 O. S., page 111
Walter Williams is appointed Overseer of the Road from the upper fork of Goose Creek to Elisha Halls in the Room of Peyton Buckner deceased. Ordered that he keep The Same in repair and Erect Stones or posts according to the directions of the Act of Assembly in such Cases made and Provided.

26 September 1750 O. S., page 112
Ordered that the Surveyors of the Roads from Sugarland Run to Goose Creek and from Goose Creek to Limestone, dig down the Banks at Goose Creek Ferry and make the same fit and convenient for taking Tobacco into the Skew, and that Francis Elzey and Richard Coleman View and direct the Performance thereof

26 September 1750 O. S., page 115
We Present the Overseer of the Road that leads from the Court house of this county to Mrs Willisses by knowledge of two of this Jury.

26 September 1750 O. S., page 117
On the Motion of Catesby Cocke Gen^t. Ordered that his Lordship the Lord Fairfax John West and William Henry Terret View a Road turned by the said Terret at his Prayer to this Court and make report to the next Court whether it be convenient to the Public to have the said Road Established.

26 September 1750 O. S., page 117
On the Presentment of the Grand jury against Vincent Lewis for failing to Perform his duty in keeping the Road in repair whereof he was Surveyor, Judgment is granted George Johnston Gen^t. as Informer against the said Defendant for fifteen Shillings Current Money and Costs.

COB 1749-1754

26 September 1750 O. S., page 117
On the Presentment of the Grand jury against Alexander Jennings for failing to Perform his duty in keeping the Road in repair whereof he was Surveyor, Judgment is granted George Johnston Gen^t. as Informer against the said Defendant for fifteen Shillings Current Money and Costs.

26 September 1750 O. S., page 117
Philip Noland is appointed Overseer of the Road. from the great Limestone Spring to Poultneys Ferry whereof David Richardson was lately Surveyor. Ordered that he keep the same in Repair and Erect Stones or Posts according to the directions of the Act of Assembly in Such Cases made and provided.

27 September 1750 O. S., page 121
A Report of the Jury on an Acre of Land Petitioned for by Edward Garret to build a Mill was returned and admitted to record.

27 December 1750 O. S., page 122
To George Mason for Occoquon Ferry till next March 1600

27 December 1750 O. S., page 124
Roger Wigginton is appointed Surveyor of the Road whereof Christopher Strother was lately Surveyor Ordered that he do keep the same in repair and Erect Stones or posts according to the directions of the Act of Assembly in such cases made and Provided.

27 December 1750 O. S., page 124
Licence is granted John Templeman to keep an Ordinary at the Falls the Ensueing Year he having Entered into and executed Bond accordingly

27 December 1750 O. S., page 124
Licence is granted John Lucas to keep an Ordinary at his house the Ensueing Year he having Entered into and Executed Bond accordingly

27 December 1750 O. S., page 125
Ordered that Anthony Russell Fielding Turner and Vincent Lewis View a Road from the Inhabitants of little River and Goose Creek down to the Road by M^r. Hutchisons and report whether it be convenient to have a Road cleared and Established there

26 March 1751 O. S., page 128
We present Thomas Bruster for not keeping his Road in Order within the Six months last Past by the Information of William Starke.

26 March 1751 O. S., page 129
Licence is granted William West to keep an Ordinary at his House the Ensueing Year he having Entered into and Executed Bond according to Law.

COB 1749-1754

27 March 1751 O. S., page 129
William Stark is appointed Overseer of the Road whereof Francis Elzey was lately Surveyor. Ordered that he do keep the same in repair and that he Erect Stones or Posts according to the Directions of the Act of Assembly in that case made and Provided.

27 March 1751 O. S., page 130
Ordered that George William Fairfax Gen^t. John Anderson Jacob Lasswell and Isaac Pennington View and mark out the most convenient Way for a Road from the Court house of this County to William's Gap and report to the next Court

27 March 1751 O. S., page 130
Robert Sanford is appointed Overseer of the Road from Hunting Creek to Dogue Run back Road. Ordered that he do keep the same in repair and Erect Stones or posts according to the directions of the Act of Assembly in that case made and provided.

27 March 1751 O. S., page 130
Edward Violet is appointed Overseer of the Road from Dogue Run to Accotinck and Ordered that he do keep the same in repair and Erect Stones or Posts according to the directions of the Act of Assembly in that case made and provided.

27 March 1751 O. S., page 132
Thomas Cockeril is appointed Overseer of the Road whereof Jacob Ramey was lately Surveyor Ordered that he do keep the same in repair and Erect Stones or Posts according to the directions of the Act of Assembly in that case made and provided.

27 March 1751 O. S., page 132
Ordered that Moses Linton Gen^t. Sampson Turley and James Lane View a way Petitioned for by the Inhabitants and report whether it be Convenient to the Publick to have a Road cleared from the new Road to Rockey Run Church.

27 March 1751 O. S., page 133
William Stark is appointed Overseer of the Road whereof Francis Elsey was lately Surveyor Ordered that he do keep the same in repair according to the directions of the Act of Assembly in such Cases made and Provided. and that he cause the banks to be dug down at Goose Creek Ferry for the conveniency of Taking Tobaccos Waggons or Carts into the Skew.

27 March 1751 O. S., page 133
Licence is granted Charles Tyler to keep an Ordinary at his house the Ensueing Year he having complyed with the Law in such Cases made and Provided.

COB 1749-1754

27 March 1751 O. S., page 134
William Peake Senior is appointed Surveyor of the Road from Herryfords Ferry to the Church of Pohick and into the main Road by William Godfreys. Ordered that he do keep the Same in repair and Erect Stones or Posts according to the directions of the Act of Assembly in Such Cases made and Provided.

28 March 1751 O. S., page 136
Richard Simpson is appointed Overseer of the Road from Bond Veales to the upper side of Difficult Run. Ordered that he do keep the same in repair and Erect Stones or Posts according to the directions of the Act of Assembly in that case made and provided.

28 March 1751 O. S., page 136
John Jackson is appointed Overseer of the Road from the Court house to the Ox Road Ordered that he do keep the same in repair and Erect Stones or Posts according to the directions of the Act of Assembly in that case made and provided.

29 March 1751 O. S., page 139
Licence is granted Nathaniel Smith to keep an Ordinary at his house in Alexandria Town the Ensueing Year he having Entered into Bond according to Law.

29 March 1751 O. S., page 139
Licence is granted Hugh West to keep an Ordinary at the Ferry in Alexandria Town the Ensueing Year he having Entered into and executed Bond according to Law.

29 March 1751 O. S., page 144
Ordered that George Mason Gen^t. keep the Ferry at Occoquon and that he be allowed for the Same fifteen hundred Pounds of Tobacco in the County Levy -

2 April 1751 O. S., page 152
Licence is granted Henry Boggess to keep an Ordinary at his house the Ensueing Year he having Entered into and Executed a Bond according to Law.

2 April 1751 O. S., page 152
Licence is granted Thomas Evans to keep an Ordinary at his house the Ensueing Year he having Entered into and Executed a Bond according to Law.

2 April 1751 O. S., page 152
John Lucas is appointed Overseer of the Road whereof William Shortridge was lately Surveyor Ordered that he do keep the same in repair and Erect Stones or Posts according to the directions of the Act of Assembly in that case made and Provided.

COB 1749-1754

25 June 1751 O. S., page 157
Robert Thomas is appointed Surveyor of the Road whereof William Payne was lately Surveyor - Ordered that he do keep the same in Repair and Erect Stones or Posts according to Law.

25 June 1751 O. S., page 158
Licence is granted Michael Regan to keep an Ordinary at his house the Ensueing Year he having Entered into and Executed Bond according to Law.

25 June 1751 O. S., page 159
On the Petition of William Stark Ordered that Anthony Russell and Vincent Lewis View the way Petitioned for and report the conveniency thereof to the next Court.

24 September 1751 O. S., page 162
Burgess Berkley is appointed Overseer of the Road whereof Fielding Turner was late Surveyor Ordered that he do keep the Same in repair and Erect Stones or Posts according to the Directions of the Act of Assembly in that case made and Provided.

24 September 1751 O. S., page 163
Samuel Mead is appointed Overseer of the Road from Edward Thompsons to Thomas Davis's and Ordered that he do keep the same in Repair and Erect Stones or Posts according to the directions of the Act of Assembly in such Cases made and Provided.

24 September 1751 O. S., page 163
Thomas Davis is appointed Overseer of the Road from his house to Goose Creek, and Ordered that he do keep the same in Repair and Erect Stones or Posts according to the directions of the Act of Assembly in such Cases made and Provided.

24 September 1751 O. S., page 163
Joseph Gardiner is appointed Overseer of the Road whereof Samuel Compton was lately Surveyor Ordered that he do keep the same in repair and Erect Stones or Posts according to the Directions of the Act of Assembly in that case made and Provided.

24 September 1751 O. S., page 164
Daniel Thomas is appointed Overseer of the Road whereof Alexander Jennings was lately Surveyor Ordered that he do keep the same in repair and Erect Stones or Posts according to the Directions of the Act of Assembly in that case made and Provided.

25 September 1751 O. S., page 164
Ordered that a Road be cleared from Little River to Ashbys Gap on the Blue Ridge and Elijah Chinn appointed Surveyor from little River to Goose Creek. and Richard Nelson from Goose Creek to the Gap, which they are Ordered to keep in Repair according to the Act of Assembly in such Cases made and Provided.

COB 1749-1754

25 September 1751 O. S., page 164
Order that a Road be Cleared from the Mountain Road by Martha Halls the most Convenient Way to the Church and Robert Stephens is appointed Surveyor thereof, and Ordered that he do keep the Same in repair and Erect Stones or Posts, according to the directions of the Act of Assembly.

25 September 1751 O. S., page 164
Ordered that a Road Viewed by Anthony Russell and Vincent Lewis be Established from Berrymans Quarter to the Church, and Vincent Lewis is appointed Surveyor thereof, Ordered that he do keep the same in repair and Erect Stones or Posts according to the directions of the Act of Assembly.

25 September 1751 O. S., page 164
Ordered that a Road be Opened from the Mountain Road by Col°. Colvills Tuskarora Quarter to the Ford of Goose Creek by Samuel Kings and William Ross is appointed Surveyor thereof. Ordered that he do keep the same in repair and Erect Stones or Posts, according to the directions of the Act of Assembly.

25 September 1751 O. S., page 165
On the Petition of the Honourable William Fairfax Esquire to Erect a Water Grist Mill on Difficult Run and to have an Acre of Land confirmed to him Convenient for that Purpose, it is Ordered that the Sherif Summon twelve Freeholders of the Vicinage to Meet upon the Land Petitioned for and to be Sworn before a Justice or the Sherif to diligently View and Examine the said Land and make report to the Court according to Law.

25 September 1751 O. S., page 165
Joshua Farguson is appointed Surveyor of the Road from the Court House to the Ox Road near Capt. Lewis Ellzeys, Ordered that he do keep the same in repair and that he Erect Stones or Posts according to the directions of the Act of Assembly in that case made and Provided.

25 September 1751 O. S., page 166
The Petition of Daniel Lyon to keep an Ordinary at his house being read and considered. Ordered the same be rejected

25 September 1751 O. S., page 166
On the Petition of Stephen Lewis Gen^t. to Erect a Water Grist Mill on Difficult Run and to have an Acre of Land confirmed to him Convenient for that Purpose, it is Ordered that the Sherif Summon twelve freeholders of the Vicinage to Meet upon the Land Petitioned for and to be Sworn before a Justice or the Sherif to diligently View and Examine the said Land and make report to the Court according to Law.

COB 1749-1754

25 September 1751 O. S., page 166
On the Petition of Edward Masterson to Erect a Water Grist Mill on Difficult Run and to have an Acre of Land confirmed to him Convenient for that Purpose, it is Ordered that the Sherif Summon twelve Freeholders of the Vicinage to meet upon the Land Petitioned for and to be Sworn before a Justice or the Sherif to diligently View and Examine the said Land and make report to the Court according to Law

26 September 1751 O. S., page 167
On the Motion of Robert Colclough Licence is granted him to keep an Ordinary at his house the Ensueing Year he having Entered into and Executed Bond according to Law -

26 September 1751 O. S., page 170
Ordered that a Road be cleared from Fitzhughs Quarter on Ravensworth to Robert Boggesss and Joseph Stephens is appointed Surveyor thereof. Ordered tht he do keep the same in repair according to the directions of the Act of Assembly.

1 January 1752 New Style, page 180
Licence is granted Thomas Sorrell to keep an Ordinary at his house the year Ensueing he having Entered into and Executed Bond according to Law.

1 January 1752 New Style, page 180
Licence is granted John Templeman to keep an Ordinary at the Falls the year Ensueing he having Entered into and Executed Bond according to Law.

1 January 1752 New Style, page 180
Licence is granted John Peake to keep an Ordinary at his house the year Ensueing he having Entered into and Executed Bond according to Law.

1 January 1752 New Style, page 180
Licence is granted James Hamilton to keep an Ordinary at his house the year Ensueing he having Entered into and Executed Bond according to Law.

1 January 1752 New Style, page 181
Licence is granted Edward Violet to keep an Ordinary at his house the Ensueing year he having Entered into and Executed Bond according to Law.

1 January 1752 New Style, page 182
Ordered that the Several and respective Landlords Build Wharfs at the Warehouses in this county where Necessary and have Sufficient Cranes for the Safety of carrying the Tobaccoes on board Vesseles

COB 1749-1754

31 March 1752 N. S., page 184
Licence is granted Robert Sandford to keep an Ordinary at his house the Ensueing Year he having Entered into and Executed Bond according to Law.

31 March 1752 N. S., page 185
Samuel Tillet is appointed Overseer of the Road from Richard Browns to little Rockey Run - Ordered that he do keep the same in repair and Erect Stones or Posts according to the directions of the Act of Assembly in Such cases made and provided.

31 March 1752 N. S., page 185
Licence is granted William West to keep an Ordinary at his house the year Ensueing he having Entered into and Executed Bond according to Law.

31 March 1752 N. S., page 185
Jacob Remy is appointed Overseer of the Road from the lower side of little Rockey Run to Cub Run. Ordered that he do keep the same in repair and Erect Stones or Posts according to the directions of the Act of Assembly in such Cases made and Provided.

31 March 1752 N. S., page 186
Ordered that William Payne James Hamilton William Duling and Sandford Remy or any three of them View the Road from James Hamiltons to the Church and lay the same off and make it convenient for the Publick

31 March 1752 N. S., page 186
William West is appointed Surveyor of the Road whereof Joseph Hutchison was lately Surveyor. Ordered that he do keep the same in repair and Erect Stones or Posts according to the directions of the Act of Assembly in that case made and Provided.

31 March 1752 N. S., page 187
On the Petition of Edward Washington to have a Road turned, It is Ordered that Joshua Ferguson John Jackson and John Higgerson View the Way Petitioned for and report to the next Court whether it will be Prejudicial to the Publick to have the same turned -

31 March 1752 N. S., page 187
On the Report of Moses Linton Sampson Turley and James Lane upon Viewing a way for a Road Petitioned for Ordered the same be cleared from a Bent in the new Road between Col°. Tayloes Quarter and James Lanes, and thence on the lower side of the said Lanes plantation and through the Woods upon a Straight Course into the Mountain Road a little above Capt. Newtons Quarter on great Rockey Run and Adam Mitchell is appointed Surveyor thereof Ordered that he do keep the same in repair according to the directions of the Acts of Assembly &c.

COB 1749-1754

31 March 1752 N. S., page 187
Licence is granted Thomas Grafford to keep an Ordinary at his house in Alexandria Town the Year Ensueing he having Entered into and Executed Bond according to Law.

31 March 1752 N. S., page 188
Daniel McCarty Gen^t. is appointed Overseer of the Road from Pohick to Accotink and the Cross Road. Ordered that he do keep the same in repair and Erect Stones or Posts according to the directions of the Act of Assembly.

1 April 1752, page 189
Ordered that a Road be Opened from Sugarland Run to the Chappel on the Beaver dam and that Samuel Jenkins do lay the same off, and Thomas Bruester is appointed Surveyor thereof and that he and his company clear the same immediately and keep it in repair

1 April 1752, page 189
Thomas Sorrell is appointed Overseer of the Road wherof Charles Broadwater Gen^t. was lately Surveyor Ordered that he do keep the same in repair and Erect Stones or Posts according to the directions of the Act of Assembly in that case made and Provided.

20 May 1752, page 191
Ordered that George William Fairfax William Ellzey Jacob Lasswell & Marquis Calmes View and mark out the most convenient way for a Road from the Court house of this County to William's Gap and make report of their Proceedings to Court -

20 May 1752, page 193
The Viewers of a Road Petitioned for by Edward Washington having made their report and the same being adjudged Useless is Ordered to be dismist

20 May 1752, page 196
Ordered that the Sherif pay the Fraction in his hands to George Mason Gen^t. towards Paying his claim in the County for Occoquon Ferry Omitted to be levied at laying the County Levy in December last

21 May 1752, page 196
On the Petition of Sundry the Inhabitants of this County it is adjudged reasonable for their benefit and Advantage to build a bridge over broad Run and Ordered John Carlyle William Ramsay and Charles Broadwater Gen^t. agree with workmen to build the said Bridge and that Francis Hague and John Hough Measure and point out the Place and report to the first day of the next Court. and that Notice be given Edmond Sands to appear then and agree about the keeping Goose Creek Ferry -

COB 1749-1754

21 May 1752, page 196
The Petition of Lewis Saunders Junr. for turning a Road is Dismist

21 May 1752, page 197
Thomas Smith is appointed Surveyor of the Road from the Gum Spring to Hunting Creek Ordered that he do keep the Same in repair and Erect Posts according to the directions of the Act of Assembly in that case made and Provided.

21 May 1752, page 199
On the Motion of Daniel M^c.Carty Gen^t. it is Ordered that the following Persons do work on the Roads whereof he is Surveyor. Robert Boggess's Titheables John Baptist Anderson Thomas Tuttle James Brown William Gilmore Mark Brown, James Brown Junr. Terence Conner and the Titheables belonging to M^r. James Steptoe.

22 May 1752, page 204
On the Petition of John Ashford. Ordered a Road be Cleared from Edward Violets to Pohick Warehouse and Edward Violet is appointed Surveyor thereof and Ordered that the hands who work on the other Road whereof he is Surveyor Clear and keep the same in repair -

22 May 1752, page 204
On the Motion of Robert Boggess Setting forth that Thomas Wadlington has in clearing Ground Stopped up the Road leading to his Mill very much to his Prejudice and disadvantage and greatly Obstructing Custom, upon consideration whereof it is Ordered the said Thomas Wadlington doe Immediately cause the same way again to be Opened and that the same for the future be deemed and Established a Publick Road for passing to and from the said Mill -

16 June 1752, page 205
On the Motion of Thomas Wadlington, It is Ordered that Daniel M^c.Carty Abraham Barnes and Samuel Conner View a way where Robert Boggess had an Order for a Road through the Land of M^r. James Steptoe and report upon Oath to the Court whether it will be Prejudicial to have the said Road Established.

16 June 1752, page 206
Licence is granted Charles Tyler to keep an Ordinary at his house the Ensueing Year he having Entered into Bond according to Law.

16 June 1752, page 206
Licence is granted Richard Brown to keep an Ordinary at his house the Ensueing Year he having Entered into Bond according to Law.

COB 1749-1754

17 June 1752, page 206
John Osborn is appointed Overseer of the Road from Edward Thompsons to the top of the Blue Ridge. Ordered that he do keep the same in repair and Erect Stones or Posts according to the directions of the Act of Assembly in that case made and Provided.

17 June 1752, page 206
Edmund Sands is appointed Overseer of the Road whereof Daniel Thomas was lately Surveyor Ordered that he do keep the same in repair according to Law and that the Persons to wit Thomas Davis Samuel Davis, Isaac Hardin John Bond Richard Canary John Colvills Tuskarora hands. John Dewit William Ross. William Mead John Cannaday - William Williams. Minors Quarter hands. John Piles William Berry John Radcliff and Benjamin Shreive. do work on the said Road under his directions -

17 June 1752, page 208
Nathaniel Smiths Licence is renewed for keeping Ordinary at his house in Alexandria Town the Ensueing Year.

17 June 1752, page 208
Licence is granted Edmund Sands to keep an Ordinary at his house at Goose Creek Ferry the Ensueing Year -

17 June 1752, page 209
Licence is granted Michael Regan to keep an Ordinary at his house the Ensueing Year he having complied with the Law -

18 June 1752, page 211
Licence is granted William Munday to keep an Ordinary at his house in Alexandria Town the Ensueing Year he having Complied with the Law.

COB 1749-1754

21 July 1752, page 213
Ordered that Thomas Smith take as many hands as are Convenient to clear and keep the Road in repair from the Post in the Main Road to the Ferry at M^r. William Cliftons.

21 July 1752, page 213
Ordered that Lewis Ellzey and Hugh West Junr. Gen^t. James Hamilton Dempsie Carroll and James Halley or any three of them View and mark the most convenient Way for a Road to be Cleared from Alexandria to Rockey Run Chappell and report to the Court

23 July 1752, page 215
On the Petition of Joseph West to Erect a Water Grist Mill on a North branch of Goose Creek and to have an Acre of Land confirmed to him Convenient for that purpose, It is Ordered that the Sherif Summon twelve Freeholders of the Vicinage to meet upon the Land Petitioned for and to be Sworn before a Justice or the Sherif to diligently View and Examine the said Land and make report to the Court according to Law.

24 July 1752, page 222
An ad quod damnum on an Acre of Land Petitioned for by Edward Masterson to Build a Mill being returned is admitted to Record.

19 August 1752, page 227
Ordered that the Titheable Persons belonging to Valinda Wade and Mr. Thomas Marshall do work on and Assist in Clearing the Road whereof George William Fairfax Gent. is Surveyor

19 August 1752, page 228
Licence is granted George West to keep an Ordinary at his house the Ensueing year he having Entered into and Executed Bond according to Law.

19 August 1752, page 229
Ordered that all the Titheables between the Falls Road and the River work on and Assist in keeping the Road in repair whereof John Lucas is Surveyor.

26 September 1752, page 239
On the Petition of Joseph Chew licence is granted him to keep an Ordinary at Alexandria the Ensueing Year, he having Complied with the Law.

17 October 1752, page 240
John Summers is appointed Overseer of the Road whereof Roger Wigginton was lately Surveyor, which he is Ordered to keep in Repair and Erect Stones or Posts according to the directions of the Act of Assembly in that case made and Provided.

COB 1749-1754

17 October 1752, page 240
On the Motion of Richard Coleman It is Ordered that Vincent Lewis Anthony Russell and William Ellzey Gen[t]. or any two of them View the most convenient way for a Road to be cleared from the Sugar Lands to the new Chappell and report their Proceedings herein to the next Court upon Oath -

18 October 1752, page 243
On the Petition of William Reardin Ordered that Catesby Cocke and John Turley Gen[t]. Edward Washington and Robert Boggess, or any two of them View the Road cleared by the Petitioner and report upon Oath to the Court if the same be Convenient for the Public -

18 October 1752, page 244
Licence is granted Henry Boggess to keep an Ordinary at his House the Ensueing Year he having Executed Bond with Hugh West and Robert Boggess his Securitys for the same -

21 November 1752, page 250
Christopher Neale is appointed Surveyor of the road from four mile run to Mason's Ferry opposite to Rock Creek -

21 November 1752, page 250
Edward Masterson Sen[r]. from the Ferry to the Churchroad -

21 November 1752, page 250
Ordered that Lewis Ellzey, James Halley, Hugh West Sen[r]. & James Hamilton or any three of them being first sworn before some Justice of the Peace for this County do View and open a road from Alexandria Town to the rockey run chappel and report the same to the next Court -

21 November 1752, page 251
Peter Wagener Gent. is appointed surveyor of the road in the room of John Graham Gent.

22 November 1752, page 252
To Edmund Sands for keeping Goose Creek Ferry & keeping a Sufficient Skew
. . . . 2000

22 November 1752, page 253
The Petition of William Thompson being read Ordered that the same be rejected -

22 November 1752, page 253
On the Petition of Thomas Baylis and others Ordered that the road from Violetts to the Warehouses of Pohick be not deemed as a Publick road for the future -

COB 1749-1754

22 November 1752, page 253
To George Mason Gent. for Occoquon Ferry. 1600
To George Mason Gent. for the Ballance of Occoquon ferry last year 791

22 November 1752, page 254
We present Thomas Bruster for not keeping the road in repair from Difficult run to Suger land run in this six months last past to the knowledge of two of this Jury -

23 November 1752, page 256
To William Ramsay John Carlyle & Charles Broadwater Gent. for the bridge over broad run 16000

23 November 1752, page 256
On the Petition of Lewis Sanders jun^r. Ordered that John Cotton James Halley, William Scutt & William Kitchen or any three of them view the road by the Petitioners plantation & report upon oath whether the same may be conveniently turned

24 November 1752, page 259
Ordered that the several Surveyors of the road (to wit) William Ramsay Gent. Thomas Smith, John Summers & William Adams, Meet with their hands at Cameron and clear and amend the road to the run of great Hunting Creek

24 November 1752, page 259
William Adams is appointed surveyor of the road from Cameron opposite to his house

19 December 1752, page 264
Daniel Ansdell is appointed Surveyor of the road in the room of Richard Stephens

19 December 1752, page 265
On the Petition of William Reardon he has liberty to turn the road by his Plantation -

21 December 1752, page 276
On the motion of Peter Wagener Ordered that Daniel M^c.Carty Gent. allot a sufficient number of Titheables to Clear the road from Occoquan Ferry to Pohick -

21 December 1752, page 276
George Chilton is appointed surveyor of the road in the room of Robert Stephens -

21 December 1752, page 278
The Petition of Hadon Edwards about turning a road is rejected

22 December 1752, page 285
Nicholas Minor is appointed surveyor of the road from Goose Creek to Thomas Davis's -

COB 1749-1754

22 December 1752, page 285
Thomas Davis is appointed surveyor of the road from Davis's Branch to Peter Lewis's mill -

22 December 1752, page 285
Owen Williams is appointed surveyor of the road from Lewis's Mill to the line of the County at the blue ridge -

23 December 1752, page 292
Licence is granted to John Anderson to keep an Ordinary George Johnston & William Ellzey Gent. enters themselves Securitys

16 January 1753, page 297
Licence is granted to Thomas Sorrell to keep an ordinary Benjamin Sebastian Gen^t. Security

20 March 1753, page 305
Ordered that William Ross & Thomas Davis Surveyors do with their hands clear the road from Goose Creek to Thomas Davis's and that Anthony Russell Gen^t. allot a sufficient number of hands to Nicholas Minor to keep the same in repair -

20 March 1753, page 305
Licence is granted to John Peake to keep an Ordinary Hugh West jun^o. Security

20 March 1753, page 305
On the motion of Henry Peyton Gent. It is Ordered that the sheriff summon a Jury & Value an acre of land belonging to John Mercer gen^t. upon little river opposite to the Petitioners land for erecting a Water Grist mill and report -

20 March 1753, page 305
Licence is granted to Samuel Jenkins to keep an Ordinary James Jenkins Security

21 March 1753, page 309
Ordered that a road be cleared from Sugarland run to the new church Richard Coleman Surveyor is appointed to clear the same with the hands to be allotted to him by Anthony Russell Gent.

21 March 1753, page 309
Ordered that a road be cleared from Broad run to the new church William Starks is appointed Surveyor to clear the same with the hands to be allotted by Anthony Russell Gent.

21 March 1753, page 309
Richard Coleman is appointed Surveyor of the road in the room of Thomas Brewster -

COB 1749-1754

21 March 1753, page 310
George Mason Gen^t. agrees to keep the Ferry at Occoquan at fifteen hundred Pounds of Tobacco per annum -

22 March 1753, page 320
Licence is granted to Robert Sanford to keep an Ordinary John Dalton Security

16 May 1753, page 343
Thomas Harrison is appointed surveyor of the road in the room of William Ramsay Gent.

16 May 1753, page 343
On the motion of George Johnston Gent. It is ordered that the Inhabitants of the Town of Alexandria be exempted from clearing the Publick roads for one year they undertaking to clear the streets and Publick Landings.

17 May 1753, page 358
On the Complaint of George Johnston Gent. William Ramsay & Peter Wagener Surveyors of the road are fined fifteen shillings Ordered that they pay the same to the churchwardens of Truro Parish for the use of the said Parish

17 May 1753, page 358
George William Fairfax Esqr. Surveyor of the road Confesses that his road is out of repair Ordered that he pay fifteen shillings to the churchwardens of Truro Parish for the use of the said Parish

17 May 1753, page 358
John West is appointed Surveyor of the road in the room of Thomas Smith -

18 May 1753, page 366
Ordered that George William Fairfax, Thomas Colvill Daniel French and Stephen Lewis - Gent or any three of them view the road turned by M^r. William Henry Terrett and report to the next Court upon oath whether it will be convenient to the Publick to have the said road Established

18 May 1753, page 367
Licence is granted to Thomas Grafford to keep an ordinary. John Ashford became security

18 May 1753, page 368
Ordered that Hugh West Gent. Proprietor of the warehouse at Hunting Creek be sumoned to appear tomorrow to shew cause why he hath not Erected a Wharf at the Publick Landing of Alexandria

COB 1749-1754

19 May 1753, page 373
M^r. Hugh West Proprietor of the warehouses at Hunting Creek having failed to build a Wharf at the Landing pursuant to the several orders heretofore made and being sumoned to appear to shew his reasons for not building the said Wharf. And he appearing and not giving sufficient reasons for failing to do the same, ordered that John Carlyle, William Ramsay George Johnston & John Dalton Gent^s. or any three of them agree with some person to Perform the same and take sufficient Security for building the said Wharf

19 June 1753, page 402
Benjamin Hutchison & John Coleman are appointed to view the most Convenient way for a road from the mountain to the road Cleared by the County of Prince William & report to the next Court

19 June 1753, page 402
Lewis Ellzey Gent. Demsie Carrol & James Halley are appointed to view the most Convenient way for a road from John Summers's to the mountain road near rocky run Chappel & report to the next Court

19 June 1753, page 402
Licence is granted to William West to keep an Ordinary William Ellzey Security

19 June 1753, page 402
Stephen Lewis, John Hough & Richard Coleman or any two of them are appointed to view the bridge built by John Trammel jun^r. & report whether it is built according to agreement & it is Ordered that William Starke Surveyor of the road do clear the road to & from the bridge

19 June 1753, page 403
Ordered that the Tithables which Cleared the old Court house road do clear the road from the black lick to the Church road whereof James Hamilton is Surveyor -

19 June 1753, page 403
Ordered that the Tithables which cleared the Church road do clear the road from Accotink to Cameron whereof John Summers is Surveyor

19 June 1753, page 405
Licence is granted to Richard Brown to keep an Ordinary William Ellzey Security

20 June 1753, page 406
Licence is granted to Newbury Lidyard to keep an Ordinary Nathaniel Smith Security.

COB 1749-1754

20 June 1753, page 407
On the Petition of Lewis Saunders for liberty to turn a road by his plantation the persons appointed to view the same report as followeth, Jany. 13th. 1753. with submission to your worships order have Viewed the road & find it may be turned without Prejudice witness our hands James Halley William Scutt, John Cotton, William Kitchen whereupon it is ordered that the said Lewis Saunders may turn the said road

20 June 1753, page 411
George William Fairfax Gent. & Peter Wagener are appointed to view the road near Pohick and report to the next Court whether the same is Damaged by Col^o. Graysons Mill dam -

20 June 1753, page 411
The order made Last Court for Viewing the road Turned by M^r. Terrett not being yet complied with it is again ordered that George William Fairfax, Thomas Colvill, Daniel French & James Hamilton or any three of them do view the said road & report to the next Court upon oath whether it will be convenient to the Publick to have the said road Established

21 June 1753, page 416
On the Petition of Nicholas Savenus Licence is Granted him to keep an Ordinary Robert Boggess Security.

18 July 1753, page 430
John Moss is appointed Surveyor of the road in the room of Robert Sanford

19 July 1753, page 431
The persons appointed to view the road turned by M^r. William Henry Terrett report as followeth In Obedience to this order we the subscribers have met at the Place therein appointed & upon one view find that the Present road is much worse & much further about than the old one & that We think where it altered & the old road again Established it would be more to the Convenience of the Publick. Given under our hands this 30th. day of June 1753. G^o. Wm. Fairfax, Thos. Colvill, Ja^s. Hamilton whereupon it is ordered that the said William Henry Terrett do open the old road as soon as the corn is Gathered

21 August 1753, page 436
Ordered that a road be cleared from the road opposite to William Wests to the road that leads to Dumfries, Daniel Hutchison is appointed surveyor of the said road. Fielding Turner Gent. is appointed to allot a sufficient number of hands to clear the same

23 August 1753, page 443
Ordered that the sheriff sell sixteen thousand Pounds of Tobacco levied for building a bridge to the highest bidder at two month's Credit & keep the Cash til he shall have further orders

COB 1749-1754

23 August 1753, page 443
Ordered that Stephen Lewis John Hough & Francis Hague view the bridge built over broad run & report whether it is built according to agreement & if it is necessary to lengthen the bridge whether the undertaker of the County ought to pay for it

18 September 1753, page 449
Licence is granted Joseph Chew to keep an Ordinary Nathaniel Smith Security

19 September 1753, page 457
Henry Boggess jun^r. is appointed Surveyor of the road in the room of Daniel M^c.Carty G^t.

17 October 1753, page 468
Licence is granted to Henry Boggess to keep an ordinary William Ellzey Security

21 November 1753, page 473
To Edmond Sands for keeping ferry over Goose Creek 2000

To George Mason Gen^t. for keeping ferry over Occoquan 1500

21 November 1753, page 473
To Tobacco to be sold for paying for building broad run bridge to remain in the Sheriffs hands til further orders 10000

21 November 1753, page 474
Ordered that William Stark surveyor of a road from Sugar land run to Goose creek assist John Tramel to clear a road over broad run bridge.

21 November 1753, page 474
Ordered that Charles Broadwater G^t. pay the money in his hands to John Tramell for building broad run bridge.

21 November 1753, page 474
Licence is granted to Nathaniel Smith to keep an ordinary John North Security.

21 November 1753, page 474
Licence is granted to Thomas Harrison jun^r. to keep an ordinary Hugh West j^r. security

21 November 1753, page 474
On the Petition of Daniel Thomas to erect a water mill upon his land on Popeshead run Ordered that the sheriff sumon a Jury and surveyor to meet on the Premises to Value an Acre of Land belonging to Leonard Dozier & return the valuation of the same & whether it will be to the prejudice of any person

COB 1749-1754

21 November 1753, page 474
John Ratcliff is appointed surveyor of the road in the room of James Hamilton

21 November 1753, page 474
John Templeman is appointed surveyor of the road in the room of Thomas Sorrell

21 November 1753, page 474
Presley Cox is appointed surveyor of the road in the room of John Moss.

21 November 1753, page 475
Ordered that William Payne j^r. Thomas Lewis, William Buckley & Demsie Carroll or any three of them do view & mark the most convenient way for a road from Accotink to rocky run Chappel.

21 November 1753, page 477
Lewis Ellzey G^t. is appointed Surveyor of the road from Accotink to Popeshead

21 November 1753, page 477
Edward Payne is appointed surveyor of the road in the room of Samuel Tillett from Popeshead on the new road to the mountain road

21 November 1753, page 477
Lewis Ellzey G^t. is appointed surveyor of the road in the room of Richard Simpson jun^r.

23 November 1753, page 485
Licence is granted to William Sewell to keep an ordinary at Alexandria William Ramsay Gen^t. Security.

17 January 1754, page 518
Ordered that James Halley, Joseph Stephens, Philip Grimes, James Hamilton & Thomas Lewis or any three of them do view & mark the most convenient way for a road from Sumers's plantation to rocky run Chappel.

17 January 1754, page 518
On the motion of Garrard Tramel, Charles Broadwater & Thomas Colvill Gen^t. are appointed to view the most convenient way for turning the road by the Petitioners Plantation -

FAIRFAX COUNTY COURT ORDER BOOK, 1754-56, PART 1

21 January 1754, page 9
On the motion of James Hamilton Ordered that John Sumers, William Duling & Robert Thomas view the road by the Petitioners plantation and report on oath whether the same may be conveniently turned.

22 January 1754, page 22
Licence is granted to Jacob Pshaw to keep an Ordinary in Alexandria Benjamin Sebastian became his Security.

19 February 1754, page 33
On the Petition of John Colvill Gen^t. for liberty to erect a water Grist mill on his Land on four mile run It is ordered that John West Gen^t. & Richard Sanford view the place & report whether it will be to the prejudice of any person -

20 February 1754, page 45
Joseph Hutchison is appointed surveyor of the road from the fork of the roads above Netherton's to little river -

20 February 1754, page 45
John Atholl is appointed surveyor of the road from Little River to the Beaver dam the fork of Goose Creek -

20 February 1754, page 45
Fielding Turner G^t. is appointed to allott the tithables to clear the roads whereof John Atholl, Joseph Hutchison, John Anderson & Thomas Kelly are Surveyors -

20 February 1754, page 45
William Ellzey G^t. John Atholl & Thomas Williams or any two of them are appointed to view the most convenient way for a road from Williams's Gap to rocky run Chappel

21 February 1754, page 48
On the Petition of Garrard Tramell for turning the road by his Plantation the viewers appointed report that it may be conveniently turned whereupon it is ordered that upon the Petitioners making a bridge & causeway over Pimmets run and clearing the road to the orchard of Templemans Plantation he may stop the old road and it is ordered that the surveyor do clear the road to M^r. Broadwater's Plantation

21 February 1754, page 48
Burgess Berkley is appointed surveyor of the road from the Post in the mountain road to Piney Branch

COB 1754-1756, PART 1

21 February 1754, page 53
Ordered that a road be cleared from Sampson Turley's plantation to the mountain road & the said Sampson is appointed surveyor thereof & the Tithables belonging to Paul Turley, Sampson Turley, William Berkley, John Berkley, James Turley, Peter Turley, Samuel Tillet, Peter Carter & Charles O'Neale are ordered to clear the same

22 February 1754, page 58
Licence is granted to John Templeman to keep an Ordinary Benjamin Sebastian Gen^t. became Security

19 March 1754, page 60
Licence is Granted to Samuel Jenkins to keep an Ordinary Franklin Perry Security

19 March 1754, page 61
Philip Grimes is appointed Surveyor of the road from Accotink as the viewers have laid it off in the room of Lewis Ellzey Gen^t.

19 March 1754, page 61
Lewis Ellzey Gen^t. is appointed to allot a sufficient number of Tithables to clear the roads whereof Philip Grimes & Edward Payne the surveyor of the Church road and himself are Surveyors -

20 March 1754, page 61
Joseph Hutchison is appointed Surveyor of the road from the Piney Branch to little river

20 March 1754, page 62
Ordered that a road be cleared from the Lower Cross street in Alexandria to the road that leads to Cameron & it is ordered that the surveyor of the Town clear the same

20 March 1754, page 62
On the Petition of Several Persons Ordered that Leonard Dozier, Thomas Ford & John Ford view and lay out the most convenient way for a road to be cleared from the mountain road to Sampson Turleys & a road out of that to the mouth of Popeshead and Leonard Dozier is appointed surveyor thereof & the Tithables mentioned in the last Courts order are ordered to work on the same -

20 March 1754, page 63
David Davis is appointed Surveyor of the road from Shanadore road to the great Lime stone & the Persons who live between the said road & the river are ordered to clear the same and that they clear the road from Nicholas Minors to the road which leads to Nolands Ferry -

COB 1754-1756, PART 1

18 April 1754, page 82
Ordered that Joseph Jacobs, William Reardon & Abraham Hargis or any two of them view and mark the most convenient way for a road from the Ox road to Occoquan Ferry & report on oath to the next Court -

18 April 1754, page 83
Licence is granted to Samuel Mead Ferry keeper to keep an Ordinary Hugh West j^r. became Security

19 April 1754, page 85
Thomas Colvill, Hugh West, Stephen Lewis Gen^t. & Richard Coleman or any three of them are appointed to view the bridge built by John Tramell over broad run and report if the same is sufficiently done & it is ordered that Charles Broadwater Gen^t. late sheriff do not pay the said Tramell til further orders -

21 May 1754, page 87
Peter Wagener, William Peake & William Simpson Surveyors of the road are ordered to clear a road from the Ox road to Occoquan Ferry the way it is marked by the viewers

21 May 1754, page 87
Fielding Turner Gen^t. is appointed surveyor of the road in the room of Burgess Berkley

21 May 1754, page 87
Thomas Connell is appointed surveyor of the road in the room of Jacob Remey

21 May 1754, page 87
Samuel Moxley is appointed surveyor of the road in the room of John West Gen^t.

21 May 1754, page 88
Ordered that Philip Grimes upon receipt of this order do clear the road from Brown's Ordinary to John Hollis's with the tithables lately appointed to clear the new road

21 May 1754, page 89
Ordered that Charles Broadwater Gen. late sheriff pay John Trammell the money in his hands levied for building broadrun bridge & that Daniel McCarty Gen. sheriff sell so much Tobacco as will be sufficient to pay the said Tramell as much money as will amount to £100 with the money in the hands of the said Broadwater -

21 May 1754, page 90
Fielding Turner Gen. is appointed to allot a sufficient number of Tithables to clear the roads whereof Thomas Connell, Francis Summers & Adam Mitchell are surveyors

COB 1754-1756, PART 1

21 May 1754, page 91
Francis Summers is appointed Surveyor of the road in the room of James Lane

21 May 1754, page 92
We Present. George W^m. Fairfax Esqr. of the Parish of Truro for not having the road between little hunting creek & Accotink run in sufficient repair to the knowledge of two of this Jury

21 May 1754, page 92
We Present; Elijah Chin of the Parish of Cameron for not having the road between little river and Goose Creek in sufficient repair to the knowledge of two of us.

21 May 1754, page 93
We Present. William Stark of Cameron Parish for not having the road between Goose Creek & broad run in sufficient repair to the knowledge of two of us.

22 May 1754, page 94
On the motion of John Alexander by his Attorney it is ordered that the Order made for clearing a road from the lower Cross street in Alexandria to the road which leads to cameron be Quashed.

22 May 1754, page 94
Anthony Russell Gen. is appointed surveyor of the road in the room of Vincent Lewis

22 May 1754, page 94
John O'Daniel is appointed surveyor of the Church road in the room of William Scutt

22 May 1754, page 98
Garrard Trammell is appointed surveyor of the road in the room of John Templeman

22 May 1754, page 98
Ordered that the several Surveyors of the roads in this County be Cont^d. in their office

18 June 1754, page 100
Licence is granted to William West to keep an Ordinary James Hamilton became Security

18 June 1754, page 101
Fielding Turner & Anthony Russell Gen^t. are appointed to agree with workmen to build a bridge over Cub run

COB 1754-1756, PART 1

18 June 1754, page 101
Thomas Owsley is appointed constable from Little River to Goose Creek to the road which leads to Broad run Chappel from thence to vestels Gap

18 June 1754, page 101
Philip Grimes is appointed constable from popes head to the falls of Occoquan a cross to Accotink at Steptoe's Mill & he having taken the oaths prescribed by Law was sworn into his said office.

18 June 1754, page 103
Thomas Self is appointed Constable from Difficult to Goose Creek between the Ox road & Potomack River

18 June 1754, page 103
James Fletcher is appointed Constable from the Falls of Occoquan along the back road to Piney run.

18 June 1754, page 103
Joseph Cash is appointed Constable from Accotink to Home's run between the back road and Ravensworth road

18 June 1754, page 103
John Ratcliff is appointed Constable from the Church road to Accotink up to difficult

18 June 1754, page 103
Nicholas Garret is appointed Constable from Difficult to the Ferry road

18 June 1754, page 104
Robert Mills is appointed Constable from the Ferry road to four mile run

18 June 1754, page 104
James Whaley is appointed Surveyor of the road in the room of John Canady

18 June 1754, page 104
George Hadon is appointed Constable between Popes head Cub run & the Ox road.

19 June 1754, page 105
Ordered that Daniel French & John Peake do view and mark the most convenient way for a road from Poseys Ferry to the main road

19 June 1754, page 107
John Peter Sarter is appointed Surveyor of the road in the room of Joseph Stephens

COB 1754-1756, PART 1

21 June 1754, page 124
Grand Jury against George William Fairfax Esqr. On a Presentment for not repairing the road This day came Hugh West Gen. deputy attorney of our sovereign Lord the King and the said George by Benjamin Sebastian his attorney and on hearing the arguments of the Parties thereon it is considered that the said George shall make a fine to our sovereign Lord the King in Fifteen shillings & pay Costs.

21 June 1754, page 124
Grand Jury against Elijah Chin On a Presentment for not repairing the road This day came Hugh West Gen. deputy Attorney of our sovereign Lord the King & the said Elijah came not altho solemnly called Therefore it is considered that the said Elijah shall make a fine to our sovereign Lord the King in the sum of fifteen shillings & Pay Costs.

21 June 1754, page 124
Grand Jury against William Stark On a Presentment for not repairing the road This day came Hugh West Gen. deputy Attorney of our sovereign Lord the King & the said William came not altho solemnly required Therefore it is considered that the said William shall make a fine to our sovereign Lord the King in the sum of Fifteen shillings & Pay costs.

16 July 1754, page 128
Licence is granted to Richard Brown to keep an Ordinary William Ellzey G^t. Security

18 July 1754, page 143
Licence is granted to Sampson Demovil to keep an Ordinary John Dalton Security.

17 September 1754, page 149
Licence is granted to Joseph Chew to keep an Ordinary in Alexandria Hugh West Security

20 November 1754, page 164
To Jacob Remey for building Cub run bridge 2650

20 November 1754, page 165
To Tobacco to be sold to pay John West Gent. for building a wharf 2000

20 November 1754, page 166
James Donaldson is appointed Surveyor of the road in the room of John Lucas

20 November 1754, page 167
George Mason Gent. agrees to keep Occoquan Ferry one year for two thousand pounds of Tobacco.

COB 1754-1756, PART 1

21 November 1754, page 185
We Present the surveyors of the lower road from goose Creek to Top of the ridge for not keeping it in repair by information of Benjn. Sebastian -

21 November 1754, page 185
We present the surveyor of the road from Doeg run to Accotinck by the information of Benjn. Sebastian for not keeping the same in repair according to Law

21 November 1754, page 185
We present the Surveyor of the road from Hunting Creek to Belhaven for not repairing the same to the knowledge of two of us -

21 November 1754, page 186
We present the Surveyor of the road from little river to the widow Adams's at Goose Creek for not keeping the said road in repair accord^g. to law to the knowledge of two of us.

21 November 1754, page 186
We present the surveyor of the road from Accotinck to Pohick run for not putting up posts of direction at the fork of said road at Boggess's & at another fork thereof leading down to M^r. Barnes's - by information of George W^m. Fairfax

17 December 1754, page 188
Anthony Russell G^t. is appointed to allot a sufficient number of Tithables to clear the road whereof Nicholas Minor is Surveyor –

17 December 1754, page 188
Thomas Colvill, Charles Broadwater & Henry Gunnell or any two of them are appointed to view the road leading from the old Court house to the mountain road and report on oath whether it may be conveniently turned.

18 December 1754, page 191
Licence is granted to Richard Moxley to keep an ordinary, Hugh West became Security

18 December 1754, page 191
Licence is granted to George Thrift to keep an Ordinary, John Hurst became Security -

21 January 1755, page 204
Licence is granted to Jacob Pshaw to keep an Ordinary Hugh West G^t. became Security

21 January 1755, page 204
Ordered that John Peake, Thomas Triplett and Samuel Johnston or any two of them view the most convenient way for a road from the Gum Spring & Doeg run to Posey's Ferry and report to the next Court.

COB 1754-1756, PART 1

21 January 1755, page 204
Ordered that the sheriff sell the Tobacco in his hands levied to pay for the building of Broad run Bridge

22 January 1755, page 205
Ordered that Hugh Caldwell, William Barry & Jacob Morris or any two of them view the most Convenient way for a road from the Glebe of Cameron Parish to Goose Creek Chappel and report to the next Court.

22 January 1755, page 207
Joseph Moxley is appointed surveyor of the road in the room of Garrard Trammel -

22 January 1755, page 207
Charles Broadwater & Thomas Colvill Gen^t. are appointed to divide the labouring tithables between Thomas Wren & James Donaldson surveyors.

18 February 1755, page 207
On the motion of John Barry it is ordered that John Peake, John Posey & Daniel French or any two of them do view the path leading thro the plantation of Verlinda Wade and report on oath whether it may be Conveniently Turned.

18 February 1755, page 207
Licence is granted to John Gladin to keep an Ordinary, Hugh West Security.

18 February 1755, page 207
The persons appointed for viewing the road leading from the old Court house to the mountain road report that the new way turned by Sybell West widow is as near as the old way whereupon she has liberty to stop the old way after she has made a bridge & cleared the new road

18 February 1755, page 208
Ordered that George William Fairfax Esqr. Surveyor do clear a road from the Gum spring and from Doeg run to Posey's Ferry the way it is marked by the viewers -

21 February 1755, page 233
John Hurst is appointed surveyor of the road in the room of Edward Masterson dec^d.

21 February 1755, page 233
Robert Mills is appointed surveyor of the road in the room of Christopher Neale

21 February 1755, page 233
William Henry Terrett Gen^t. is appointed surveyor of the road from the old School House to the road at Cameron

COB 1754-1756, PART 1

21 February 1755, page 233
Daniel Talbut is appointed surveyor of the road in the room of John Somers from Accotinck to the old school house. Ordered that John West G^t. do allot the labouring Tithables to work thereon

21 February 1755, page 233
Robert Boggess is appointed surveyor of the road in the room of Henry Boggess J^r.

FAIRFAX COUNTY COURT ORDER BOOK, 1754-56, PART 2

18 March 1755, page 286
The persons appointed for viewing the path leading thro the plantation of Virlinda Wade report on oath that it may be conveniently turned, whereupon she has liberty to stop the old way after the new one is cleared.

18 March 1755, page 287
Ordered that William West, Elijah Chinn & John Roberts do view the road from Mark Cantons near little river to Peytons mill path and report on oath to the next Court whether it may be conveniently turned.

18 March 1755, page 287
John West Junr. is appointed surveyor the road in the room of William Adams -

18 March 1755, page 287
On the motion of John Gest ordered that John West, Daniel M^c.Carty G^t. & John Posey do view the most convenient way for a road from his house to the road by William Tramell's

18 March 1755, page 287
Licence is Granted to Henry Awbrey Ferry-keeper to keep an Ordinary Oneas Campbell became his security -

19 March 1755, page 288
Licence is granted to William Ashford to keep an Ordinary Michael Reagan Security

19 March 1755, page 288
On the motion of William Ashford Ordered that William Gunnell Jun. Joseph Moxley & Owen Williams or any two of them view the most convenient way for a road, from the main road to the said Ashfords house and report on oath to the next Court

19 March 1755, page 290
John Rhodes is appointed surveyor of the road in the room of William Henry Terrett

19 March 1755, page 293
Licence is granted Richard Coleman to keep an Ordinary, William Ellzey G^t. Security -

19 March 1755, page 293
Licence is granted Samuel Jenkins to keep an Ordinary, James Spurr and Charles Broadwater Securitys -

20 May 1755, page 298
Christopher Neale is appointed surveyor of the road in the room of Thomas Connell

COB 1754-1756, PART 2

20 May 1755, page 298
Garrard Boling is appointed surveyor of the road in the room of Benjamin Sebastian G^t.

20 May 1755, page 299
Mark [torn—Canton?] is appointed surveyor of the road in the room of Elisha Chinn

17 June 1755, page 303
John Smarr is appointed Surveyor of the road in the room of Joseph Hutchison

18 June 1755, page 309
Ordered that Daniel M^c.Carty Gent. sheriff pay John Trammell forty two pounds fourteen shillings & one half penny the ballance due to him for building broad run bridge

19 June 1755, page 314
Abednego Adams is appointed Surveyor of the road in the room of Samuel Moxley

19 June 1755, page 314
John Ford is appointed Surveyor of the road in the room of Lewis Ellzey Gent.

19 June 1755, page 314
Sampson Turley is appointed surveyor of the mountain road in the room of Edward Payne.

19 June 1755, page 315
Francis Padget is appointed surveyor of the road in the room of Fielding Turner Gent.

19 June 1755, page 315
William Reardon is appointed surveyor of the road from Occoquan Ferry road to the Ox road & Lewis Ellzey Gent. is appointed to divide the Tithables who clear the roads whereof William Peake, William Simpson & Peter Wagener are Surveyors -

19 June 1755, page 318
On the Petition of John Hough he has liberty to build a water Mill on his leased Land. lying in the Gap of the short hills, he Complying with the Act of assembly -

20 June 1755, page 322
On the motion of William Ashford it is ordered that George Thrift do open the old road from the said Ashfords house to the main road.

15 July 1755, page 328
John Anderson is Cont^d. Surveyor of the road.

COB 1754-1756, PART 2

15 July 1755, page 329
Ordered that the prison bounds extend from the river bank to the back line of Alexandria & to the north side of Cameron Street & to the south side of Kings Street.

17 July 1755, page 346
Ordered that the sheriff sell the Tobacco on Saturday next Levied for paying John West G^t. for building a wharf & that he pay him ninety eight pounds current money

24 July 1755, page 375
On the motion of William Ashford Ordered that an Attachment be issued against George Thrift for not opening a road from Ashfords house to the main road

19 August 1755, page 383
Licence is granted James Pendleton to keep an Ordinary Peter Wagener Security

19 August 1755, page 384
James Lane Junr. is appointed surveyor of the road in the room of Christopher Neale

19 August 1755, page 384-B
Ordered that John Poultney, Oneas Campbell & David Richardson or any two of them do view the most convenient way for a road from the lower part of the Island on Potomack river belonging to John Trammel to the place where he is erecting his house & that they make report thereof to the next Court.

21 August 1755, page 400
On the motion of Sampson Turley Gent. Ordered that the persons between wolf run & Popes head work on the mountain road between Richard Browns & little Rockey run

16 September 1755, page 401
On the Petition of Charles Thrift for turning a road. Ordered that George William Fairfax & Thomas Colvill Gent. view the said road & report on oath whether it may be conveniently turned

18 November 1755, page 425
To Richard Coleman for repairing Difficult Bridge 80

18 November 1755, page 426
We present the several Surveyors of the road from the old Court house to Difficult run & from Difficult run to sugar Land run & from sugar Land run to Broad run for not keeping the same in repair by information of William Ramsay G^t.   Townshend Dade foreman

COB 1754-1756, PART 2

19 November 1755, page 426
Blanch Flower Duncan is appointed Surveyor of the road in the room of James Donaldson

19 November 1755, page 428
To George Mason Gent. for keeping Occoquan Ferry 2000

To Henry Awbrey for keeping Ferry over Goose Creek 2 years 2400

20 November 1755, page 436
Ordered that Sybill West do give Bond for keeping Ferry at Hunting Creek

16 December 1755, page 441
Licence is granted to Paul Irmill to keep an Ordinary in Alexandria Peter Wagener & Benjamin Sebastian Gen^t. Securitys

20 January 1756, page 445
Licence is granted to Jacob Pshaw to keep an ordinary at the Ferry in Alexandria Hugh West Gent. security

20 January 1756, page 445
John Shelton is appointed surveyor of the road in the room of Thomas Davis

21 January 1756, page 448
Licence is granted to William Dodd to keep an Ordinary at his house James Hamilton Gent. security

22 January 1756, page 450
Licence is granted to Jacob Pshaw to keep an ordinary on M^r. Sebastians Lot in Alexandria John Dalton Security

22 January 1756, page 450
Grand Jury against Thomas Wren On a Presentment for not repairing the road The said Thomas being sumoned and failing to appear it is ordered that he be fined fifteen shillings & that he pay Costs

22 January 1756, page 451
Grand Jury against Richard Coleman On Presentment for not repairing the road It is ordered that the said Richard be fined fifteen shilling & that he pay Costs

17 February 1756, page 453
Licence is granted to Richard Brown to keep an ordinary at his house Hugh West and William Ellzey securitys

COB 1754-1756, PART 2

17 January 1756, page 453
Licence is granted James Ingo Dozer to keep an ordinary at his house John Hampton & Leonard Dozer securities

18 March 1756, page 496
A Petition signed by sundry persons praying that an act may be made for Establishing a Ferry upon the land of Josias Clapham on Potomac River which is ordered to be Certified

20 April 1756, page 499
Thomas Shaw is appointed Surveyor of the road in the room of John West Junr. Gent. and it is ordered that the Tithables belonging to the said Shaw and West do assist to clear the said road

20 April 1756, page 499
Robert Adam is appointed surveyor of the road in the room of Thomas Harrison dec^d.

20 April 1756, page 500
Licence is granted to Ezekiel Hickman to keep an ordinary at his house William Ellzey security

18 May 1756, page 501
Samuel Johnston is appointed surveyor of the road in the room of Abednego Adams

18 May 1756, page 501
Ordered that the several Surveyors of the roads be Continued in their office

18 May 1756, page 501
Licence is granted Richard Coleman to keep an Ordinary at his house William Ellzey Gent. security

19 May 1756, page 507
It is again Ordered that the surveyors of the road from the old Court house to Difficult run & from Difficult to sugarland run & from sugarland run to Broad run be sumoned to appear at the next Court to answer a presentment of the Grand Jury against them for not keeping the said road in repair

19 May 1756, page 509
We present George William Fairfax for not keeping in repair the road from Dogue run to John Pozeys Ferry landing and from thence to little hunting creek to the knowledge of two of us

COB 1754-1756, PART 2

21 May 1756, page 519
The persons appointed to view a way for a road for John Trammell having return'd their report it is ordered that the said Trammell may clear the same as it is mark'd by the Viewers & keep the same open

15 June 1756, page 522
William Parle is appointed surveyor of the road in the room of John Smar

15 June 1756, page 524
William Wadlington is appointed surveyor of the road in the room of John Atholl

15 June 1756, page 524
Samuel Duncan is appointed surveyor of the road in the room of John Anderson

15 June 1756, page 524
Licence is granted Charles Mason to keep an ordinary at his house George Johnston Gent. security

16 June 1756, page 527
John Posey is appointed surveyor of the road from Little Hunting Creek to Accotinck

16 June 1756, page 527
Licence is granted to George Thrift to keep an Ordinary at his house Joseph Stephens & James Donaldson securities

18 June 1756, page 553
On the Petition of George Mull Licence is granted him to keep an ordinary at his house George Johnston Gent. security

FAIRFAX COUNTY MINUTE BOOK 1756-1763, PART 1

1756, page 12A (pt. 2)
On the motion of John Carlyle Gent. Ordered that John West jun^r. Thomas Shaw, Richard Sanford & Robert Sanford or any three of them view a way for a road between the plantations of Baldwin Dade & Townshend Dade from Alexandria to the plantation of John Carlyle and report on oath whether the same may be Conveniently Cleared

17 August 1756, page 14A (pt. 2)
John Athell is appointed surveyor of the road from the Beaver dam of Goose Creek to the School house above Andersons & that Samuel Duncan be surveyor from thence to the top of the blue ridge

21 July 1756, page 16A (pt. 2)
Licence is granted to John Watson to keep an Ordinary in Alexandria Benjamin Sebastian Security

18 August 1756, page 20A (pt. 2)
It appearing to the Court that James Ingoe Dozer suffered in his Ordinary Unlawfull Gaming, Ordered that his Licence be suspended

19 August 1756, page 24
Licence is granted to John Gladin to keep an Ordinary at his house Hugh West Gent. Security

21 September 1756, page 25
Licence is granted to William West to keep an Ordinary at his house Peter Wagener and Hugh West Security

19 October 1756, page 30
Joseph Cockeril is appointed surveyor of the road in the room of William Grimes dec^d.

(no date) - 1756 (pt. 2), page 31A
Benjamin Grayson of Cameron is appointed Constable above Cub Run from the Ox road up Cub run to Benjamin Hutchinson's

(no date) - 1756 (pt. 2), page 31A
Ordered that a road be Cleared from Thomas Kelleys road below William Ellzey's plantation to Cross the Creek at Dawson's Ford & be Continued to Nicholas Minor's John Davis is appointed surveyor from Kelleys road to Segalon run and William Mead is appointed surveyor from thence to Nicholas Minors and that James Hamilton Gent. allot the Tithables to clear the said road

16 November 1756, page 37
Charles Thrift is appointed surveyor of the road in the room of Blanch Flower Duncan and it is ordered that he remove all the gates from the said road

FCMB 1756-1763, PART 1

16 November 1756, page 37
Joseph Jones is appointed surveyor of the road in the room of Owen Williams

16 November 1756, page 37
To George Mason Gent. for keeping Occoquan Ferry .... [torn]

To John Moss Junr. for keeping Ferry over Gooscreek 1200

17 November 1756, page 37
John Ford is appointed surveyor of the road in the room of Lewis Ellzey Gent.

17 November 1756, page 37
William Bucklew is appointed surveyor of the road in the room of Edward Payne

17 November 1756, page 38
We present William Tramel of the parish of Truro for keeping a Disorderly house profaning the Sabath within one month last past by the information of Joseph Cash

17 November 1756, page 38
We present the surveyor of the Highway between little rockey run & Browns Ordinary for not keeping the road in sufficient repair to the knowledge of two of us

17 November 1756, page 38
We present the surveyor of the Highway between Accotink & the old school house for not keeping the road in sufficient repair to the knowledge of two of us

17 November 1756, page 39
Sampson Trammell is appointed surveyor of the road in the room of Richard Coleman

17 November 1756, page 39
Ordered that Nicholas Minor turn the road near Tuscarora run according to the direction of James Hamilton & Oneas Campbell Gent.

17 November 1756, page 39
Franklin Perry is appointed surveyor of the road in the room of James Spur

17 November 1756, page 40
The Petition of William Tramell for keeping an Ordinary is rejected

18 January 1757, page 43
Licence is granted Charles Tyler to keep an Ordinary at Colchester Hugh West and Peter Wagener security's

FCMB 1756-1763, PART 1

15 March 1757, page 47
Benjamin Grayson of Cameron is appointed surveyor of the road in the room of Francis Padget

15 March 1757, page 48
Ordered that Robert Boggess John Peake & John Barry view the most Convenient way for a road from Edward Violets to Pohick Warehouse and report to next Court

15 March 1757, page 49
Stephen Emzie is appointed surveyor of the road in the room of John Shelton

21 March 1757, page 80
Licence is granted to Edward Willit to keep an Ordinary in Alexandria William Gladin & James Lane security's

22 March 1757, page 81
Licence is granted James Ingo Dozer to keep an ordinary at his house who with James Lane acknowledged a bond for the same

17 May 1757, page 100
On the motion of Francis Summers he is discharged from being surveyor of the road

17 May 1757, page 100
Richard Moxley is appointed surveyor of the road in the room of Samuel Johnston

17 May 1757, page 100
Joseph Reid is appointed surveyor of the road in the room of William Buckley & it is ordered that the tithables belonging to Edward Payne assist in clearing the said road

17 May 1757, page 101
William Dodd is appointed surveyor of the road in the room of Nicholas Minor

18 May 1757, page 101
On the Petition of Josias Clapham he has liberty to Erect a water Grist Mill on his land whereon he now lives

18 May 1757, page 101
Richard Stephens is appointed surveyor of the road from the back lick road to the county road

18 May 1757, page 106
John Peak is appointed surveyor of the road from Accotink to Little hunting Creek

FCMB 1756-1763, PART 1

18 May 1757, page 106
Thomas Triplet is appointed surveyor of the road from Dogue run to Posey's ferry & from thence to little hunting Creek and it is ordered that all the Tithabels on the East side of the County road do clear the same

22 June 1757, page 131
Benjamin Davis is appointed surveyor of the road in the room of Charles Thrift

19 July 1757, page 137
Charles Thrift is appointed surveyor of the road from the falls warehouse up to the falls road leading to Difficult

19 July 1757, page 137
Benjamin Davis is appointed surveyor of the road from the Falls warehouse to the falls landing and it is ordered that he clear the same with the hands of Colonel Lee & Stephen Vinayard & his son and they are Exempted from clearing the other part of the said road

18 October 1757, page 167
William Talbot is appointed surveyor of the road in the room of Daniel Talbot

15 November 1757, page 169
We present John Ford Surveyor of the road from Dificult to James Doziers for not keeping it in repair according to Law to the knowledge of two of this Jury

15 November 1757, page 170
To George Mason Gent. for keeping Occoquan Ferry 2000

To John Jenkins for keeping Goose Creek Ferry 1200

16 November 1757, page 171
Ordered that Mercy Chew & Elinor Jackson Ann Mason John Gladin George Thrift William Ashford and George Mull be sumoned to appear at the next Court to answer their retailing liquors without licence

16 November 1757, page 171
On the motion of George Mason Gent. it is ordered that Every person who passes the Ferry over Occoquan after nine a Clock and before day break pay their ferriage if demanded and the said George Mason agrees to keep the said Ferry the Ensuing year for two thousand pounds of Tobacco

16 November 1757, page 174
Benjamin Sebastian is appointed surveyor of the road in the room of Robert Adam

FCMB 1756-1763, PART 1

18 November 1757, page 181
Licence is granted to Ann Mason to keep an ordinary in Alexandria Peter Wagener security

18 November 1757, page 181
Licence is granted to John Gladin to keep an ordinary at Cameron John Hunter security

22 February 1758, page 194
On the motion of William Bayley Licence is granted him to keep an ordinary at his house in Colchester Benjamin Sebastian security

22 February 1758, page 194
Licence is granted to Charles Tyler to keep an ordinary in Colchester Peter Wagener security

22 February 1758, page 195
Sanford Remy is appointed surveyor of the road in the room of John Ratliff

22 February 1758, page 196
Ordered that the male labouring tithables belonging to Henry Fitzhugh & Ann Neale do assist in clearing the road whereof William Talbot is surveyor

21 March 1758, page 205
Licence is granted to Eleanor Jackson to keep an ordinary George Johnston Security

21 March 1758, page 205
Licence is granted to John Hereford to keep an ordinary at his Ferry John Posey security

22 March 1758, page 210
Licence is granted to Edward Willet to keep an ordinary Grafton Kirk and William Trammell security's

18 April 1758, page 224
John Hereford & David Piper having viewed the road leading to Hereford's ferry report on oath that the same may be conveniently turned from Mr. Ferguson's cleared ground opposite to M^{rs}. Bronaughs corn field ordered that George Mason Gent. have liberty to stop the old road when he has sufficiently cleared the new one

18 April 1758, page 224
Leonard Dozer is appointed surveyor of the road in the room of Sampson Turley

16 May 1758, page 225
Licence is granted to Robert Boggess to keep an ordinary at his house William Ellzey security

FCMB 1756-1763, PART 1

16 May 1758, page 225
Licence is granted to John Farrow to keep an ordinary William Ellzey security

16 May 1758, page 225
Licence is granted to James Ingo Dozer to keep an ordinary security to be given in the Clerk's office

16 May 1758, page 227
We present the surveyor of the road from the Town of Alexandria to Cameron for not keeping the same in repair

16 May 1758, page 227
We present the overseer of the road from the Bridge at Cameron to John Summers for want of a direction post to the knowledge of two of us

16 May 1758, page 227
We present the overseer of the road leading from Alexandria to four mile run for want of direction posts to the knowledge of two of us

16 May 1758, page 227
We present the overseer of the road from Hunting Creek to the Dogue run for want of a direction post to the knowledge of two of us

16 May 1758, page 227
We present the overseer of the road from Accotink to the back lick for want of a direction post to the knowledge of two of us

16 May 1758, page 227
We present the overseer of the road from Robert Boggess's to Alexandria for want of a direction post to the knowledge of two of us

16 May 1758, page 227
We present the Overseer of the road from Pohick Church to Hereford's Ferry for want of a direction post to the knowledge of two of us

20 June 1758, page 251
Ordered that William Payne Junr. Townshend Dade & Henry Gunnell or any two of them view the road through the plantation of James Wren & report on oath whether it is not prejudicial to the inhabitants by being stopped and whether as convenient a way cannot be found

20 June 1758, page 255
David Piper is appointed surveyor of the road in the room of William Peake

FCMB 1756-1763, PART 1

15 August 1758, page 273
Licence is granted to John Holless to keep an ordinary Hugh West Gent. security

18 August 1758, page 287
William Self an evidence for Andrew Riddle against John Gladin having attended fifteen days ordered that he pay him for the same and five times coming and going from Westmoreland one hundred & five miles one thousand eight hundred & forty two pounds of Tobacco & four shillings & six pence for ferriages over Occoquan

18 August 1758, pasge 289
William Douglass is appointed surveyor of the road in the room of Charles Thrift

20 September 1758, page 308
Licence is granted to Joseph Stevens to keep an ordinary William Ellzey, Henry Boggess and Francis Dade security's

20 September 1758, page 310
Ordered that Charles Broadwater & Henry Gunnell Gent. view the most Convenient way for a road from the old Court house to Turley's mill & report thereof on oath to the next Court

22 November 1758, page 314
To John Rhodes for setting up a post with Direction 25

23 November 1758, page 316
On the motion of John West Junr. Gent. ordered that Thomas Shaw, Presley Cox, Richard Sanford & Robert Sanford, or any two of them view the road from Cameron to the top of Cameron hill & report on oath to the next Court whether the same may be conveniently turned

23 November 1758, page 317
On the motion of Daniel M^c.Carty Gent. ordered that Charles Green Clerk, George William Fairfax Gent. Peter Wagener & William Payne or any two of them view the road from Accotinck to the lower end of M^r. Barns's plantation and report on oath to the next Court whether the same may be Conveniently turned

19 December 1758, page 320
Thomas Shaw and Presley Cox having viewed the road from Cameron pursuant to an Order of this Court returned a report on oath that the same may be Conveniently turned ordered that the surveyor of the said road clear the same accordingly

19 December 1758, page 320
Licence is granted Thomas Boylston to keep an ordinary at the Falls who with William Douglas acknowledged a bond for the same

FCMB 1756-1763, PART 1

19 December 1758, page 320
Ordered that Townshend Dade, Robert Adam, Lee Massey & Benjamin Sebastian, Gent. view the most convenient way for a road from Alexandria to the Falls warehouse and make report thereof on oath to the next Court

21 February 1759, page 329
Three Petitions signed by sundrie persons praying that twenty two thousand pounds of Tobacco levied by the Justices on the inhabitants of this county for building a warf at Alexandria may be repaid by the Justices were presented and ordered to be certified

17 April 1759, page 330
On the motion of William Bayley ordered that Daniel M^c.Carty, William Bronaugh, Benjamin Grayson Gent. and William Peake or any three of them, view the bridle way from Colchester to Occoquan Warehouse and report on oath to the next Court whether Peter Wagener has turned the said way to the prejudice of the Public

18 April 1759, page 334
William Bronaugh Gent. is appointed to view the most Convenient way for a road from the Ox road to Colchester in the room of William Peake

18 April 1759, page 334
Licence is granted to Elizabeth Fallon to keep an ordinary in Colchester Peter Wagener security

18 April 1759, page 338
Joseph Stevens being convicted for suffering unlawfull Gaming in his ordinary before John Carlyle & Charles Broadwater Gent. two of his Majesty's Justices of the peace is disabled from keeping ordinary by virtue of his licence

18 April 1759, page 338
Charles Broadwater and Henry Gunnell Gent. having viewed the road from Bull run to the old Court house pursuant to an order of this Court report that they cannot find a better than the old road to that place ordered that Henry Gunnell, William Gunnell, Joseph Moxley & Owen Williams or any three of them view the most Convenient way for a road from the old Court house to the Falls Warehouse and make report thereof to the next Court

18 April 1759, page 338
Licence is granted to Eleanor Jackson to keep an ordinary George Johnston G^t. security

18 April 1759, page 338
Licence is granted to Edward Willett to keep an ordinary Grafton Kirk & Samuel Conner security's

FCMB 1756-1763, PART 1

18 April 1759, page 338
Ordered that the old road from the Fall's Warehouse to Alexandria be cleared by William Frizell the Surveyor.

15 May 1759, page 340
The persons appointed to view a road from the old Court house to the Falls warehouse report as follows we find the nearest way to answer the other part of the road that comes from Bull run will go through the Court house corn field and through M^r. Scotts pasture and some part of his Corn field through or near John Urtons and into the old falls road at the head of a branch known by the name of Sasse Branch then along the old road a little below Thomas Ashburys to an old former road that goes through the Honourable Philip Ludwell Lees plantation and then along the old road to the Falls Henry Gunnell, Owen Williams, William Gunnell, Joseph Moxley, Charles Broadwater Gent. is appointed Surveyor & it is ordered that he with the male labouring tithables belonging to the roads whereof Thomas Wren and Charles Thrift are surveyors clear the said road according to the report except through the Corn fields

15 May 1759, page 340
The persons appointed to view a Bridle way from Colchester to Occoquan warehouse returned a report as follows in obedience to an Order of the County Court of Fai [torn] the subscribers have viewed the Bridle way from Colchester to Occoquan Wareh [torn] find that the way Peter Wagener has turned is no ways prejudicial to the pub[l]ick a[torn] the Complaint made to this Court about the same is frivolous given under our [torn] 2^d. May 1759. William Bronaugh, William Peake Benjn. Grayson where [torn] ordered that the new way be Established

16 May 1759, page 342
[torn] resent the surveyor of the road from the main road nigh the Falls Church leading to the [torn] d for want of a Sign or direction post to the knowledge of two of us

16 May 1759, page 342
[torn] esent the Surveyor of the road leading from William Ashfords to John Hollis for want [torn]n or direction post to the knowledge of two of us

16 May 1759, page 342
[torn] ent the surveyor of the road leading from the main road to rock creek for want [torn]ection post to the knowledge of two of us

16 May 1759, page 342
Ordered that Thomas Hampton, John Hampton, Edward Washington, and James Ingo Dozer or any three of them view the most Convenient way for a road from the Ox road to M^r. John Balendines Mill & make report thereof to the next Court

FCMB 1756-1763, PART 1

17 May 1759, page 347
Licence is granted Pierce Bayley to keep an ordinary David Piper Security

19 June 1759, page 361
Ordered that Charles Broadwater & Henry Gunnell Gent. Join the Commission.^{rs} of Loudoun in agreeing with workmen to repair the bridge over Difficult run

19 June 1759, page 347
The persons appointed to view a road from the ox road to Colchester report that there may be a better way than the old road made by leaving the County road at the plantation of John Fords and going through the said Plantation along a path formerly a mill road into the Ox road at the Wolf pit hill Alexander Henderson is appointed Surveyor and it is ordered that he clear the same with the labouring tithables belonging to the roads whereof William Simpson & Peter Wagener are Surveyors

19 June 1759, page 347
Ordered that William Simpson surveyor of the road clear a road from the Ox road to John Ballendines Ford

20 June 1759, page 368
Grand Jury against John Hurt On Presentment for not setting up a post with directions for Travellers The Defendant is fined fifteen shillings and ordered to pay Costs

20 June 1759, page 368
Grand Jury against Joseph Cockerill On Presentment for not setting up a post with direction for Travellers The Defendant is fined fifteen shillings & ordered to pay Costs

22 June 1759, page 387
Grand Jury against William Talbot On Presentment for not setting up a Post with directions for travellers The said William appearing & making no defence is fined fifteen shillings

18 July 1759, page 394
Henry Cavener a witness for Andrew Riddle against John Gladin having attended five days ordered that the said Andrew pay him for the same and three times coming and going from Westmoreland ninety seven miles and for his ferriages over Occoquan three shillings and nine hundred and ninety eight pounds of Tobacco

18 July 1759, page 394
Williams Self an evidence for Andrew Riddle against John Gladin having attended five days Ordered that the said Andrew pay him for the same and three times coming and going from Westmoreland one hundred & five miles and for his ferriages over Occoquan three shillings & one thousand & seventy pounds of Tobacco

FCMB 1756-1763, PART 1

18 July 1759, page 394
John Marmaduke an evidence for Andrew Riddle against John Gladin having attended four days Ordered that the said Andrew do pay him for the same and three times coming & going from Westmoreland ninety six miles and for his ferriages over Occoquan two shillings & nine hundred & sixty four pounds of Tobacco

19 July 1759, page 396
Sampson Dorrell Gent is appointed Surveyor of the roads in the room of Rich^d. Moxley deceased

19 July 1759, page 398
Licence is granted to Benjamin Sebastian to keep an Ordinary John West Gent security

21 August 1759, page 399
Licence is granted to Robert Boggess to keep an Ordinary Hugh West Security

21 August 1759, page 402
On the motion of James Donaldson it is ordered that Charles Broadwater Henry Gunnell and Edward Blackburn Gent view the most Convenient way for a road from his house into the main road that leads to Alexandria and report on oath to the next Court.

21 August 1759, page 403
Licence is granted to Michael Grater to keep an Ordinary at his House in Alexandria George Johnston Gent Security

23 August 1759, page 410
On the motion of Thomas Simons Ordered that the Sheriff Sumon a Jury to meet & value an acre of Land for a Mill belonging to the Heirs of Stephen Lewis dec^d. according to Law and return a report thereof to the next Court

18 September 1759, page 411
Licence is granted to James Ingo Dozer to keep an Ordinary at his House William Bayley security

20 September 1759, page 419
Licence is granted to John Hollis to keep an Ordinary Hugh West security

20 September 1759, page 419
Ordered that the Tithables belonging to George Washington Esqr. & his Tenants assist in clearing the road whereof John Peak is Surveyor

FCMB 1756-1763, PART 1

20 November 1759, page 420
Licence is granted to Seth Johnson to keep an Ordinary Hugh West security

20 November 1759, page 420
Daniel French is appointed Surveyor of the road in the room of Thomas Triplett and it is order.d that the Tithables belonging to George Washington Esqr. at the home plantation assist in clearing the same

21 November 1759, page 422
To George Mason Gent. [torn—Occoq-?] uan Ferry 2000

21 November 1759, page 423
To John Hurst for setting up Posts with directions for Travellers 50

To William Frizell for D^o 25

21 November 1759, page 423
To Joseph Cockerill for setting up a Post with directions for Travellers 10

21 November 1759, page 424
We present the Overseer of the road from Alexandria to the four mile run for not keeping a Sign Post to the knowledge of two of us.

21 November 1759, page 424
We present the Overseer of the ox road for not setting up a sign Post at the fork of the mountain road to the knowledge of two of us.

21 November 1759, page 424
We present the Overseer of the road over the other side of Holme's run for not setting up a sign post & not keeping the road in good repair to Dogue run to the knowledge of two of us.

21 November 1759, page 424
We present the overseer of the Falls road for not setting up a Sign Post which leads to the Sugar Land road to the Knowledge of two of us.

21 November 1759, page 424
We present the overseer of the road for not setting up a Sign Post to Clifton's ferry to the knowledge of two of us.

21 November 1759, page 424
We Present the Overseer of the road from Accotinck run to Dogue run for not keeping it in good repair and not setting up a Sign Post to the knowledge of two of us.

FCMB 1756-1763, PART 1

21 November 1759, page 424
We present the overseer of the road from the Falls church to the back lick for not setting up a Sign Post to the knowledge of two of us.

21 November 1759, page 425
John Summers jun^r. is appointed Surveyor of the road in the room of Joseph Cockerill

21 November 1759, page 425
Ordered that John West Gent late Sheriff pay to Charles Broadwater Gent. the Sum of fourteen pounds five shillings and four pence for this County's proportion for building Difficult bridge when the bridge is received.

21 November 1759, page 425
William Adams Gent. is appointed Surveyor of the road in the room of Thomas Shaw

21 November 1759, page 425
Ordered that Gilbert Simpson sen^r. be Surveyor of the road leading to Clifton's ferry and the Tithables belonging to M^r. Clifton, Harry Taylor, Edward Williams, & Samuel King are order.d to clear the same.

22 November 1759, page 427
The order made in August Court last for viewing a road on the motion of James Donaldson not being complied with it is again Ordered that Charles Broadwater Henry Gunnell & Edward Blackburn Gent. view the most convenient way for a road from his House into the main road which leads to Alexandria and report on oath to the next Court.

18 December 1759, page 428
Licence is granted to Nathan Hughes to Keep an Ordinary John Dalton security

19 December 1759, page 431
Ordered that the Surveyor of the mountain road be sumoned to appear at the next Court to answer the Presentment of the Grand Jury for not setting up posts with directions for Travellers from the Ox road and that John Ford be discharged

19 December 1759, page 431
James Moor is appointed Surveyor of the road in the room of John Ford

19 December 1759, page 432
Grand Jury against Edward Violett On a Presentment for not repairing the road The Sheriff having return.d that he left a Copy of the Sumons at the Deft^s. house and he failing to appear it is Consider.d that the said Defend^t. be fined fifteen shillings according to Law.

FCMB 1756-1763, PART 1

19 December 1759, page 432
Grand Jury against William Douglas On a Presentment for not repairing the road. The suit is Dismiss.d.

19 December 1759, page 433
Presley Cox is appointed Surveyor of the road from Hunting Creek to Dogue run.

19 February 1760, page 438
On the motion of James Donaldson it is Order.d that Charles Broadwater, Henry Gunnell, & Edward Blackburn Gent view the most convenient way for a road from Trammell's old Field to the Falls road & report to the next Court on oath

20 February 1760, page 440
Licence is granted to Jacob Shilling to keep an Ordinary at the Ferry in Alexandria John West jun^r Esqr. security

15 April 1760, page 455
On the Motion of Thomas Monroe he has liberty to erect a water grist mill on his own land at Quarter Branch

15 April 1760, page 455
Gerard Trammell is appointed Surveyor of the road from William Ashfords [torn] old Court house and it is ordered that the Tithables which formerly worked o[torn] road clear the same.

15 April 1760, page 455
Ordered that William Payne jun^r. & John Summers view the roa[torn] William Donaldson's to Summer's old School House and repor[torn] convenient way.

20 May 1760, page 458
Licence is granted to Charles Tyler to keep an Ordinary at Colchester Hector Ross & Peter Wagener securities

20 May 1760, page 458
[torn] that Bryan Fairfax, Henry Gunnell & Edward Blackburn Gent or [torn] them view the Bridge built by Charles Broadwater Gent over Difficult [torn] the next Court if the same is built according to Agreement.

21 May 1760, page 460
[torn]am Payne is appointed Surveyor of the road view.d & mark.d by William [torn] jun^r. & John Summers from above William Donaldson's to Summer's [torn]chool House William Adam Gent. is appointed to allot the Tithables to clear [torn]he same

FCMB 1756-1763, PART 1

22 May 1760, page 462
Licence is granted to John Plummer to keep an Ordinary in Alexandria George Johnston security

22 May 1760, page 465
On the motion of Thomas Colvill Gent. Order.d that Thomas Wren, James Donaldson James Wren & John Hurst or any three of them view the road whereof Garrard Trammell is Surveyor and report on oath to the next Court whether it will be inconvenient to the public to turn the road from the Plantation of the said Colvill

22 May 1760, page 466
[torn]ce is granted William Bayley to keep an Ordinary Joseph Stephens secu[torn]

22 May 1760, page 467
Licence is granted to Eleanor Jackson to keep an Ordinary John Dalton security.

22 May 1760, page 467
Licence is granted to Ann Mason to keep an Ordinary John Dalton security

22 May 1760, page 467
Licence is granted to David Young to keep an Ordinary John Dalton security

17 June 1760, page 472
Order.d that James Donaldson, Thomas Wren, Abraham Lay and James Wren or any three of them view the road from the old Court House to the fork of the road by the plantation of John Ratcliff and report on oath the most convenient way

17 June 1760, page 473
Grand Jury against Alexander Henderson On a Presentment for not setting up a Post with directions for Travellers The said Alexander appearing and making no defence it is Considered that he make a Fine to our Sovereign Lord the King in the sum of fifteen shillings & that he pay Costs.

17 June 1760, page 474
Joseph Davis is appointed Surveyor of the road in the room of William Payne

18 June 1760, page 477
Order.d that John West Gent. late Sheriff pay to Charles Broadwater Gent. £14.[torn] for building a Bridge over Difficult.

FAIRFAX COUNTY COURT MINUTE BOOK 1756-1763, PART 2

16 September 1760, page 514
Licence is granted to Robert Boggess to keep an Ordinary Hugh West Security

16 September 1760, page 514
Licence is granted Benjamin Sebastian to keep an Ordinary at Alexandria Hugh West security

21 October 1760, page 523
Licence is granted to John Hollis to keep an Ordinary for one year who with Hugh West his Security entered into Bond

21 October 1760, page 525
Seth Johnston is appointed Surveyor of the road in the room of Charles Broadwater Gen^t. Henry Gunnell Gen^t. is appointed to allot the Titheables to clear the said Road

21 October 1760, page 526
Humphrey Peake is appointed Surveyor of the road in the room of Sampson Darrell Gen^t.

22 October 1760, page 529
Franklin Perry is appointed Surveyor of the road in the room of William Douglas

18 November 1760, page 532
Nicholas Garrett is appointed Surveyor of the Road in the room of Thomas Wren

18 November 1760, page 532
The persons appointed to View a Road from the old Courthouse To the fork of the Road by John Ratcliffs having returned their Report it is ordered that Gerrard Trammell surveyor of the road do clear the same the way it is marked by the Viewers

18 November 1760, page 533
William Gunnell is appointed Surveyor of the Road in the Room of Joseph Moxley

18 November 1760, page 533
Ordered that James Donaldson William Shortridge and Charles Thrift view the most Convenient way to turn the road from Mr Turbevilles plantation and Report on oath to the next Court

18 November 1760, page 534
To George Mason Esqr. for keeping Occoquan Ferry. 2000

(December 1760), page 549
Licence is granted to Seth Johnston to keep an ordinary Hugh West became his security

FCMB 1756-1763, PART 2

(December 1760), page 550
On the motion of Simon Pearson ordered that William Payne Junr. Sanford Ramy & Joseph Moxley veiw the road from the back Lick to the to the church which goes through his plantation and report on oath to the next Court whether the same may be conveniently turned

(December 1760), page 550
Samuel Talbut is appointed surveyor of the road in the room of William Talbut

(December 1760), page 550
Daniel McCarty Gent appointed Surveyor of the Road in the room of Edward Violett

(December 1760), page 550
Licence is granted to Michael Gratter to keep an Ordinary at Alexandria who with George Johnston entered into Bond

(December 1760), page 551
Ordered that William Payne Junr. Gent. allot a Sufficient number of Titheables to Clear the road whereof Garrard Trammell is Surveyor

17 February 1761, page 552
James Moore is appointed Surveyor of the Road from Difficult branch to Capt. Lewis Ellzey's

17 February 1761, page 552
Thomas Withers Coffer appointed Surveyor the Road from Capt. Lewis Ellzeys to James Ingo Dozers

17 February 1761, page 552
Ordered that Franklin Perry surveyor turn the road through M^r. Turbe.vills old field the way it is marked by the Viewers

17 February 1761, page 553
The persons appointed to veiw aroad which goes through the Plantation of Simon Pearson returned a report and it is ordered that the said Simon may turn the said Road round the corner of Edmund Butlers fence the way it is marked to the church through his old field

17 February 1761, page 553
Ordered that the way by M^r. Wrens plantation and through Gerrard Trammell's Old field be established a bridle way for James Donaldson Gen^t. and that he have leave to clear the same

17 February 1761, page 553
Edward Blackburn Gen^t. is appointed Surveyor of the road from the old Courthouse to the ox road Charles Broadwater is appointed to allot the Titheables to work on the same

FCMB 1756-1763, PART 2

17 February 1761, page 553
Benjamin Talbott appointed Surveyor of the Road from the ox Road to Bull Run & Charles Broadwater Gen^t. is appointed to allot the Tytheables to work on the same

17 March 1761, page 563
On the Petition of Samuel Littlejohn ordered that Lewis Ellzey Thomas Ford and Thomas Withers Coffer view the road from the ox road to Hollis's old feild and report on oath whether it will be Inconvenient to the publick to turn the said road round his fence and whether it is necessary for the said road to be Continued

17 March 1761, page 566
Licence is granted to Nathan Hughes to keep an Ordinary who with George Johnston his security entered into bond

19 March 1761, page 577
Edward Payne is appointed Surveyor of the road from Popeshead to the County line

19 March 1761, page 577
William Kinchloe is appointed surveyor of the road from the County line to the fork of the road by Waters's Ordinary Lewis Ellzey is appointed to allot the Tytheables to work on the same,

21 April 1761, page 578
On the motion of John Berkley ordered that Lewis Ellzey Benjamin Talbott Sarshall Grasty and Thomas Ford or any three of them view the most Convenient way for a Road from the ox road to Bull run and report on oath to the next court and that the road lately opened be shut up till further order

19 May 1761, page 579
Licence is granted Michael Seelig to keep an ordinary John Carlyle Gen^t. Security

19 May 1761, page 580
Licence is granted to James Brown to keep an Ordinary at his house who with Samuel Canterbury his security entered into bond

19 May 1761, page 581
We present Elizabeth Waters of Truro parish for keeping a Tipling house to the Knowledge of two of us within six months

22 May 1761, page 594
William Adams Gen^t. is appointed to allot a sufficient Number of Tytheables to clear the Road whereof John Rhodes is Surveyor

FCMB 1756-1763, PART 2

16 June 1761, page 595
John Barry appointed Surveyor of the Road from the fork of the Road by Robert Boggess's to Dougue run and it is ordered that the Tytheables below dogue Run clear the same

16 June 1761, page 595
George Washington Esquire is appointed Surveyor of the road from Douge Run to the Gum Spring and the Road leading to Posey's ferry and that the Tytheables above Dougue Run & his Tytheables at Dougue Run Quarter clear the same

16 June 1761, page 597
Licence is granted to Charles Tyler to keep an ordinary at Colchester who with Hector Ross his security entered into Bond

18 June 1761, page 614
John Lister is appointed Surveyor of the Road in the room of William Adams Gen^t.

21 July 1761, page 617
Licence is granted to William Linton to keep an Ordinary at Colchester who with Hugh West his security entered into Bond

22 July 1761, page 625
Licence is granted to David Young to keep an Ordinary who with George Johnston his security entered into bond

15 September 1761, page 633
Henry Gunnell Gen^t. is appointed to allot the Tytheables to clear the Road whereof Seth Johnston is Surveyor

16 September 1761, page 635
Samuel Grigsby is appointed Surveyor of the Road in the room of Seth Johnston

16 September 1761, page 635
William Stone is appointed Surveyor of the Road in the Room of John Daniel

20 October 1761, page 662
George Simpson is appointed Surveyor of the Road from Dozers to the head of Giles's Run and Benjamin Grayson Gen^t. is appointed to allot the Tytheables to clear the same

20 October 1761, page 662
John Hampton is appointed Surveyor of the Road from Giles's Run to occoquan Warehouse and that Benjamin Grayson gen^t. allott the Tytheables to clear the same

FCMB 1756-1763, PART 2

20 October 1761, page 663
Licence is granted to George Simpson to keep an Ordinary who with Samuel Canterbury his Security entered into bond

17 November 1761, page 664
Owen Williams is appointed Surveyor of the road in the room of William Gunnell

17 November 1761, page 664
Licence is granted to Robert Boggess to keep an Ordinary who with Hugh West his security entered into Bond

17 November 1761, page 664
Samuel Tillett appointed Surveyor of the Road in the room of Benjamin Talbot

17 November 1761, page 664
On the motion of Benjamin Talbot ordered that Charles Broadwater William Kitchen Sampson Demoville and Joseph Bennett or any three of them view The most convenient way to turn the Road from his Plantation and report on oath to the next Court

17 November 1761, page 665
Thomas Beach is appointed Surveyor of the Road in the room of Samuel Talbot

18 November 1761, page 667
We present Paul Irmell of Truro parish for retailing spirituous Liquors contrary to Law by information of Benjamin Gilping within twelve months last past

18 November 1761, page 667
We present Elizabeth Waters of Truro parish for retailing spirituous Liquors contrary to Law within twelve months this last past to the Knowledg of two of us

18 November 1761, page 667
We present the Surveyor of the Road from Accotinck to the old school house for not keeping the said Road in repair to the Knowlege of two of us within one month

18 November 1761, page 667
We present the Surveyors of the Road from the Falls church to the ox Road for not keeping the said Road in repair to the knowledge of two of us within one month last past

18 November 1761, page 668
We present the Surveyor of the Road from Pohick Church to Colchester for not keeping the said Road in repair to the knowledge of two of us within one month last past

FCMB 1756-1763, PART 2

18 November 1761, page 668
We present the Surveyor of the back Road from dogue Run to great hunting Creek for not keeping the said Road in repair for one month last past to the knowledge of two of us

18 November 1761, page 668
We present Philip Trammell of Truro parish for retailing spirituous Liquors within twelve months last past to the knowledge of two of us

18 November 1761, page 669
Licence is granted to Benjamin Sebastian to keep an Ordinary who with Hugh West his security entered into Bond

13 December 1761, page 669
Ordered that William Kitchen Sampson D'Moville and Joseph Bennet be summoned to appear at the next Court to explain and amend their report about a Road

13 December 1761, page 670
To George Mason Gen^t. for keeping Occoquan Ferry 2000

16 February 1762, page 673
Licence is granted John Minor to keep an Ordinary who with Thomas Shaw entered into Bond

16 February 1762, page 674
On the Motion of Daniel McCarty Gen^t. ordered that Charles Green William Payne Junr. John Barry and Peter Wagener or any three of them view the Road from the ford over Accotinck to the upper end of Barn's quarter and report on oath if it will be convenient to turn the same

16 February 1762, page 674
Ordered that Lewis Ellzey, James Halley William Turner, & James D'Neale or any three of them view whether it be convenient to stop the Road from the Fork of the road that leads to Prince William into the Mountain Road and report on oath to the next Court

16 February 1762, page 674
Ordered that Charles Broadwater and Edward Blackburn Gen^t. allot the Titheables to clear the Road whereof William Stone is Surveyor

16 February 1762, page 675
John Sarter is appointed surveyor of the Road in the room of John Peter Sarter

16 February 1762, page 675
William Payne Junr. Esquire is appointed Surveyor of the Road in the room of Gerrard Trammell

FCMB 1756-1763, PART 2

16 February 1762, page 675
Licence is granted to John Seal to keep an Ordinary who with Robert Douglas his security entered into Bond

17 February 1762, page 684
Licence is granted to William Garner to keep an Ordinary Security to be given in the Clerks Office

16 March 1762, page 692
Joseph Cash is appointed Surveyor of the road in the room of William Williams

16 March 1762, page 692
On the motion of John Urton ordered that Henry Gunnell Charles Thrift Thomas Wren and William Shortridge or any three of them view the road by the said Urton's Plantation and report on oath if it may be conveniently turned

18 March 1762, page 695
Ordered that Edward Blackburn, John Monroe, Benoni Halley and Samuel Talbot or any three of them view and mark the most convenient way for a road from the old Courthouse road near John Monroes, to intersect with the road to Alexandria near Balty Stearns and report to the next Court

18 March 1762, page 695
John Barry is appointed surveyor of the road from Doeg run to Accotink

18 March 1762, page 695
Robert Boggess is appointed Surveyor of the roads from Pohick to Accotink

20 April 1762, page 704
On the motion of John Seale Ordered that William Douglass Thomas Wren, James Wren and William Shortridge or any three of them view the road from below the old Court house to Difficult and report on oath if it may be conveniently turned

20 April 1762, page 704
On the Petition of John Carlyle Gent ordered that the Sheriff summon a Jury to meet on his land at four mile run and view if it will be to the prejudice of any person if the Petitioner build a Mill thereon and report to the next Court

20 April 1762, page 704
Ordered that Alexander Henderson, Daniel McCarty, John Ford & William Linton or any three of them view the road from Pohick to Colchester and report on oath the most convenient way to shun the miry places in the said Road

FCMB 1756-1763, PART 2

18 May 1762, page 705
Licence is granted to Patrick Byrn to keep an Ordinary who with Hugh West his security entered into Bond

19 May 1762, page 709
We present the Surveyor of the Road from Cameron to Mr William Adams's Plantation for not having the Road in proper repair to the knowledge of two of us within one Month last past

19 May 1762, page 709
We present the Surveyor of the road from Cameron to Mr Daniel French's Plantation for not having the Road in proper repair to the knowledge of two of us within one Month last past

19 May 1762, page 710
We present the Surveyor of the Road from the Falls Church to Rock Creek for not having a Sign post erected at the Forks of the Road to the knowledge of two of us within one Month last past

19 May 1762, page 710
We present the Surveyor of the Road from Alexandria to the four Mile run for not having the road in proper repair to the knowledge of two of us within one Month last past

19 May 1762, page 710
We present Michael Grater of the County of Fairfax for retailing spiritous Liquor without Licence to the knowledge of two of us

19 May 1762, page 710
We present John Plumer of the County of Fairfax for retailing spiritous Liquor without Licence by information of Spencer Neale

19 May 1762, page 711
We present the Surveyor of the Streets & publick Landings for not keeping them in repair within one Month last past by the information of George Johnston

19 May 1762, page 712
William Summers is appointed Surveyor of the road in the room of John Rhodes

19 May 1762, page 713
Ordered that a road be cleared from the old Court house road near John Monroes to intersect with the road to Alexandria near Balty Stearns the way it is viewed and marked John Monroe is appointed Surveyor it is ordered that Charles Broadwater and Edward Blackburn Gen^t. allot the Titheables to clear the same

FCMB 1756-1763, PART 2

19 May 1762, page 713
John Seal is appointed Surveyor of the road in the room of Edward Blackburn G^t.

19 May 1762, page 713
On the Petition of John Carlyle Gent for liberty to erect a Water Grist Mill the Jurors return a report which is ordered to be recorded & leave is given to the said John Carlyle to erect the same according to the Prayer of his Petition

19 May 1762, page 713
Licence is granted to Michael Grater to keep an Ordinary who with George Johnston his Security entered into bond

15 June 1762, page 716
The persons appointed to view the road from the fork of the road that leads to Prince William into the mountain road returned a report whereupon it is ordered that the new road may be stopped

15 June 1762, page 717
John Ashford is appointed Surveyor of the road in the room of Presley Cox

18 June 1762, page 742
Grand Jury against Samuel Talbott Surveyor of a Road On Presentment The Defendant having been duly summoned and failing to appear it is considered that the Defendant be fined fifteen shillings and that he pay costs

18 June 1762, page 743
Grand Jury against John Summers Surveyor of a Road On Presentment The Sheriff having returned that he left a Copy of the Summons at the Defendants usual place of abode and the said Defendant failing to appear it is considered that the Defendant be fined fifteen shillings and that he pay costs Mem^o fine is remitted at July Court

18 June 1762, page 743
Grand Jury against Peter Wagener Surveyor of a Road On Presentment On hearing it is ordered that the Defendant be fined fifteen shillings and that he pay Costs

18 June 1762, page 743
Grand Jury against Presley Cox Surveyor of a Road On Presentment On hearing it is ordered that the Defendant be fined fifteen shillings and that he pay Costs

18 June 1762, page 746
Grand Jury against Gerrard Boling Surveyor of a Road On Presentment The Defendant having been duly summoned & failing to appear it is considered that the said Defendant be fined fifteen shillings and that he pay costs &c

FCMB 1756-1763, PART 2

18 June 1762, page 747
John Minor is appointed Surveyor of the Road in the room of John Lister

18 June 1762, page 747
Leonard Pratt is appointed Surveyor of the Road in the room of John Hurst

18 June 1762, page 747
Edward Dulin is appointed Surveyor of the Road in the room of William Payne Gen^t.

20 July 1762, page 751
John Summers being fined last Court for not keeping his road in repair whereof he is Surveyor it appearing to the Court that he had not received a Copy of the order appointing him Surveyor the fine is remitted

17 August 1762, page 769
Licence is granted Daniel Sexton to keep an Ordinary at Colchester who with Peter Wagener and William Bayly his security entered into Bond

22 September 1762, page 788
Licence is granted to George Parrott to keep an Ordinary who with Gerrard Bolling his Security entered into Bond

19 October 1762, page 790
Gerrard Trammill Junr. appointed Surveyor of the lower part of the Road from pimmet's run to the upper part of Nicholas Garrett's Plantation. Edward Blackburn Gen^t. is appointed to allot the Titheables to clear the same

16 November 1762, page 792
Joseph Moxley appointed Surveyor of the road in the room of Owen Williams

17 November 1762, page 792
Alexander Henderson Merchant is appointed Surveyor of the road in the room of Peter Wagener

17 November 1762, page 793
To George Mason Gen^t. for keeping Occoquan Ferry 2000

17 November 1762, page 794
We present Elizabeth Waters for keeping a Tipling house and retailing spirituous Liquors without Licence within six months last past to the knowledge of two of us

FCMB 1756-1763, PART 2

17 November 1762, page 794
We present Thomas Grafford for retailing Spirituous Liquors not having Licence within six months last past to the knowledge of two of us

17 November 1762, page 795
We present the Surveyor of the road from Cameron to William Adams Gen^t. for not being in proper repair to the knowledge of two of us

17 November 1762, page 795
We present the Surveyor of the road from Alexandria to Cameron for the road being in bad order to the knowledge of two of us

17 November 1762, page 795
We present the Surveyor the road from the Falls Church to Maj^r. Broadwater's Spring branch for the road being in bad order to the knowledge of two of us

17 November 1762, page 795
We present the Surveyor of the road from Maj^r. Broadwater's Spring Branch to the Ox road for the road being in bad Order to the knowledge of two of us

17 November 1762, page 795
We present the Surveyor of the road from Cameron to M^r. French's Quarter for the road being in bad order to the knowledge of two of us

17 November 1762, page 795
We present the Surveyor of the road from Cameron to the old School house for the road being in bad order to the knowledge of two of us

17 November 1762, page 795
We present the Surveyor of the road from the old School House to Fitzhugh Quarter for the road being in bad order to the knowledge of two of us

17 November 1762, page 795
We present the surveyor of the road from William Adams Gent. to Gerrard Trammells for the road being in bad order to the knowledge of two of us

17 November 1762, page 795
We present the surveyor of the road from Difficult to the old Court house for the road being in bad order to the knowledge of two of us

17 November 1762, page 795
We present John Plummer for retailing Spirituous Liquors without Licence within six months last past to the knowledge of two of us

FCMB 1756-1763, PART 2

21 December 1762, page 798
Licence is granted to Samuel Johnston Johnston Ferry keeper to keep an Ordinary who with Benjamin Sebastian his Security entered into Bond

22 December 1762, page 799
John Seale is appointed Surveyor of the road in the room of Samuel Grigsby from the old Court house to the road leading from the Chesnut tree to the Falls Warehouse

15 February 1763, page 801
Ordered that Edward Washington, John Ford, Abraham Hargiss and William Simpson or any three of them view the most convenient way for a road from the ox road to Colchester and report to the next Court on oath

15 February 1763, page 801
On the motion of Benjamin Talbott it is ordered that Charles Broadwater Edward Blackburn, Joseph Bennett and John Monroe or any three of them view the road that goes thro his Plantation from the old Court house to Bull run and report on oath to the next Court the the most convenient way to turn the same

15 February 1763, page 801
Licence is granted to Samuel Bayly to keep an Ordinary at Colchester who with George Johnston his Security entered into Bond

15 February 1763, page 801
Abednego Adams is appointed Surveyor of the road in the room of Humphry Peak

15 February 1763, page 802
Licence is granted to William Linton to keep an Ordinary at Colchester who with Hugh West his Security entered into Bond

15 February 1763, page 802
Licence is granted to Benjamin Sebastian to keep an Ordinary at Alexandria who with Hugh West his Security entered into Bond

16 February 1763, page 805
On the motion of John Seale it is ordered that William Douglass, Thomas Wren, James Wren, and Henry Gunnell, or any three of them view the road by the Petitioners Plantation and report on oath to the next Court whether it is convenient to turn the same

16 February 1763, page 808
Licence is granted to David Young to keep an Ordinary Bond to be given in the Clerk's Office

FCMB 1756-1763, PART 2

16 February 1763, page 808
Licence is granted to Robert Lindsey to keep an Ordinary Bond to be given in the Clerk's Office

16 March 1763, page 828
Dorson Vallandingham is appointed Surveyor of the road in the room of John Seale from the old Court house to the Falls road

16 March 1763, page 829
Licence is granted to Mathew Sparrow to keep an Ordinary who with John Ward his Security entered into Bond

16 March 1763, page 829
Ordered that the Tytheabbles belonging to the road whereof George Simpson and John Hampton are Surveyors clear the road from the ox Road to Colchester according to the report of the Viewers Peter Wagener Gen^t. is appointed Surveyor

16 March 1763, page 830
The persons appointed to view the road by John Seals returned their report whereupon he has liberty to stop the old road when he has sufficiently cleared the new way and made a Bridge

19 April 1763, page 852
Licence is granted to John Linton to keep an Ordinary a Colchester who with George Johnston his Security entered into Bond

19 April 1763, page 852
Ordered that Benjamin Sebastian, Robert Alexander, William Adams & James Buckman or any three of them view the most convenient way for a road from the rock hole on four mile Creek to Mills now erecting by John Carlyle & Robert Adam Gen^t. and report on oath to the next Court

19 April 1763, page 852
John West Junr. is appointed Surveyor of the road in the room of Benjamin Sebastian

17 May 1763, page 855
Licence is granted to Thomas Sanders to keep an Ordinary at the ferry in Alexandria who with Hugh West his Security entered into Bond

18 May 1763, page 856
The persons appointed to view the road from the rock hole on four mile Creek to Mills now erecting by Messrs. Carlyle and Adam returned their report on oath and it is ordered that James Buckman clear the said Road according to the said report

FCMB 1756-1763, PART 2

18 May 1763, page 856
James Buckman is appointed Surveyor of the road from four mile Run to rock Creek

18 May 1763, page 856
Daniel McCarty, Alexander Henderson, Peter Wagener and John Ford are appointed to view the most convenient way to turn the road to shun the Hill below Boggesses and report on oath to the next Court

18 May 1763, page 858
We present the Surveyer of the road from Cameron to William Adam's path by the Knowledge of two of us within one month last past

18 May 1763, page 859
We present the Surveyor of the road from Dogue run to little Hunting Creek by the knowledge of two of us for not keeping the road in repair within one month last past

18 May 1763, page 862
John Carlyle Gen^t. is appointed Surveyor of the road from Alexandria to Cameron in the room of John West Gen^t.

19 May 1763, page 869
On the Petition of Robert Adam Gen^t. setting forth that he owns the Land on one side of four mile Run and intends to build a Water Mill on the said Run but that John Carlyle Gen^t. owns the Land opposite and praying to have one acre of the said Land viewed and valued according to Law, It is ordered that the Sheriff summon a Jury of twelve free holders of his Bailiwic to meet upon the Land Petitioned for who being met and duly sworn are diligently to view and examine the said Land and the Land adjacent thereto on both sides the said Run which may be affected or laid under Water by building the said Mill together with the Timber and other conveniences thereon, and report the same with the true of value of the acre petitioned for and of the damages to the said John Carlyle or to any other person or persons under their hands & Seals and that the Sheriff return the said report to the next Court and it is ordered that the Surveyor of this County survey and lay off the said Acre of Land

21 June 1763, page 884
The Petition of John West Junr. for leave to build a Mill over Poultneys run is continued untill the next Court for the Court to be advised whether it will be to the prejudice of any person

21 June 1763, page 884
Ordered that William Ramsay, John West Junr. Robert Adam & William Adams Gen^t. or any two of them meet at four mile run and allott the Labouring Tithables to the several Surveyors near that place

FCMB 1756-1763, PART 2

21 June 1763, page 885
Dorson Valandingham is appointed Surveyor of the road in the room of John Seale

21 June 1763, page 885
Charles Broadwater and Edward Blackburn Gen^ت. are appointed to allot the hands to clear the same

21 July 1763, page 903
Thomas Saunders is appointed Ferry keeper at Hunting Creek who together with John Carlyle his security entered into Bond

FAIRFAX COUNTY MINUTE BOOK 1763-1765

16 August 1763, page 43
Licence is grt Daniel Bush to keep an Ordinary in Alexandria Hugh West Security

20 September 1763, page 53
On the Pet^n. of John Trammell ordered that the Sheriff summon a Jury to value an acre of Land on difficult run opposite to the Petitioner Land on which he intends to erect a Water Grist Mill and that the Surveyor survey the same

18 October 1763, page 55
On the Pet^n. of Thos. Pollard Ord^d. that Thos. Ford Leonard Dozer Ed. Payne & Jno. Turley or any 3 view the road that goes thro' the Pet^r. plantation & report on oath if the same may be conveniently turned

18 October 1763, page 55
Charles Alexander app^d. Surv^r. of the road in the room of Gerard Bo[torn]

18 October 1763, page 55
Thos. Shaw app^d. Surv^r. of the road in the room of Jno. Minor

18 October 1763, page 55
Licence is granted John Conner to keep an Ord^y. at Cameron H. West Secy.

20 December 1763, page 56
To Geo. Mason Gent for keep^g Occ^o. Ferry 2000

20 December 1763, page 57
To Geo. Mason G^t. for keep^g Occ^o. Ferry one month 161

21 December 1763, page 58
To Robt. Speake for Timber for a Bridge 60

21 December 1763, page 58
George Hereford Appointed Surveyor of the Road in Room of David Piper

21 December 1763, page 58
Ordered that John Hurst Joseph Moxley William Gunnell & James Wren to Veiw the Most Convenient way for the road from M^r. Sebastians Quarter to Maj^r. Chas. Broadwaters Quarter & make a Report

21 December 1763, page 60
Ord. that Jno. West j^r. Thomas Shaw Wm Ramsay Benjn. Sebastian or any three view if there cannot be a nearer & better way from the Falls Church to Alexandria & report on oath

FCMB 1763-1765

21 December 1763, page 61
Ord^d. that Hector Ross, Alex^r. Henderson & Peter Wagener view the most conv^t. way to turn the road by Carpenter's Spring

18 January 1764, page 70
John Hurst app^d. Surv^r. of the road in the room of Joseph Moxley

19 January 1764, page 71
On the motion of Benjamin Sebastian Ord that Henry Gunnell, W^m. Douglas, Edwd. Blackburn & Thos. Wren or any three of them view the most Convenient way if necessary to turn the road from M^r. Sebastian's Quarter to Maj^r. Broadwaters Qr & report on oath

21 February 1764, page 74
Lewis Ellzey Ed. Payne Moses Simpson & James Halley or any 3 to view the most conv^t. way for a road from W^m. Garner's to Ravensworth road & report

21 February 1764, page 74
Ord^d. that W^m. Payne, W^m. Payne jun^r. Benjn. Talbott & James Halley or any three view the most conv^t. way for a road out of the old Court house road near Benoni Halley's to the Ravensworth road near Jno. Hollis's

21 February 1764, page 74
James D'neal app^d. Surv^r. of the road in the room of W^m. Kincheloe

21 February 1764, page 74
The persons app^d. to view the most conv^t. way to turn the road by Carpenter's Spring report on oath that it is the best way as the fence is lately made thro' M^r. Geo. Johnston's plantation whereupon it is ordered that the road may be turned accordingly

21 February 1764, page 74
On the motion of James Donaldson Ord^d. that Ch^s. Broadwater Henry Gunnell, W^m. Shortridge & W^m. Darnes or any three view the most conv^t. way to turn the road from the old Court h^s. to the Falls Wareh^s. & rep^t. on oath

22 February 1764, page 75
John Canaday app^d. Surv^r. of the road in the room of Ch^s. Alexander

22 February 1764, page 75
The persons app^d. to view the most convenient way for turn^g the road to shun the hill below Boggess's report on oath that it ought to be turn.d to the left hand going from Colchester to Alexandria Ord^d. that Robt. Boggess Surveyor of the road clear the same accordingly

FCMB 1763-1765

22 February 1764, page 77
Licence is granted R^d. Arell to keep an Ord^y. George Johnston Sec^y

22 February 1764, page 78
On the motion of Ch^s. Broadwater Ord^d. that W^m. Ramsay, Robt. Adam, Jno. Dalton & Jno. Muir or any three view the most conv^t. way if necessary to turn the road from M^r. Sebastian's Quarter to M^r. Broadwater's Quarter at the lower end where the old road came in & report the former viewers having returned an imperfect report

20 March 1764, page 83
Thomas Hardin is appointed Surveyor of the road in the room of Joseph Cash and it is ordered that Bryan Fairfax & William Payne Junr. allott the Tyth^s. to clear the said road between the said Hardin & Daniel McCarty Surv^{rs}.

20 March 1764, page 83
The persons appointed to view the road from William Gardners to Ravensworth road returned a report Thomas Windsor is appointed Surveyor and ordered to clear the same road from the Ox Road to the pond William Keen appointed Surv^r. of the road from the Pond to Steptoes Quarter and that Thos. Ford direct the Surveyors in clearing the same Bryan Fairfax & Wm. Payne appointed to allott the Tyth^s. to clear the same

20 March 1764, page 83
The persons appointed to view the most convenient way to turn the road by Joseph Pollards plantation report that it may be conveniently turned along the Line between the Petitioners plantation & M^r. Beckwiths thereupon it is ordered that Edwd. Payne mark the way & that Thos. Pollard may open the said road according to the said Report

20 March 1764, page 83
On the motion of John Posey it is ordered that Daniel McCarty, Thomas Triplett, Abednego Adams, & W^m. Triplett; or any three view the most convenient way to turn the road from above M^{rs}. Manley's plantation to Poseys ferry and report on oath

20 March 1764, page 84
The persons appointed to view the Most convenient way to turn the road from M^r. Broadwaters Quarter to M^r. Sebastian's Quarter returned a report Ordered that the Surv^r. of the road clear the same accordingly after the last day of November next - Gerard Tramell having no notice of the application for the order nor the time of the Execution of it tho he was informed there was such an order and being now present and opposing the s^d. Order prays leave to appeal and leave is given to Tramell to enter into Bond at the next Court

FCMB 1763-1765

21 March 1764, page 85
The persons appointed to view the road on the motion of James Donaldson returned a report and it is ordered that James Donaldson may turn the road from the top of the hill by John Mason's fence along the Line of the Land of the Rev.^{d} M^r. Scott to the Falls Road.

21 March 1764, page 85
Daniel McCarty Gen^t. is appointed Surv^r. of a road from Pohick to Accotinck in the room of Robert Boggess. Ordered that George Mason Junr. W^m. Simpson, Vincent Boggess & Richard Lightfoot assist to clear the said Road

17 April 1764, page 97
Licence is granted to Micajah Powell to keep an Ordinary John Minor Sec^y

17 April 1764, page 98
The person appointed to view the road from above M^{rs}. Manleys Plantation to Poseys ferry returned a report whereupon it ord^d. that the s^d. Posey clear the same accordingly

15 May 1764, page 100
Licence is granted to James Bell to keep an Ordinary in Alexandria who with William Ramsay his Security entered into Bond

15 May 1764, page 100
George Simpson is appointed Survey^r. of the road from William Gardners to the place where the road to Colchester leaves the Ox road John Ford is appointed Sur^r of the road from that place to Colchester Ware house Lewis Ellzey and H Ross Gen^t. are appointed to allott the Tithables

17 May 1764, page 104
William Bayly is appointed Surv^r. of the road in the room of Alexander Henderson

19 June 1764, page 124
Daniel French is appointed Surv^r. of the road in the room of John Ashford

23 August 1764, page 147
On the motion of John Ballendine Ord^d. that C. Broadwater Ja^s. Jenkins Franklin Perry & W^m. Shortridge view the most conv^t. way for a road from his mill at the Falls Landing to or near the mouth of Difficult run & report

19 September 1764, page 160
Licence is granted to Benjn. Grayson Gen^t to keep an Ord^y. in Alexandria he having entered into Bond at the last Court

FCMB 1763-1765

16 October 1764, page 168
Ordered that Lund Washington with his hands at the Quarter sufficiently clear the new Road that he has opened to the satisfaction of neighbors or immediately open the old Road

20 November 1764, page 169
Licence is granted to Pearse Bayly to keep an Ordinary at Colchester W^m. Bayly & Benjn. Sebastian Sec^y

21 November 1764, page 171
Joseph Cash appointed Surveyor of the Road from the fork of the road below W^m. Johnston to Doeg Run in the room of Daniel McCarty Gen^t.

20 March 1765, page 199
Licence is granted to Elijah Williams to keep an Ordinary at the place called Browns Ordinary W^m. Ellzey Security

15 April 1765, page 202
Ordered that George Simpson & John Ford be Surv^s of the Road from Browns Ordinary to Colchester & that Daniel McCarty Hector Ross & Alexander Henderson Gen^{ts}. or any two of them allot the Tith^s. to clear the same

15 April 1765, page 203
Licence is granted to W^m. Gardner to keep an Ordinary at his House H. West Security

16 April 1765, page 208
Licence is granted to Richard Arrell to keep an Ordinary in Alexandria. George Johnston Gen^t. Security

17 April 1765, page 209
Thomas Darnes is apointed Surv^r. of the road in the room of Jam^s Buckman

FAIRFAX COUNTY MINUTE BOOK 1765-1766

20 May 1765, page 2
William Gunnell is appointed Surveyor of the road in the room of John Hurst

20 May 1765, page 2
John Frizzell is appointed Surveyor of the road in the room of Leonard Pratt

20 May 1765, page 3
John Dalton is appointed Surveyor of the road in the room of John Carlyle Gen^t. John Carlyle & & Robert Adam are appointed to allott the hands to clear the same

19 June 1765, page 11
Ordered that Lewis Ellzey Ed^d. Payne W^m. Turner & Ja^s. D^{e}Neale or any three view the most convenient way for a road from the ford of Piney Branch into the Ox Road above the Pohic Road & report to the next Court

21 June 1765, page 20
W^m. Turner is appointed Surv^r. of the road in the room of Edwd. Payne

19 August 1765, page 31
Peter Mauzy is app^d. Surv^r. of the road in the room of Ja^s. D'Neal

19 August 1765, page 32
A Report of the Jury for valuing an acre of Land for John Trammell opposite to the place where he intends to erect a Water grist mill being returned and the s^d. Trammell tendering in Court 25/ the Valuation & Damages not knowing to whom the same ought to be Paid Ord. that the rep^t. be rec^d. & the acre vested in the s^d. Tramell according to Law G. Johnston on behalf of W^m. Douglas & Eliz. his wife who is guardian to Eliz. Lewis an Infant objected to the s^d. report being recorded, for that the s^d. Infant hath Lands lying on Difficult Run which probably may be affected by this Inquisition & that the s^d. Infant or Guardian had not notice of the Execution of the same

16 September 1765, page 58
William Lister appointed Surveyor of the road in the room of Nich^s. Garrett

16 September 1765, page 58
Hugh West is appointed Surveyor of the road in the room of Dorson Vallandigham

21 October 1765, page 62
William Richards is appointed Surveyor of the road in the room of Joseph Davis

17 March 1766, page 77
Edwd. Blackburn, Ja^s. Jenkins Franklin Perry & John Shortridge are app^d. to view a Road that leads from Alexandria to the Falls Ware house & report to the next Court

FCMB 1765-1766

17 March 1766, page 78
Ordered that Ge^o. W^m. Fairfax Gen^t. apply to the Court of Loudoun to Join in making an order to repair Difficult Bridge

17 March 1766, page 78
Bennett Hill is app^d. Sur^r. of the road in the room of Danl. McCarty G^t.

17 March 1766, page 78
A Report for a Road from the Falls Mill to the mouth of difficult Run is returned & ordered to be cleared John Seale app^d. Sur^r. of the same Chas. Broadwater & Edwd. Blackburn Gen^t. are app^d. to allott the Tith^s. to clear the same

18 March 1766, page 78
Peter Wagener Junr. is appointed Surveyor of the road in the room of William Bayly

18 March 1766, page 78
John Hampton is app^d. Sur^r. of the road in the room of John Ford Hector Ross Alex. Henderson & Peter Wagener Gen^t. are app^d. to allott the District & the hands between him & [torn] Simpson

18 March 1766, page 79
George William Fairfax Daniel McCarty Hector Ross & Alexander Henderson Gen^t. or any three view the most convenient way for a Road from the Ford from Piney Branch into the Ox Road above the Pohic Road and report to the next Court the former viewers having returned an imperfect Report

19 May 1766, page 80
John Hurst is app^d. Surv^r. of the Road in the room of W^m. Gunnell

19 May 1766, page 80
Thomas Triplett is app^d. Surv^r. of the road in the room of Abednego Adams

19 May 1766, page 80
Isaac Rose app^d. Surv^r. of the road from the forks of the road to four mile run leading to the Falls

19 May 1766, page 81
Daniel Talbot app^d. Surv^r. of the road in the room of D^l. French

19 May 1766, page 81
James Steward app^d. Surv^r. of the road in the room of Hugh West-Ch^s. Broadwater G^t. app^d. to allot the Tith^s. to clear the same

FCMB 1765-1766

19 May 1766, page 81
Ordered that Thos. Palmer be Surv^r. of the road in the room of James Buckman and that the Tithables belonging to Robt. Alexander, W^m. Russell, Philip Alexander W^m. Greenwood, John Butcher, Philip Ellis, Owen Brayley James Vineyard, Benjn. Molds, John Rhodes, Jno. Williams Samuel Homes, John Talbott James Gordon & John Flint clear the same

19 May 1766, page 81
Charles Broadwater G^t. app^d. Surv^r. of the road in the room of John Seal

19 May 1766, page 81
John Alexander j^r. app^d. Surv^r. of the road in the room of Wm Frizell

FAIRFAX COUNTY COURT MINUTE BOOK 1768-1770

15 August 1768, page 1
James Harper is appointed Surveyor of the road in the room of William Lister

15 August 1768, page 1
William Triplett is appointed Surveyor of the road in the room of Joseph Cash

20 September 1768, page 41
Henry Moore is appointed Surveyor of the room of William Linton

21 September 1768, page 53
Licence is granted Richard Arell to keep an Ordinary in Alexandria who together with Charles Broadwater Gen^t. acknowledged a Bond for the same

17 October 1768, page 55
Licence is granted Peirce Bayly to keep an Ordinary at Colchester who together with Robert Harrison his Security acknowledged a Bond

21 November 1768, page 56
William Moore is appointed Surveyor of the Road in the room of William Courts

21 November 1768, page 57
Licence is granted David Young to keep an Ordinary in Alexandria who with John Dalton His Security acknowledged a Bond for the same

22 November 1768, page 62
Ordered that a Road be opened from the Falls Church to the road leading from Alexandria to the Falls Warehouse, William Darnes is appointed Surveyor of the same - John Hurst & Joseph Moxley are appointed to pilot the way - William Adams & Henry Gunnell Gen^t. to allott such & so many hands as they shall think necessary to open the road and direct what hands shall work upon it for the future

22 November 1768, page 62
William Clifton is appointed Surveyor of the road in the room of Thomas Triplett

22 November 1768, page 63
Solomon Nicholas is appointed Surveyor of the Road in the room of James Keen

23 November 1768, page 64
Edward Ford is appointed Surveyor of the Road in the room of Peter Mauzy

23 November 1768, page 65
Licence is granted Michael Gretter to keep an Ordinary in Alexandria

FCMB 1768-1770

23 November 1768, page 66
Licence is granted Richard Leake to keep an Ordinary in Alexandria

23 November 1768, page 69
We present the Overseer of the road from Cameron to M^r. William Adam's for not keeping the road in repair agreeable to the act of Assembly in that case made and provided in this present year 1768 within two months last past to the knowledge of two of us

23 November 1768, page 71
We present Bennett Hill of Truro Parish and County of Fairfax for retailing spiritous Liquors within six months last past contrary to act of Assembly in that case made and provided to the knowledge of two of us

23 November 1768, page 71
We present the Overseer of the road from M^r. William Adam's to the old Court house for not keeping the road in repair agreeable to act of Assembly in this present year 1768 within two months last past by Information of M^r. George West

23 November 1768, page 71
We present the Overseer of the road from Cameron to Alexandria for not keeping the road in repair agreeable to act of Assembly in this present year 1768 within two months last past by Information of Peter Wagener Senr.

24 November 1768, page 74
Ordered that William Adams Gen^t. devide the hands that work upon the Road from Cameron to M^r. Adam's' path and also the hands that work from Cameron to the old Court house road and give the Surveyors of the said road a List of the hands that shall work upon the same

20 February 1769, page 76
It appearing to the Court that William Clifton is unable to serve as Surveyor of the road, Humphrey Peak is appointed in the room of Thomas Triplett

20 February 1769, page 76
On the motion of Edward Shelvin it is ordered that Francis Whiting William Talbott, & William Williams view the road that leads from Accotinck run to the back Lick and report on oath whether it is convenient for the said Road to be turned

21 February 1769, page 80
George Harley is appointed Surveyor of the road in the room of Solomon Nicholas

FCMB 1768-1770

21 February 1769, page 86
On the motion of Robert Boggess Jun'. It is ordered that Daniel M°Carty Alexander Henderson and Peter Wagener Jun'. view the most convenient way for a road to be opened from Colchester to the said Boggess' mill and report to the next Court

21 February 1769, page 86
On the motion of William Ramsay it is ordered that James Jenkins Gerrard Trammell, & Franklin Perry view the most convenient way for a road from young Gerrard Trammell's to Tolston mill upon Difficult mark out the same and report to the next Court

20 March 1769, page 96
George Grasty is appointed Surveyor of the road from Hunters old Store house to the little Falls in the room of John Ballendine

22 March 1769, page 100
Henry Burnum is appointed Surveyor of the Road in the room of James Stuart

22 March 1769, page 110
Licence is granted James Rhodes to keep an Ordinary at Hunting Creek Ferry who with John Carlyle Gen'. acknowledged a Bond for the same

23 March 1769, page 115
On the motion of William Ramsay Gen'. It is ordered that James Jenkins Gerrard Trammell & Franklin Perry view the most convenient way for a road from John Hursts plantation above the Falls Church to Tolston Mill upon Difficult mark out the same and report on oath to the next Court

17 April 1769, page 123
Licence is granted William Gardner to keep an Ordinary who with Augustus Darrell acknowledged a Bond for the same

18 April 1769, page 126
On the Petition of John Hooe setting forth that he holds land on one side of bull run and intends to build a water Grist mill on the said run and prays that one acre of Land opposite belonging to Susanna Wickliff widow may be viewed and valued according to Law Whereupon it is ordered that the Sheriff summon a Jury of twelve Free holders of his Bailiwic to meet upon the Land petitioned for who being met and duly sworn are diligently to view and examine the said Land and the Lands adjacent thereto on both sides the said Run which may be affected or laid under water by building the said Mill together with the Timber and other Conveniences thereon and report the same with the true value of the acre petitioned for and of the Damage to the said Susanna Wickliff or to any other person or persons under their hands & seals and that the Sheriff return the said report to the next Court and it is ordered that the Surveyor of this County survey and lay of the said acre of Land

FCMB 1768-1770

18 April 1769, page 126
Licence is granted Francis Summers Junr. to keep an Ordinary who with William Payne Gen^t. acknowledged a Bond for the same

15 May 1769, page 129
We present William Linton of the parish of Truro & County of Fairfax for retailing spirituous Liquors without a Licence within six months last past to the knowledge of two of us

15 May 1769, page 130
We present the Surveyor of the road from opposite against William Adams to Pimmets run for being in bad repair contrary to the act of Assembly in that case made and provided to the knowledge of two of us

15 May 1769, page 130
We present the Surveyor or Overseer of the Road from Cameron to opposite to William Adam's path for being in bad repair to the knowledge of two of us contrary to the act of Assembly in that case made & provided

15 May 1769, page 130
We present the Overseer or Surveyor of the road from the school house near John Summer's to Accotinck Run for being in bad repair to the knowledge of two of us contrary to the act of Assembly in that case made and provided

15 May 1769, page 130
We present the Overseer of the road from the fork of the road leading to John Summer's to the old Court house road for being in bad repair to the knowledge of two of us contrary to the act of Assembly in that case made and provided

15 May 1769, page 130
We present John Lomax & Rachel his wife for retailing spirituous Liquors without a Licence to the knowledge of two of us as appears by the Clerk's certificate within one month last past

15 May 1769, page 130
We present William Sanford of the County of Fairfax for retailing spirituous Liquors without a Licence as appears by the Clerks certificate within one month last past to the knowledge of two of us

15 May 1769, page 130
Licence is granted to Charles Turner to keep an Ordinary in Alexandria who with John West Junr. acknowledged a Bond for the same

FCMB 1768-1770

15 May 1769, page 131
Moses Ball is appointed Surveyor of the road from Cameron to M^r. William Adam's in the room of Thomas Shaw

16 May 1769, page 133
Licence is granted Elizabeth Sanford to keep an Ordinary at Cameron who with William Sanford her Security acknowledged a Bond for the same

17 May 1769, page 142
On the motion of the Rev^d. M^r. Lee Massey It is ordered that the road from Pohic Church to the Ox road at Keens hill be cleared - William Reardon is appointed Surveyor who is ordered to clear the same with his own hands and the hands belonging to M^r. Massey

19 June 1769, page 145
Benjamin Moody is appointed Surveyor of the road in the room of Edward Dulin

17 July 1769, page 195
On the Petition of Thomas Pollard setting forth that he holds Land on one side of Popeshead and intends to build a water grist mill on the said Run & prays that an acre of Land opposite belonging to James Keith may be viewed and valued according to Law Whereupon it is ordered that the Sheriff summon a Jury of twelve free holders of his Bailiwic to meet upon the Land petitioned for who being met and duly sworn are diligently to view and examine the said Land and the Land adjacent thereto on both sides of the said Run which may be affected or laid under water by building the said Mill together with the Timber and other conveniences thereon and report the same with the true value of the acre petitioned for and of the Damages to the said James Keith or to any other person or persons under their hands & seals and that the Sheriff return the said Report to the next Court and it is ordered that the Surveyor of this County Survey and lay off the said acre of Land

16 October 1769, page 240
Francis Whiting is appointed Surveyor of the road in the room of Thomas Beach

16 October 1769, page 242
Licence is granted Henry Burnum to keep an Ordinary at the old Court house who with Benjamin Moody his Security acknowledged a Bond for the same

17 October 1769, page 243
Licence is granted William Courts to keep an Ordinary at Colchester who with Peter Wagener & Peirce Bayly acknowledged a Bond for the same

17 October 1769, page 243
Licence is granted Richard Arell to Keep an Ordinary at Alexandria who with Charles Broadwater Gen^t. acknowledged a Bond for the same

FCMB 1768-1770

17 October 1769, page 246
Licence is granted John Lomax to keep an Ordinary in Alexandria who with David Young acknowledged a Bond for the same

20 November 1769, page 248
Licence is granted Bennett Hill to keep an Ordinary who with Peirce Bayly acknowledged a Bond for the same

22 November 1769, page 255
Robert Lindsey is appointed Surveyor of the road from Pimmitts Run to M^r. William Adams and it is ordered that he repair the Bridge over the said run

24 November 1769, page 269
Licence is granted James Bell to keep an Ordinary at Alexandria who with Michael Gretter acknowledged a Bond for the same

18 December 1769, page 279
Martin Cockburn is appointed Surveyor of the road from Court's Ferry to the road from Alexandria to Colchester and also from the said Ferry to Pohic Church

18 December 1769, page 279
Peter Wagener is appointed Surveyor of the road from Pohic Run to Occoquan Ferry in the room of Henry Moore

19 February 1770, page 285
On the motion of Thomas Sangster Ordered that Thomas Withers Coffer Edward Ford, William Gardner and James D^{e}Neale or any Three of them view the road that leads from Samuel Littlejohns to Occoquan new road and report on oath whether the same may be conveniently turned

20 March 1770, page 345
Ordered that the road that leads from Hollis' old field to Sangsters be opened, Thomas Withers Coffer appointed Surveyor, William Payne & Edward Payne Gen^t. are appointed to allott the hands to clear the same

23 March 1770, page 355
Licence is granted Andrew Wailes to keep an Ordinary in Alexandria who with Charles Turner acknowledged a Bond for the same

FAIRFAX COUNTY COURT ORDER BOOK 1770-72

16 April 1770, page 1
Ordered that William Hardin Surveyor turn the Road by James Green's Plantation the way it is already marked by Daniel Jennings and John Williams

16 April 1770, page 3
Thomas Sangster is appointed Surveyor of the road from Hollis' in the room of Thomas Coffer

16 April 1770, page 3
William Payne & Edward Payne or either of them are appointed to allot the Tithables to clear the road from Thomas Sangsters to the wolf Run Shoals on Occoquan

16 April 1770, page 5
The persons appointed to view a road from Littlejohns to Occoquan new road having returned a Report It is ordered that Thomas Sangster clear the same the way it is marked by the viewers

17 April 1770, page 13
John Jackson is appointed Surveyor of the Road from the old Road at the upper side of Ballendine's Farm to Difficult Charles Broadwater Gentleman is appointed to allott the Tithables according to a former Order

21 May 1770, page 15
Licence is granted Michael Gretter to keep an Ordinary in Alexandria who with Abraham Barnes acknowledged a Bond for the same

21 May 1770, page 17
We present William Hawkins of the parish of Fairfax & County of Fairfax for retailing Spirituous Liquors in the parish and County aforesaid within three months last past contrary to an act of Assembly in that case made and provided by Information of Benjamin Boswell

21 May 1770, page 17
We present the Surveyor of the Road from Pohick to Accotinck in the back Road for not being in Order agreeable to the act of Assembly in that case made and provided by Information of William Linton

21 May 1770, page 17
We present the Surveyor of the Road from Accotink to Dogue Run on the back Road for not being in Order agreeable to the act of Assembly in that case made and provided by Information of John Reardon & William Linton

COB 1770-1772

21 May 1770, page 17
We present the Surveyor of the Road from Dogue Run on the back Road to Hunting Creek for not being in Order agreeable to an act of Assembly in that case made & provided by Information of John Reardon & William Linton

21 May 1770, page 18
We present the Surveyors of the Road and each of them from Hunting Creek to Alexandria for not being in Order agreeable to an act of Assembly in that case made and provided by Information of John Reardon & William Linton

21 May 1770, page 18
We present the Surveyors of the Road from Thomas Sangsters to Occoquan for not being in Order agreeable to an act of Assembly in that case made and provided to the knowledge of two of us

21 May 1770, page 18
We present the Surveyor of the Road from the old Court house to William Shortridge's for not being in Order agreeable to an act of Assembly in that case made and provided to the knowledge of two of us

21 May 1770, page 18
We present Jacob Hubbard of the parish of Fairfax and County of Fairfax for retailing Spirituous Liquors in the parish & County af^d. within three months last past contrary to an act of Assembly in that case made and provided to the knowledge of two of us

21 May 1770, page 20
Licence is granted Dorotha Young to keep an Ordinary in Alexandria who with Michael Gretter acknowledged a Bond for the same

21 May 1770, page 20
Ordered that Charles Broadwater Henry Gunnell & John Wren or any two of them view the road from the fork of the old Road near Benjamin Talbott's to the Church Road near Lewis Sander's and report on oath whether the same ought to be opened

21 May 1770, page 26
Edward Shelvin is appointed Surveyor of the Road from Accotink run to the back Lick in the room of Thomas Harden

22 May 1770, page 27
Licence is granted Joseph Gord to keep an Ordinary in Alexandria who with John West Junr. acknowledged a Bond for the same

COB 1770-1772

22 May 1770, page 31
Robert Alexander is appointed Surveyor of the Road from four mile Run to George Town Ferry in the room of Thomas Palmer

23 May 1770, page 34
On the motion of Lee Massey Cl. It is ordered that Martin Cockburn and Bennett Hill Surveyors assist William Reardon with their hands to clear a Road from Pohic Church to Keen's Hill

23 May 1770, page 35
Licence is granted Joseph China to keep an Ordinary at Cameron who with Robert Boggess & Gerrard Bolling acknowledged a Bond for the same

18 June 1770, page 37
Benjamin Southerd is appointed Surveyor of the Road from the Ponds to the Ox Road in the room of Thomas Windsor

18 June 1770, page 38
Thomas Lewis is appointed Surveyor of the road from Accotink to Popes head in the room of Philip Grymes

19 June 1770, page 40
The persons appointed to view the Road from the fork of the old Road near Benjamin Talbotts to the Church Road by Lewis Sanders returned the following report Viz^t. "In obedience to the above Order we the Subscribers have met and have viewed the Road Required and find it as followeth Viz^t. (first) We find it will be usefull & necessary to have a Road from the forks of the old Court house Road near Benjamin Talbotts to the Church Road near Lewis Sanders for the benefitt of the people in the upper part of both parishes to go to the new Church built by M^r. Edward Payne also to Mill, also a nearer and most Convenient way for the People to Colchester & Dumfries, secondly we find that the old former Road to be the best way that leads through part of the Plantation of Benjamin Talbotts, William Kitchens, Edward Washingtons and John Cottons, Thirdly we find that there might be a road got something streighter through a small Corner of Benoni Hallys Plantation, William Kitchens, M^{rs}. Mary Ferguson, Edward Washingtons and John Cottons but with much more Labour and Trouble and we believe it will not be so good a way given under our hands this 18°. day of June 1770 Chas Broadwater, Henry Gunnell, John Wren Whereupon it is ordered that a Road be opened according to the above Report

19 June 1770, page 41
Charles Broadwater & Bryan Fairfax Gen^t. are appointed to agree with Workmen to repair Difficult Bridge and that they also apply to the Court of Loudoun to appoint Comissioners to assist them in the same

COB 1770-1772

19 June 1770, page 42
Licence is granted Francis Summers to keep an Ordinary who together with Peter Wise acknowledged a Bond for the same

21 August 1770, page 60
John Bowling is appointed Surveyor of the road from the Falls Church to back lick Run in the room of Elisha Powell

22 August 1770, page 66
Licence is granted John Lomax to keep an Ordinary at Alexandria at Hunting Creek Ferry who with John Carlyle Gen^t. acknowledged a Bond for the same

23 August 1770, page 77
Ordered that George Mason & George Washington Gen^t. be summoned to appear at the next Court to give Security according to Law for keeping the Ferrys at their Respective Landings in this County

16 October 1770, page 101
Benoni Halley is appointed Surveyor of the road from the fork of the old Road near Benjamin Talbott's to the Church road near Lewis Sanders Charles Broadwater Gen^t. is appointed to alott the hands to clear the same

16 October 1770, page 108
Jervis Hammond is appointed Surveyor of the road from Popes head to Accotink Run

19 October 1770, page 128
Licence is granted Richard Arell to keep an Ordinary in Alexandria who with Robert Harrison acknowledged a Bond for the same

19 November 1770, page 130
Ordered that Edward Payne Lewis Elzey & Thomas Coffer view the Road from John Pinkstones to the mountain road and report the Condition thereof

19 November 1770, page 131
Licence is granted David Loofborrow to keep an Ordinary who with Thomas Sangster acknowledged a Bond for the same

19 November 1770, page 131
Samuel Jenkins is appointed Surveyor of the road from the Bridge above the old Court house to Difficult in the room of James Harper

COB 1770-1772

19 November 1770, page 131
On the motion of Sampson Turley It is ordered that Marmaduke Beckwith, Thomas Pollard & Sanford Payne go upon the Land belonging to the said Sampson on Johnimore & report whether a Road is necessary and the most convenient way from the said Turley's plantation into the mountain Road

20 November 1770, page 135
John House & Michael Gretter acknowledged a Bond for keeping Rock Creek Ferry

22 November 1770, page 145
Licence is granted William Courts to keep an Ordinary at Colchester William Ellzey & Peter Wagener acknowledged a Bond for the same

22 November 1770, page 153
We present Sarah Mills of the parish of Truro & County of Fairfax for retailing spirituous Liquors in the said County without Licence within three months last past contrary to an act of assembly in that case made & provided by Information of William Payne Baylis & Bennett Hill

22 November 1770, page 153
We present Benjamin Buckley of the parish of Truro & County of Fairfax for retailing spiritous Liquors in the said County without Licence within three months last past contrary to an act of assembly in that case made & provided by Information of William Payne Baylis

22 November 1770, page 153
We present Edward Wathen of the parish of Truro & County of Fairfax for retailing spirituous Liquors in the County af^d. without Licence within three months last past contrary to an act of assembly in that case made and provided by Information of William Payne Baylis & Bennett Hill

22 November 1770, page 153
We present Samuel Gates of the parish of Truro & County of Fairfax for retailing spirituous Liquors in the County af^d. without Licence within three months last contrary to an act of assembly in that case made & provided by Information of William Payne Baylis

22 November 1770, page 153
We present Isaac Gates of the parish of Truro & County of Fairfax for retailing spirituous Liquors in the said County without Licence within three months last past contrary to an act of assembly in that case made & provided by Information of William Payne Baylis

COB 1770-1772

22 November 1770, page 154
We present William Johnson Junr. of the parish of Truro & County of Fairfax for retailing spirituous Liquors without Licence in the said County within three months last past contrary to an Act of Assembly in that case made & provided by Information of William Payne Baylis

22 November 1770, page 154
We present Edward Violett of the parish of Truro & County of Fairfax for retailing spirituous Liquors in the said County without Licence within three months last past contrary to an act of assembly in that case made & provided by Information of William Payne Baylis

22 November 1770, page 154
We present James Brown of the Parish of Truro & County of Fairfax for retailing Spirituous Liquors in the said County without Licence within three months last past contrary to an act of Assembly in that case made & provided by Information of William Payne Baylis, Bennett Hill and Peirce Bayly

22 November 1770, page 155
We present Henry Burnum of the parish of Fairfax & County of Fairfax for retailing Spirituous Liquors in the County af^d. without Licence within three months last past contrary to an act of assembly in that case made and provided to the knowledge of two of us

22 November 1770, page 155
We present William Gardner of the parish of Truro & County of Fairfax for retailing spirituous Liquors in the County af^d. without Licence within six months last past contrary to an act of assembly in that case made and provided to the knowledge of two of us

22 November 1770, page 155
We present Francis Summers of the Parish of Fairfax & County of Fairfax for retailing spirituous Liquors without Licence in the said County within three months last past contrary to an act of assembly in that case made & provided to the knowledge of two of us

19 December 1770, page 167
Ordered that Thomas Shaw, Presley Cox Daniel Talbott & Thos. Monroe or any three of them view the road proposed to be altered by Daniel French and Report on oath to the next Court

19 December 1770, page 167
John Ward is appointed Surveyor of the road in the room of Daniel Talbott from Douge Run to Hunting Creek

21 January 1771, page 169
On the motion of Benjamin Talbott It is ordered that Chas. Broadwater Henry Gunnell & John Wren view the road opened through his plantation and report on oath to the next Court

COB 1770-1772

21 January 1771, page 170
William Brummitt is appointed Surveyor of the road in the room of William Triplett

21 January 1771, page 170
James Green is appointed Surveyor of the road in the room of William Hardin

21 January 1771, page 170
Nathaniel Popejoy is appointed Surveyor of the road in the room of William Summers

18 March 1771, page 173
William Bowling is appointed Surveyor of the road from Alexandria to four mile Run in the room of Isaac Rose

18 March 1771, page 173
Robert Boggess Junr. is appointed Surveyor of the roads in the room of Bennett Hill

19 March 1771, page 176
Hanah Johnson Samuel Bayly & Thomas Triplett entered into Bond for keeping a Ferry from the said Johnson's Plantation to the opposite Landing in Maryland according to Law

20 March 1771, page 194
Franklin Perry this day appearing on his Recognizance and charged with taking a Shovel full of Meal out of the meal Trough at Tolson Mill whereof George Ogelvay is miller on hearing the Witnesses against him and all circumstances relating to the said fact It is ordered that he give Security himself in the sum of Twenty pounds and two Securities each in ten pounds for his appearance at the next Grand Jury Court to answer a Bill of Indictment to be preferred against him Whereupon the said Franklin & John Rhodes & Samuel Talbott severally acknowledged themselves indebted to our Sovereign Lord King George the third his Heirs and Successors Viz^t. the said Franklin in the sum of Twenty pounds and the said John & Samuel in ten pounds each to be levied of their respective Goods & Chattels Lands and Tenements on Condition that the said Franklin do personally appear at the next Grand Jury Court to answer a Bill of Indictment to be preferred against him that then the above Recognizance to be void otherwise to remain in full force

20 May 1771, page 208
Licence is granted Francis Summers to keep an Ordinary who with Daniel Jenkins acknowledged a Bond for the same

20 May 1771, page 208
Licence is granted Andrew Wales to keep an Ordinary who with Charles Turner acknowledged a Bond for the same

COB 1770-1772

20 May 1771, page 208
Licence is granted John Steel to keep an Ordinary who with Alex: Black acknowledged a Bond for the same

21 May 1771, page 211
Licence is granted James Brown to keep an Ordinary at Pohic

21 May 1771, page 211
Licence is granted William Hawkins to keep an Ordinary who with Joseph Cash & Butler Stone Street Ashford acknowledged a Bond for the same

21 May 1771, page 216
William Turner is appointed Surveyor of the road from the Loudoun Line to Popeshead Run in the room of James Hally

21 May 1771, page 217
Henry Gunnell Junr. is appointed Surveyor of the road from Pimitt Run to M^r. William Adams' in the room of Robert Lindsey

21 May 1771, page 218
On the Petition of George Lampkin setting forth that he holds land on one side of wolf Run and is desirous of building a water Grist Mill on the said Run and prays to have an opposite acre of Land belonging to Richard Clark viewed and valued according to Law Whereupon it is ordered that the Sheriff summon a Jury of twelve freeholders of his Bailiwic to meet upon the Lands Petitioned for who being met and duly sworn are diligently to view and examine the said Land adjacent thereto on both sides the said Run which may be affected or laid under water by building the said Mill together with the timber & other conveniences thereon and report the same with the true value of the acre Petitioned for and of the damages to the said Richard Clark or to any other person or persons under their hands & seals and that the Sheriff return the Report to the next Court and it is ordered that the Surveyor of this County survey and lay off the said acre of Land

19 August 1771, page 231
William Adams Gen^t. is appointed Surveyor of the Road in the room of Moses Ball

21 August 1771, page 249
On the Petition of Charles Alexander leave is granted him to build a water grist Mill on his Land upon four mile Run

27 September 1771, page 278
Ordered that Daniel McCarty Alexander Henderson Peter Wagener & Peirce Bayly or any three of them view the road on the upper side of Accotink and report on oath the most convenient way to turn the same

COB 1770-1772

27 September 1771, page 278
Daniel McCarty & Peter Wagener are appointed to allott the hands to clear the road from Loofborrows to Colchester from the upper Church to the new Church & from Hollis's to the new Church from the back Lick to accotink and from Accotink to Pohic

28 September 1771, page 287
Licence is granted Elizabeth Sanford to keep an Ordinary at Cammeron who with William Sanford & Peter Gullatt acknowledged a Bond for the same

23 October 1771, page 303
Licence is granted Richard Arell to keep an Ordinary in Alexandria who with Philip Alexander acknowledged a Bond for the same

23 October 1771, page 303
Licence is granted Edward Harden to keep an Ordinary at Rock Creek who with Gerrard Bowling acknowledged a Bond for the same

18 November 1771, page 304
Licence is granted Dorothy Young to keep an Ordinary in Alexandria who with John Ratcliff her Security acknowledged a Bond for the same

18 November 1771, page 304
Licence is granted Michael Gretter to keep an Ordinary in Alexandria who with Charles Turner his Security acknowledged a Bond for the same

18 November 1771, page 305
Licence is granted David Loofburrow to keep an Ordinary who with John Tillett his Security acknowledged a Bond for the same

19 November 1771, page 307
Licence is granted John Lomax to keep an Ordinary at Hunting Creek ferry who with Michael Gretter his Security acknowledged a Bond for the same

19 November 1771, page 312
To William Triplett for Timber to Repair the Road . . . .£1..5..6

20 November 1771, page 315
Richard Arell is appointed Surveyor of the Road from Alexandria to Cameron Run

21 November 1771, page 320
We present James McDonald of the County of Fairfax & Parish of Truro for retailing spirituous Liquors within six months last past by Information of James Brown

COB 1770-1772

21 November 1771, page 320
We present John House and Margaret his wife of the County of Fairfax and Parish of Fairfax for retailing spirituous Liquors without Licence within six months last past by information of Edward Harding

16 December 1771, page 324
John Shepperd is appointed Surveyor of the Road from Difficult bridge to the dirt Bridge near the old Court house in the room of Samuel Jenkins

16 December 1771, page 324
Ordered that a Road from the fork of the old Court house Road to the old Church Road near Lewis Sanders be opened according to the viewer's Report Benoni Hally is appointed Surveyor and Charles Broadwater Gen^t. is appointed to allott the hands to clear the same

17 December 1771, page 327
On the motion of Charles Broadwater Gen^t. Ordered that Henry Gunnell William Gunnell Junr. & George Williams or any two of them view to most proper way for a Road from Edward Blackburns to Difficult Run at the said Broadwater's Mill to intersect with the old Lawyer's Road and report on oath to the next Court

17 December 1771, page 327
Ordered that Daniel Jenkins, John Waller and Benoni Hally or any two of them view the most convenient way for a road from Benjamin Talbotts to M^r. Thomas Lewis' Mill on Difficult and report on oath to the next Court

20 January 1772, page 329
Shadrach Green is appointed Surveyor of the Road in the room of Martin Cockburn

20 January 1772, page 329
Robert Moss is appointed Surveyor of the road from Accotinck to the back Lick in the room of Edward Shelton

20 January 1772, page 329
Ordered that Townshend Dade Richard Sanford, Thomas Shaw and Robert Alexander or any three of them view a way from the termination of Queen's Street to the Road leading from Alexandria to George Town and Report on oath to the next Court whether a Road be necessary

FAIRFAX COUNTY COURT ORDER BOOK 1772-1774

16 March 1772, page 3
On the motion of Thomas Throgmorton Ordered that Thomas Pollard Paul Turley Marmaduke Beckwith and Samuel Tillett or any three of them view the most convenient way for a Road from Thomas Sangsters to the said Throgmorton's Mill on Bull Run and report on oath to the next Court

16 March 1772, page 4
Bryan Fairfax Gen^t. is appointed Surveyor of the Road from Difficult bridge to the Bridge branch near the old Court house in the room of Samuel Jenkins

17 March 1772, page 10
The viewers appointed to view the most convenient way for a Road from Edward Blackburns to M^r. Charles Broadwaters Mill on difficult returned the following Report Viz^t. "January 15th 1772 In obedience to the within Order we the Subscribers have met and viewed the within mentioned way from M^r. Blackburns to Major Broadwater's Mill and find there may be a road made convenient from the said Blackburn's fence to Walter Garretts Plantation and thro' his Plantation by his consent to line made by Peter Harley thro' the Land of Col^o. Thomas Ludwell Lee then along that Line and near it to the said Mill Henry Gunnell, George Williams William Gunnell Whereupon it is ordered that Henry Gunnell who is appointed Surveyor do clear the same according to the said Report Bryan Fairfax Gen^t. is appointed to allott the hands to clear the said Road

17 March 1772, page 11
The persons appointed to view the most convenient way for a Road from Benjamin Talbotts to Thomas Lewis' Mill on Difficult returned the following Report Viz^t. "January 18°. 1772 In obedience to the within Order we the Subscribers have met and viewed the way for the road required and find that it will be convenient for to open it from the plantation of Benjamin Talbotts then along the main Road to the end of Samuel Talbotts fence then thro' the plantation of M^r. John Wallers then along a blazed Row of Trees and Saplings and the old Mill path to M^r. Thomas Lewis' Mill on Difficult Run it being the best and most convenient way Benoni Hally Daniel Jenkins, John Waller Whereupon it is ordered that Thomas Lister who is appointed Sureyor do clear the same according to the said Report, Henry Gunnell Gen^t. is appointed to allott the hands to clear the said Road

17 March 1772, page 13
Licence is granted Henry Burnum to keep an Ordinary who with George Kilgore acknowledged a Bond for the same

20 April 1772, page 26
Jeremiah Williams is appointed Surveyor of the Road from the Cross Road where Ricketts lives to the Maddams branch in the room of John Sumers

COB 1772-1774

18 May 1772, page 28
Licence is granted Thomas Forbes to keep an Ordinary at the Falls of Potomack who with Richard Thompson & Thomas Blackburn ack^d. a Bond for the same

19 May 1772, page 34
The Viewers appointed to view a way for a Road from the termination of Queen's Street to the Road leading from George Town to Alexandria and returned the following report; Viz^t. "May 18th 1772 Agreeable to an Order of Court dated January 1772 we have viewed the way from the terminationof Queen's Street to the Road leading from George Town to Alexandria and Report upon Oath that we believe a Road to be necessary and that the said Road shall go along M^r. Carlyle's Ditch which appears to be upon a paralel line with Queen Street to the County Road from George Town to Alexandria Townshend Dade, Richard Sanford, Thomas Shaw Robt. Alexander Whereupon it is ordered that a Road be opened according to the above Report and that the Surveyor of Road from four mile Run to Alexandria clear the same

20 May 1772, page 38
The Grand Jury sworn yesterday returned into Court with the following presentments Viz^t.

We present George Mason the Younger of the County of Fairfax and parish of Truro owner of a Mill on Pohic Creek for not having the Bridge on the Road over the said Mill race in good Order and agreeable to an act of Assembly in that case made and provided within six months last past to the knowledge of two of us

20 May 1772, page 38
We present the Surveyor of the back road from Hunting Creek to Dogue Run for the said Road not being in good Repair agreeable to an act of Assembly in that case made and provided within six months last past to the knowledge of two of us

20 May 1772, page 38
We present the Surveyor of the Road from Alexandria to Cameron for the said Road not being in good Repair agreeable to an Act of Assembly in that Case made and provided within six months last past to the knowledge of two of us

20 May 1772, page 38
We present the Surveyor of the road from William Adams Gen^t. to Pimets Run for the said Road not being in good Repair agreeable to an act of Assembly in that case made and provided within six months last past to the knowledge of two of us

COB 1772-1774

20 May 1772, page 38
We present the Surveyor of the road from William Adams Gen^t. to Pimmetts run for not keeping a Stone or post with plain inscriptions thereon where the road leads from the Alexandria Road to the Ferry at George Town agreeable to an act of Assembly in that case made & provided within six months last past to the knowledge of two of us

20 May 1772, page 38
We present the Surveyor of the Road where the Road divides at Boggess's for not having sign Posts for directions to each road that is the River side and back Roads to Alexandria agreeable to an act of Assembly in that case made and provided within six months last past to the knowledge of two of us

20 May 1772, page 39
We present the Surveyor of the Road from Accotink to the back Lick for not having a sign post for directions at the angle of the road leading to the back Lick and Alexandria agreeable to an act of Assembly in that case made and provided within six months last past to the knowledge of two of us

20 May 1772, page 39
We present the Surveyor of the road from Hunting Creek to Dogue Run on the back road for not having a Sign post for directions down the said Road at the angle of the Roads leading from Hunting Creek agreeable to an act of Assembly in that case made and provided within six months last past to the knowledge of two of us

20 May 1772, page 39
We present the Surveyor of the Road from Hunting Creek to the Gum Spring for not having a Sign post for directions down the said Road at the angle of the back and river side roads agreeable to an act of Assembly in that case made and provided within six months last past to the knowledge of two of us

20 May 1772, page 39
We present the Surveyor of the road from Cameron to the Falls Church for not having a Sign post for directions up the Road from Cameron agreeable to an act of Assembly in that case made and provided within six months last past to the knowledge of two of us

20 May 1772, page 39
We present the Surveyor of the Road from Cameron leading to John Summer's for not having a sign post for directions at the angle of the road leading from Cameron agreeable to an act of Assembly in that case made and provided within six months last past to the knowledge of two of us

COB 1772-1774

20 May 1772, page 39
We present the Surveyor of the Road from Colchester up the Ox Road for not having a Sign post for directions at the angle of the Road up the said Ox Road agreeable to an act of Assembly in that case made and provided within six months last past to the knowledge of two of us

20 May 1772, page 39
We present the Surveyor of the Road from the ox road to the Forge for not having a sign post & directions at the angle of the Road leading to the said Forge & Colchester agreeable to an act of Assembly in that case made & provided within six months last past to the knowledge of two of us

20 May 1772, page 40
We present the Surveyor of the Ox Road where it intersects at Thomas Sangsters for not having a Sign post & directions at the said intersection leading from Colchester to the Mountains agreeable to an act of Assembly in that case made and provided within six months last past to the knowledge of two of us

20 May 1772, page 40
We present the Surveyor of the road from Thomas Sangsters to Frying Pan for not having a Sign post with directions where the said Road leads out from the Mountain road agreeable to an act of Assembly in that case made and provided within six months last past to the knowledge of two of us

20 May 1772, page 40
We present the Surveyor of the Road from Thomas Sangsters to Occoquan for not having a Sign post for directions where the said road leads out from the mountain road agreeable to an act of Assembly in that case made and provided within six months last past to the knowledge of two of us

20 May 1772, page 40
We present the Surveyor of the road from the Loudoun Line for not having a sign post for directions at Thomas Sangsters leading to Colchester agreeable to an act of Assembly in that case made and provided within six months last past to the knowledge of two of us

20 May 1772, page 40
We present the Surveyor of the Road from Alexandria to Cameron for not having a Sign post at Cameron with the directions for the Roads leading up and down the Country agreeable to an act of Assembly in that case made and provided within six months last past to the knowledge of two of us

COB 1772-1774

20 May 1772, page 40
We present the Surveyor of the Road from the Alexandria Road to the four mile Run for not having a Sign post with directions where the said Road leads out as well as where the Road forks leading to George Town and the Falls Warehouse agreeable to an act of Assembly in that case made and provided within six months last past to the knowledge of two of us

20 May 1772, page 40
We present the Surveyor of the Road from the back lick to the Falls Church for the said Road being in bad Order, and for having several Gates across the same road contrary to an act of Assembly in that case made and provided within six months last past to the knowledge of two of us

20 May 1772, page 41
We present the Surveyor of the road from the back Lick to the Falls Church for not having a sign post for directions at the intersection of the said Road at Prices leading from thence to the said Church agreeable to an act of Assembly in that case made and provided within six months last past to the knowledge of two of us

20 May 1772, page 41
We present the Surveyor of the Road from Accotink to the School house nigh John Summers for not having a sign post with directions at the Intersection of the Road at Prices leading from Accotink to Alexandria agreeable to an act of Assembly in that case made and provided within six months last past to the knowledge of two of us

20 May 1772, page 41
We present the Surveyor of the Road from the old School house nigh Francis Summers to the old Court house for not having a sign Board with directions at the angle of the said Road also where it intersects the Road leading to the old Court house Road agreeable to an act of Assembly in that case made and provided within six months last past to the knowledge of two of us

20 May 1772, page 41
We present William Courts of the Parish of Truro and County of Fairfax for selling spirituous Liquors without Licence within six months last past as appears to us by a List of Ordinary Licences delivered to us by the Clerk and to the knowledge of two of us

20 May 1772, page 41
We present Philip Daws Copper Smith of the parish and County of Fairfax for selling Rum without Licence within six months last past as appears to us by a List of Ordinary Licences delivered to us by the Clerk and by the Information of Agness Stonemier

COB 1772-1774

20 May 1772, page 41
We present Hanah Perry of the Parish and County of Fairfax for selling Rum without Licence within six months last past as appears by a List of Ordinary Licences delivered to us by the Clerk and by the Information of Agness Stonemier

20 May 1772, page 42
We present John Ward of the Parish and County of Fairfax for retailing Spirituous Liquors at his Mill in the Parish of Truro by his hireling man Thomas Parrot without Licence within six months last past as appears to us by a List of Ordinary Licences delivered to us by the Clerk and by the Information of Thomas Devaund

22 May 1772, page 55
Licence is granted William Courts to keep an Ordinary at Colchester who with Peirce Bayly his Security acknowledged a Bond for the same

15 June 1772, page 62
George McCormuck is appointed Surveyor of the road from Pimmetts run to M^r. William Adams in the room of Henry Gunnell

15 June 1772, page 62
William Follin is appointed Surveyor of the road from the forks of the road at the falls Church to the old Road from Rock Creek to Alexandria in the room of John Frizell

15 June 1772, page 62
John Bowling is appointed Surveyor of the Road from the back Lick to the Falls Church

15 June 1772, page 62
James D^{e}Neale is appointed Surveyor of the Road from the Alexandria Road to the Ox Road near the Piney branch in the room of Edward Payne

15 June 1772, page 62
Ordered that Charles Broadwater William Payne Benjamin Moody & James Wren or any three of them view the most convenient way for a Road from Edward Blackburns to the Church Road near M^r. Benjamin Moody's and report on oath to the next Court

15 June 1772, page 63
Samuel Tillett is appointed Surveyor of the Road from the Loudoun Line to Thomas Sangsters in the room of Edward Ford

15 June 1772, page 63
Robert Moss is appointed Surveyor of the Road from Accotinck to the back Lick in the room of Edward Shelton

COB 1772-1774

**2	1772	114**
On the Petition of William Carlin setting forth that he has Land on one side of four mile Run and intends to build a water Mill on the said Run and prays to have an acre of Land opposite viewed and valued according to Law whereupon it is ordered that the Sheriff summon a Jury of twelve free holders of his bailiwic to meet upon the Land petitioned for who being met and duly sworn are diligently to view and examine the said Land adjacent thereto on both sides the said Run which may be affected or laid under water by building the said Mill together with the timber and other conveniences thereon and Report the same with the true value of the Acre Petioned for and of the damages to the Proprietor or to any other person or persons under their hands and Seals and that the Sheriff return the Report to the next Court and it is ordered that the Surveyor of this County survey and lay off the said Acre of Land

23 September 1772, page 123
John West Junr. Henry Gunnell William Payne & William Adams Gen^t. or any three of them are appointed to view the main Road from Alexandria to Difficult Bridge and from Cameron to the Mountain Road and make out a List of Titheables appointed to work on the said main Roads and the contiguous Roads according to the best Information that can obtain from the several Surveyors as well as the List of Titheables and divide the said two main Roads into such districts as think proper particularizing the Titheables proper to work on each such district to which they are to add such Titheables as they think can be conveniently spared from the contiguous Roads and attend with their Report thereof at the next Court

23 September 1772, page 124
George Mason Daniel McCarty and William Payne Gen^t. or any two of them are appointed to view the main Roads from Colchester to Loudoun County and the Ox Road from Loofburrows to Difficult and make out a List of Titheables appointed to work on the said main Roads & the contiguous Roads according to the best Information they can obtain from the several Surveyors as well as the List of Titheables and divide the said two main Roads into such districts as they think proper particularizing the Titheables proper to work on each such district to which they are to add such Titheables as they think can be conveniently spared from the contiguous Roads and attend with their Report thereof at the next Court

24 September 1772, page 131
Peter Wagener Junr. is appointed Surveyor of the road from Colchester to Pohic in the room of Peter Wagener

19 October 1772, page 133
Licence is granted to John Hill to keep an Ordinary at Cameron who with Robert Boggess acknowledged a Bond for the same

20 October 1772, page 139
Licence is granted Joseph Courts to keep an Ordinary in Alexandria who with William Courts acknowledged a Bond for the same

COB 1772-1774

16 November 1772, page 140
Licence is granted Dorothy Young to keep an Ordinary at Alexandria who with John Dalton acknowledged a Bond for the same

16 November 1772, page 140
Licence is granted Michael Gretter to keep an Ordinary at Alexandria who with John Ratcliff acknowledged a Bond for the same

17 November 1772, page 141
The viewers appointed to view a way for a Road from Colchester to Boggesses Mill returned the following Report Viz "Pursuant to an Order of the worshipfull Court of Fairfax dated the [blank] day of [blank] 177 [blank] we the Subscribers viewed a way for a Road from Colchester to Boggess' Mill on Pohic and find that a Road may be had to leave the main Road from Colchester to Alexandria above Gile's Run to go along an old Road or path through M^r. Bayly's Land and thro' George Mason's Land above the plantation where Joseph Garberry lives then to cross the Church Road at the Widow Atcheson's plantation and to go through her plantation into the old Fields which leads down to the Mill Daniel McCarty Alex Henderson Pet Wagener Junr. whereupon it is ordered that Robert Boggess open a Road agreeable to the above Report

17 November 1772, page 141
Licence is granted Richard Arell to keep an Ordinary in Alexandria who with Philip Alexander acknowledged a Bond for the same

17 November 1772, page 141
Licence is granted John Lomax to keep an Ordinary at Hunting Creek ferry who with Peter Wise acknowledged a Bond for the same

17 November 1772, page 145
William Cash is appointed Surveyor of the Road from Accotink to Dogue Run in the room of William Brummitt

17 November 1772, page 146
To the Trustees of the Road from Vestal's & William's Gaps leading to the Towns of Alexandria & Colchester £45

17 November 1772, page 147
Ordered that the Several Surveyors of the Roads in this County be summoned to appear before the Commissioners appointed to lay off the several districts on the main Roads leading from the Vestals & William's Gaps to Alexandria & Colchester at such day as they shall appoint

COB 1772-1774

17 November 1772, page 147
Edward Payne Sampson Turley, Thomas Ford & Thomas Withers Coffer are appointed to view the main Road from the Loudoun Line to Loofburrows and report whether it is necessary to alter or straighten the said Road

17 November 1772, page 147
Alexander Henderson, Lee Massey, Henry Moore & William Courts are appointed to view the main Road from Loofburrows to Colchester and Report whether it is necessary to alter or straighten the said Road

18 November 1772, page 148
Licence is granted Joseph Jones to keep an Ordinary in Alexandria who with John Dalton acknowledged a Bond for the same

21 December 1772, page 158
On the motion of George Washington Gen^t. Proprietor of a Mill in this County It is ordered to be recorded that he marks his Flour with a Brand having the following Letters G: WaSHINGTON

21 December 1772, page 158
Bryan Fairfax and Henry Gunnell Gen^t. are appointed to view the work performed at Difficult Bridge and report in what manner it is executed

22 December 1772, page 161
To Bryan Fairfax Gen^t. for Timber to repair the Road . . . . [0].8.6

22 December 1772, page 163
Charles Craig is appointed Surveyor of the Road from bricken's branch to the little Falls in the room of Daniel Jennings

22 December 1772, page 164
John West Junr. & William Adams Gen^t. are appointed to view the main Road from Alexandria to Difficult Bridge and Henry Gunnell and William Payne Gen^t. from Cameron to the Mountain Road and make out a List of Titheables appointed to work on the said main Roads and the contiguous Roads according to the best Information they can obtain from the several Surveyors as well as the List of Titheables and divide the said two main Roads into such districts as they shall think proper particularizing the Titheables proper to work on each such district to which they are to add such Titheables as they think can be conveniently spared from the contiguous Roads and attend with their Report thereof at the next Court

22 December 1772, page 165
Benjamin Boylstone is appointed Surveyor of the Road from the Falls Warehouse to the Red House in the room of George Grasty

COB 1772-1774

15 February 1773, page 167
Licence is granted William McDaniel to keep an Ordinary at Colchester who together with Cleon Moore Peirce Bayly & Samuel Bayly acknowledged a Bond for the same

15 February 1773, page 168
On the motion of Samuel Talbott Ordered that Charles Broadwater John Wren, John Monroe, and Joseph Bennett or any three of them view the road from the upper end of the said Talbotts Plantation to the lower end report on oath whether it is necessary to turn the same

15 February 1773, page 168
On the motion of Joseph Thompson Ordered that William Adams Simon Pearson and John Hurst and John Bolling or any three of them view whether it is necessary for the said Thompson to have a Road from his Plantation into the County Road and report on oath to the next Court

15 February 1773, page 168
On the motion of Alexander Henderson Gen[t]. Ordered that William Moon Thomas Lucas, Thomas Sangster and John Tillett or any three of them view whether it is necessary for the said Henderson to have a Road from his Plantation on Bull Run into the main Road and Report on Oath to the next Court

15 February 1773, page 168
Licence is granted John Graham to keep an Ordinary in Alexandria John Dalton Security

16 March 1773, page 173
On the motion of Thomas Lewis Gen[t]. leave is granted him to build a Water Grist Mill on wolf Trap Run and it is ordered that the Sheriff summon a Jury to assertain the damage which may be done to any person holding Lands adjoining thereto and report on Oath to the next Court

16 March 1773, page 178
Licence is granted James Brown to keep an Ordinary at Pohic Robert Boggess Security

18 March 1773, page 182
Peter Wagener is appointed Surveyor of the road from Pohic to Colchester in the room of Peter Wagener Jun[r].

17 May 1773, page 193
Licence is granted Francis Summers to keep an Ordinary who with William Summers his Security acknowledged a Bond for the same

COB 1772-1774

18 May 1773, page 200
Licence is granted John Brawner to keep an Ordinary who with Moses Ball his Security entered into Bond

18 May 1773, page 202
We present George McCormick of the Parish of Fairfax & County of Fairfax Surveyor of the Road leading from Gerrard Trammell's Bridge towards Willm. Adams for suffering the said Road to be bad and not in repair agreeable to the act of Assembly in that case made and provided within two months last past to the knowledge of two of us

18 May 1773, page 202
We present Thomas Wren of the Parish of Fairfax and County of Fairfax Surveyor of the Road leading from the Falls Church to Joseph Cockerills for neglecting to keep the said Road in necessary Repair agreeable to the act of Assembly in that case made and provided within two months last past to the knowledge of two of us

18 May 1773, page 202
We present Thomas Lucas of the Parish of Truro and County of Fairfax Surveyor of the Road leading from Thomas Sangsters to Occoquan for neglecting to keep the said Road in necessary Repair agreeable to act of Assembly in that case made and provided within two months last past to the knowledge of two of us

18 May 1773, page 202
We present John Waller of the Parish of Truro and County of Fairfax Surveyor of the Road leading from the Ox Road to the old Court House for neglecting to keep the said Road in necessary Repair agreeable to act of Assembly in that case made and provided within two months last past to the knowledge of two of us

18 May 1773, page 202
We present the Surveyor of the Road leading from Dogue Run to Cameron for neglecting to keep the said Road in necessary Repair agreeable to act of Assembly in that case made and provided within two months last past to the knowledge of two of us

18 May 1773, page 203
We present the Surveyor of the river side Road leading from Pohic to Accotink for neglecting to keep the said Road in necessary Repair agreeable to act of Assembly in that case made and provided within two months last past to the knowledge of two of us

18 May 1773, page 203
We present the Overseer of the Road leading from Alexandria to Cameron for neglecting to keep the said Road in necessary Repair agreeable to act of Assembly in that case made and provided within two months last past to the knowledge of two of us

COB 1772-1774

18 May 1773, page 203
We present the Overseer of the Road leading from where John Hollis formerly lived to Pohic Warehouse for neglecting to keep the said Road in necessary Repair agreeable to act of Assembly in that case made and provided within two months last past to the knowledge of two of us

18 May 1773, page 203
We present the Surveyor of the Road leading from the Falls Warehouse to the fork of the Road by Turberville's Quarter for neglecting to keep the said Road in necessary Repair agreeable to act of Assembly in that case made and provided within two months last past to the knowledge of two of us

18 May 1773, page 203
We present the Surveyor of the Road leading from the Forks of the Road near Francis Summers Junr. to the foot of Cameron Hill for neglecting to keep the said Road in necessary Repair agreeable to act of Assembly in that case made and provided within two months last past to the knowledge of two of us

18 May 1773, page 203
We present the Surveyor of the Road from the fork of the Road near Francis Summers Junr. to Accotink Run for neglecting to keep the said Road in necessary Repair agreeable to act of Assembly in that case made & provided within two months last past to the knowledge of two of us

18 May 1773, page 203
We present the Surveyor of the Road leading from the wolf Pit Hill to Pohic Church for neglecting to keep the said Road in necessary Repair agreeable to act of Assembly in that Case made and provided within two months last past to the knowledge of two of us

18 May 1773, page 204
We present Elizabeth Sanford of the Parish of Fairfax & County of Fairfax for retailing spirituous Liquors in the parish and County aforesaid Contrary to the act of Assembly in that case made and provided within six months last past to the knowledge of two of us

18 May 1773, page 204
We present Mary Bell of the parish of Fairfax in the County of Fairfax for retailing spirituous Liquors in the parish & County aforesaid Contrary to the act of Assembly in that case made and provided within six months last past to the knowledge of two of us

COB 1772-1774

19 May 1773, page 208
On the motion of Alexander Henderson Gen^t. It is ordered that John Hampton John Cotton Junr. Thomas Lucas and John Read or any three of them view a way for a Road from the said Henderson's Plantation on Bull Run into the main Road and report on Oath to the next Court the former viewers having returned a Report that a Road is necessary

21 June 1773, page 227
John Pettuth is appointed Surveyor of the Road from Dogue Run to Accotink on the lower Road in the room of John Barry

22 June 1773, page 228
Alexander Henderson Gen^t. is appointed to view the several Roads and to allot the Titheables to work on the said Roads with Daniel McCarty and William Payne Gen^t. agreeable to an Order of Court made 23 September 1772 in the room of George Mason Gen^t.

22 June 1773, page 228
Philip Adams is appointed Surveyor of the Road from the Falls Church to X Road by Joseph Cockerill in the room of Thomas Wrenn

19 July 1773, page 234
John Carlyle William Ramsay Bryan Fairfax an William Ramsay Gen^t. or any two of them are appointed to view the Road at the Falls Warehouse down to the Landing and report what is necessary to be done to the said Road and it is ordered that John Seale keep the said Road in Repair with John Baptist Tuttle, Richard Hurdle Zachariah Kennett and negroe Jack belonging to Thompson & Magruder and John Alverson and Alexander Jackson until the Commissioners return their Report

21 July 1773, page 243
Licence is granted William Courts to keep an Ordinary in Colchester who with William Thompson acknowledged a Bond for the same

22 July 1773, page 246
John Lock Williams is appointed Surveyor of the Road from Dogue Run to Hunting Creek in the room of John Ward

22 July 1773, page 251
Samuel Talbott is appointed Surveyor of the Road from the old Court house to the Ox Road in the room of John Waller

18 October 1773, page 269
Richard Clark is appointed Surveyor of the road from Thomas Sangsters to Occoquan in the room of Thomas Lucas

COB 1772-1774

18 October 1773, page 270
Licence is granted David Loofburrow to keep an Ordinary who with Moses Simpson acknowledged a Bond for the same

21 October 1774, page 284
Ordered that John West Gen^t. pay James Coleman the proportion due from this County for rebuilding Difficult Bridge agreeable to a Report returned by Bryan Fairfax & Henry Gunnell Gen^t.

15 November 1773, page 286
Licence is granted Andrew Wailes to keep an Ordinary in Alexandria who with Francis Summers acknowledged a Bond for the same

15 November 1773, page 287
The Persons appointed to view a way for a Road from Alexander Henderson's Plantation on Bull Run into the main Road returned the following Report, Viz^t. "In obedience to an order of the worshipfull Court of Fairfax County We the Subscribers have viewed a way for a Road from the Plantation of Alexander Henderson on Bull run to the main Road, and we find the way most convenient for the Neighbourhood to be from the lower part of the said Henderson's Plantation by the Tobacco Ground of Yelverton Reardon, the Plantation of William Moon, through the Lane between the Plantations of Marmaduke Beckwith and James Keen through the plantation of Sampson Turley / to which we are told he consents on Condition that the Surveyor of the road is obliged to keep up two Gates) and from thence with the Line of William Simpson by the Plantation of Peter Ryley to the main Road but if M^r. Turley does not consent to the road going thro' his Plantation then we find the most convenient way to be round the same towards Occoquan John Reed, John Cotton Junr. Thomas Lucas Whereupon it is ordered that John Cotton Junr. who is appointed Surveyor clear the same according to the above Report and that he affix two Gates on the Road through Sampson Turley's Plantation and it is also ordered that the following persons with their Titheables assist to clear the said Road Viz^t. Alexander Henderson's at Bull Run, John Cotton, Junior William Powell, John Powell, Sarah Reeds William Moon, Thos. Lucas James Keen, Marmaduke Beckwith Yelverton Reardon, Richard Clark John Chappell, James Williams, Joseph Williams, George Lampkin Benjamin King Hargiss King, Benjamin Jacobs Sampson Turley William Simpson, Baxter Simpson, and Thomas Harris.

16 November 1773, page 291
Licence is granted Richard Arell to keep an Ordinary in Alexandria who with Philip Alexander acknowledged a Bond for the same

16 November 1773, page 291
Licence is granted Sopia Beall to keep an Ordinary who with John West Junr. acknowledged a Bond for the same

COB 1772-1774

17 November 1773, page 299
On the motion of George Mason Junr. It is ordered that Daniel McCarty & Peter Wagener Junr. view the Road at the said Mason's Mill and report whether the same can be conveniently turned

18 November 1773, page 305
William Triplett is app^d. Surveyor of the Road in the Room of William Cash

18 November 1773, page 307
We present Susanna Pattinson Widow of the Parish of Truro & County af^d. for retailing spirituous Liquors without Licence in the parish & County aforesaid contrary to an Act of Assembly in that case made and provided within six months last past by the Information of Robert Doughlas and Thomas Davis

18 November 1773, page 308
We present James McDaniel yeoman, of the Parish of Truro & County aforesaid for retailing Spirituous Liquors without Licence in the parish & County af^d. contrary to an Act of Assembly in that case made and provided within six months last past by the Information of Robert Doughlas & Thomas Davis

18 November 1773, page 308
We present Richard Lightfoot Yeoman of the parish of Truro & County aforesaid for retailing spirituous Liquors without Licence in the parish & County af^d. contrary to an act of Assembly in that case made and provided within six months last past by the Information of Robert Doughlass

18 November 1773, page 308
We present Sarah Mills Widow of the Parish of Truro & County af^d. for retailing Spirituous Liquors without Licence in the Parish & County af^d. contrary to an act of Assembly in that case made and provided by the Information of Robert Doughlass and Thomas Davis

18 November 1773, page 308
We present George Mason Junr. Yeoman of the parish of Truro & County af^d. owner of the Mill upon Pohic in the Parish & County af^d. for the Bridge on the main road over his Mill Race being in bad Order within two months last past contrary to an Act of Assembly in that case made & provided by the information of Robert Boggess and Robert Doughlass

18 November 1773, page 310
We present Shadrack Green yeoman of the parish of Truro & County af^d. Overseer of the Road from Herefords Ferry to the Road leading from Colchester to Alexandria also from the said Ferry to Pohic Church for the said Roads being in bad Order contrary to an Act of Assembly in that case made and provided within two months last past by the information of Peter Wagener Junior Gen^t.

COB 1772-1774

18 November 1773, page 310
We present John Seale yeoman of the Parish of Fairfax & County af^d. Overseer of the Road from the Falls Warehouse to the Landing at the same place for the said Road being in bad Order contrary to an Act of Assembly in that case made and provided within two months last past by the Information of Nathan Frizell

18 November 1773, page 311
We present Thomas Forbers yeoman of the Parish of Fairfax & County af^d. for retailing spirituous Liquors without Licence in the parish and County af^d. contrary to an act of Assembly in that case made and provided within six months last past by the Information of Nathan Frizell & Joseph Woodward

18 November 1773, page 311
We present the Overseer of the Road from the old Courthouse to the Ox Road in the County af^d. for the said Road being in bad Order within two months last past contrary to an act of Assembly in that case made provided by the Information of Benjamin Talbott

18 November 1773, page 311
We present the Overseer of the Road from the old Storehouse to the forks of the Road in the County af^d. for the said Road being in bad Order within two months last past contrary to an Act of Assembly in that case made and provided by the Information of Peter Wagener Junior Gentleman

18 November 1773, page 312
Charles Broadwater, Bryan Fairfax & William Payne Gen^t. or any two of them are appointed to view the way or Road from the Falls Warehouse to the Landing and that they allott such hands to clear the same as they may think necessary

18 November 1773, page 313
On the motion of Moses Simpson setting forth that he he owns Land on one side of Sandy Run and intends to build a water grist Mill on the said Run and prays to have an Acre of Land belonging to Robert Carter Esqr. opposite viewed and valued according to Law Whereupon it is ordered that the Sheriff summon a Jury of twelve freeholders of his Bailiwic to meet upon the said Land of Robert Carter's who being met and duly sworn are diligently to view and examine the said Land adjacent thereto on both sides the said Run which may be affected or laid under Water by building the said Mill together with the Timber and other Conveniences thereon and report the same with the true value of the acre Petitioned for and of the damages to the said Robert Carter or to any other person or Persons under their hands and Seals and that the Sheriff return the Report to the next Court and it is ordered that the Surveyor of this County Survey and lay off the said acre of land

22 February 1774, page 319
To John Deakins & Henry Burnum for Timber to repair Difficult Bridge . . . . £[0].7.3/4

COB 1772-1774

22 February 1774, page 319
To the Trustees of the Roads from Vestals & William's Gaps leading to Alexandria & Colchester £45.-.-

FAIRFAX COUNTY COURT ORDER BOOK 1783-88

18 November 1783, page 52
On the motion of James Collins It is ordered that Thomas Magruder, William Scott and Edward Blackburn or any two of them view the road from the run above the old Courthouse to where intersects the road below the said Courthouse and report on oath to the next Court the most convenient way to alter the same

16 December 1783, page 55
The viewers appointed to view the road by the old Courthouse returned the following Report "In obediance to an order of Fairfax Court we the Subscribers being duly sworn have viewed the way proposed by James Collins and find that a road from bridge branch to the S.E.^{t} end of the said Collin's old field and then to strike the old road may be made a better way for the publick then the old road and the distance in the difference but small if any Ed. Blackburn Thos. Magruder" Whereupon it is ordered that the road be opened agreeable to the said Report

21 June 1784, page 72
Ordered that William Stone, John Cotton, Thomas Palmer & Benjamin Talbutt or any three of them view a way for a road from Benjamin Talbutt's plantation to the Loudoun line at Amos Fox's mill on Difficult and report on oath to the next Court

16 August 1784, page 75
The viewers appointed to view a way for a road from Benjamin Talbutt's plantation to the Loudoun Line at Amos Fox's mill on Difficult, returned the following Report Viz^t. "July 15th. 1784 In obedicnce to the above Order of Court we the Subscribers being first sworn have viewed the way required and marked it out to the best of our Judgments William Stone, John Cotton, Thomas Palmer Whereupon it is ordered that the way be opened agreeable to the above report

16 August 1784, page 75
Ordered that Henry Gunnell, Thomas Gunnell and George Smith view a way for a road from the road leading out of the Alexandria road about three miles above the old Court house to the great Falls of Potomack and report on oath to the next Court the most convenient way for the same

19 November 1784, page 103
Robert Adam, William Adams, James Wren & William Henry Terrett are appointed to view the road from the Ponds, /on the road from Cameron to the Falls Church) to Alexandria and report on oath to the next Court the most convenient way to turn the same

17 January 1785, page 106
Ordered that George Gilpin, William Henry Terrett, William Sanford and Thomas West or any three of them view the road from buck horn bridge to the top of the hill on the west side of Holmes Run and report on oath the most convenient way to turn the same

COB 1783-1788

21 February 1785, page 108
On the motion of Joseph Bennett ordered that Charles Broadwater Edward Blackburn John Coats & John Wren or any three of them view the road from the mouth of John Hurst's lane to the upper part of Talbut's fence and report on oath whether it is necessary to turn the said road

20 April 1785, page 119
On the motion of Daniel Mills Ordered that Jeremiah Williams Daniel Mills William Summers and John Summers or any three of them view the most convenient way to turn the road from the great road by said Mills's to William Paynes old Mill on Accotinck run & report on oath to the next Court

20 April 1785, page 120
John Dowdle is appointed Overseer of the Road leading from the little falls to four mile run in the room of John Ball.

20 April 1785, page 120
John Mills is appointed Overseer of the Road leading from the Falls church to Fitzhugh's gate in the room of Elisha Powell.

17 May 1785, page 129
On the Petition of sundry inhabitants of this County, Ordered that Charles Little, William Payne, Lund Washington and Robert Powell Gent. or any three of them view the most convenient way for opening a road from the falls church to Alexandria into the Town of Alexandria, and report on oath with an accurate survey of the way to be so viewed to the next Court.

18 May 1785, page 137
The viewers appointed to view a way for a Road from Daniel Mills's to William Paynes old Mill on Accotink returned the following report, to wit. "In Obedience to the above Order we the subscribers being sworn have viewed and are of opinion that the new way from the main road below Daniel Mill's to the old Mill on Accotinck is or may be made as convenient to the neighbouring people as the way thro' William Payne's Plantation. William Summers, Jeremiah Williams, John Summers." Whereupon it is Ordered that the said Road be opened agreeable to the said report

20 May 1785, page 145
Edward Harden is appointed Overseer of the road from George Town Ferry to Adams's Mill in the room of George Chapman

COB 1783-1788

22 September 1785, page 170
On the motion of Alexander Henderson, Ordered that Sampson Turley, Richard Clark, Hargiss King and William Moon or any three of them view the most convenient way for a road from Occoquan above the Wolf run shoals to the said Alexander Hendersons Mill, and report on oath to the next Court.

18 October 1785, page 174
On the motion of sundry inhabitants: Ordered that Thomas Gunnell, George Smith, John Jackson & John Shortridge or any three of them view the road from Scott's run to Allen Davis's lane and report on oath to the next Court the most convenient way to alter the same.

19 October 1785, page 177
Robert Adam is appointed surveyor of the road from Alexandria to four mile run in the room of Benjamin Dulany.

20 October 1785, page 178
William Darne is appointed Surveyor of the road from the Falls Church to the Road leading from Alexandria to the falls ware-house.

20 October 1785, page 178
John Jackson is appointed Overseer of the Road from Difficult Mill to the forks of the road at the farm.

20 October 1785, page 178
George Hunter is appointed Surveyor of the road from the Red house to the falls-landing.

20 October 1785, page 178
William Halley is appointed Surveyor of the road from the forks of the road below Sangsters to the mountain road by Hally's.

20 October 1785, page 178
James Moore is appointed Overseer of the road from Sangsters shop to difficult.

20 October 1785, page 178
Edward Ford is appointed Surveyor of the road from Loofboroughs to the Loudoun line.

20 October 1785, page 178
Edward Blackburn is appointed Surveyor of the road from the Ox Road to the old Courthouse.

20 October 1785, page 178
John Fowler is appointed Surveyor of the Road from Pohic Church to the ponds.

COB 1783-1788

20 October 1785, page 178
James Halley is appointed Surveyor of the Road from the Ponds to the Ox Road.

20 October 1785, page 178
George Smith is appointed Surveyor of the road from the forks at difficult to the red house.

20 October 1785, page 178
Phillip Alexander is appointed Surveyor of the road from Four mile run to Rock creek Ferry.

20 October 1785, page 178
Robert Adam is appointed Surveyor of the road from the bridge at Alexandria to four mile run.

20 October 1785, page 178
Sampson Cockerill is appointed Surveyor of the road from Ricketts to the Alexandria Road.

20 October 1785, page 179
William Payne is app[d]. Surveyor of the road from the forks of the road above Dickens to Ricketts.

20 October 1785, page 179
Gerrard Spinks is appointed Surveyor of the road from the back line to the forks of the road above Dickens.

20 October 1785, page 179
Edward Davis is appointed Overseer of the Road from the forks of the road above Summers's to the road that leads to the old Courthouse.

20 October 1785, page 179
Joseph Bennitt is appointed Surveyor of the road from Accotink to the Ox Road.

20 October 1785, page 179
Simon Summers is app[d]. Surveyor of the road from Ricketts to Accotink.

20 October 1785, page 179
Joseph Cockerill is app[d]. Surveyor of the road from Ricketts to the Leesburgh road.

20 October 1785, page 179
John Shepherd is app[d]. Surveyor of the road from Gullatt's fence to Difficult.

20 October 1785, page 179
William Wren is app[d]. Surveyor of the road from Gerr[d]. Tramells to Gullatts fence.

COB 1783-1788

20 October 1785, page 179
Thomas Lindsay is app^d. Surveyor of the road from the falls church to Trammells.

20 October 1785, page 179
Robert Powell is app^d. Surveyor of the road from the mile tree to the church.

20 October 1785, page 179
William Adams is appointed Surveyor of the road from the forks of the road at Cameron hill to the mile tree.

20 October 1785, page 179
William Deneale is appointed Surveyor of the road from the Ox road to the Loudoun line, and from Piney run to the main road.

20 October 1785, page 179
Richard Simpson is app^d. surveyor of the road from Accotink above Prices to the Ox road.

20 October 1785, page 179
Benjamin Gwin is app^d. surveyor of the road from Prices to Accotinck.

20 October 1785, page 179
John Moss is appointed surveyor of the road from the forks of the road above Summers's to Prices.

20 October 1785, page 179
Francis Summers is appointed surveyor of the road from the north run of hunting creek to the forks of the road above Summers's.

20 October 1785, page 179
George Gilpin is appointed surveyor of the road from the forks of the road above Herberts to the north run of hunting creek.

20 October 1785, page 179
Martin Cockburn is appointed surveyor of the road from Hereford's ferry to the old church and to the road leading from Colchester to Alexandria.

20 October 1785, page 179
Thomas Lucas is app^d. surveyor of the road from the Wolf run shoals to Sangsters.

20 October 1785, page 179
William Simpson is appointed surveyor of the road from the upper part of Woolf Pit hill to Loofboroughs.

COB 1783-1788

20 October 1785, page 179
William Thompson is appointed surveyor of the road from Colchester Ware-house to the upper part of Wolf Pitt hill.

20 October 1785, page 179
Peter Wagener is app^d. Surveyor of the road from Occoquan Ferry to Pohic run.

20 October 1785, page 179
Gilbert Simpson is appointed Surveyor of the road from Pohic run to Accotink on both roads, and from the new church to the river side road.

20 October 1785, page 180
Charles Smith is app^d. Surveyor of the road from Accotink to back lick.

20 October 1785, page 180
Richard Chichester is appointed surveyor of the road from the forks of the road below Pattersons to dogue run.

20 October 1785, page 180
Charles Little is app^d. surveyor of the road from dogue run to Cameron.

20 October 1785, page 180
Daniel McCarty Junr. is appointed surveyor of the road from the lower side of Accotink run on the river side road to dogue run.

20 October 1785, page 180
Lund Washington is appointed surveyor of the road from dogue run to the Gum spring, and from the spring to Poseys ferry.

20 October 1785, page 180
William Dorrell is app^d. surveyor of the road from the Gum spring to Cameron

20 October 1785, page 180
William Moon is appointed surveyor of the road from Alexander Hendersons plantation on bull run by Sampson Turleys to the Ox road.

20 October 1785, page 180
Henry Dogan is appointed surveyor of the road from the mountain road to Bull run near Hooe's mill.

20 October 1785, page 180
John Cotten is appointed surveyor of the road from the forks of the old Courthouse road to the old church road near Saunders's.

COB 1783-1788

20 October 1785, page 180
Thomas Herbert is appointed surveyor of the road from Alexandria to Cameron run also to the forks of the road at Cameron hill.

20 October 1785, page 180
George Gilpin, James Wren & William Payne Gent. or any two of them are appointed to allot the hands to work on the several roads in Fairfax Parish.

22 November 1785, page 181
Licence is granted William Lindsay to keep an Ordinary at Colchester who together with Cleon Moore his security acknowledged a bond for the same.

19 December 1785, page 183
On the motion of Samuel Adams, It is Ordered that John Jackson, George Smith, Charles Thrift and William Darne or any three of them view a way for a road from the red house to the said Adams's Mill on Pimmets run, and from thence to the road leading from the little falls to Alexandria and report on oath to the next Court the most convenient way for the same.

20 March 1786, page 186
On the motion of William Deneale, Ordered that Charles Broadwater, John Hunter, Edward Blackburn and John Coats or any three of them view the most convenient way to turn the road from Hunter's fence to Collins's fence and report on oath to the next Court.

23 March 1786, page 200
Ordered that the Sheriff collect the sum of Sixty pounds from the several persons in this County in proportion to their property assessed under the revenue Law agreeable to an act of Assembly "Entitled an act for keeping certain Roads in repair." And that he pay the same to George Gilpin and Charles Little Gent. Commissioners of this County, and it is Ordered that the Commissioners of the Land Tax furnish the said Sheriff with a list of the said proportion agreeable to the said Revenue Law.

18 April 1786, page 204
The viewers appointed to view a way for a road from Hunter's fence to Collins's fence returned the following report. Viz^t. "In Obedience to the within order of the worshipful Court of Fairfax being first sworn, we have viewed the way proposed by M^r. William Deneale to be as near as the present road to begin at the said Deneales corner on wolf-trap and thence with Lewis's line to his corner, & thence along the line of said Deneale & Fairfax to their corner near Collins's fence, and we think the way he wants the road removed will answer the publick as well as the road now goes. Charles Broadwater, John Hunter, Edward Blackburn." Whereupon it is ordered that the said Road be opened agreeable to the s^d. Report.

COB 1783-1788

18 April 1786, page 204
Ordered that Thomas Herbert, William Adams, James Wren, William Henry Terrett, George Hunter, Thomas Lindsay, Jeremiah Moore, John Sheppard & William Payne or any five of them view the Road from Alexandria to Difficult bridge, also from Alexandria to Lewis Ellzeys meadow branch and also from Alexandria to George Town ferry, and report on Oath to the next Court what alterations are necessary towards straitening the said Roads.

19 June 1786, page 244
George Gilpin, Charles Little & William Deneale Gent. are appointed to assist in viewing the roads and reporting agreable to an order made at April Court last.

19 June 1786, page 245
William Adams tendered in Court Forty three shillings being the value of an Acre of Land condemned opposite his mill on Pimmets run.

19 June 1786, page 245
The viewers appointed to view a way for a Road from the red house to Adams's Mill, on Pimmetts run, and from thence to the road leading from the little falls to Alexandria returned the following report: "In Obedience to an Order of the Worshipful Court of Fairfax County to us directed dated December Court 1785, we the subscribers have viewed the Road from the Red house to Samuel Adams's Mill, and find the most convenient and best way to be beginning at a white oak about one hundred and fifty yards below the red house, thence to a parcel of Rocks, and along a parcel of broken bushes to Pimmets run, thence crossing the said run along a parcel of broken bushes to the said Mill, thence from the said Mill thro' the said Adams's Plantation to the long branch of Pimmetts thence along John Butchers fence to Mathew Earp's, and thence crossing the Church road to the Alexandria road Given under our hands this 26°. day of January 1786. George Smith, Charles Thrift, William Darne." Whereupon it is Ordered that the said Road be opened agreeable to the said report.

COB 1783-1788

19 July 1786, page 251
The persons appointed to View the Roads from Alexandria to difficult bridge, also to Lewis Ellzeys meadow branch returned the following report - Viz^t.
"In Obedience to an order of the Worshipful Court of Fairfax, We the subscribers have viewed the road leading from Alexandria to Lewis Ellzey's meadow branch as far as the ford over Holmes's run and are of opinion the nearest and Straitest way is from Duke Street crossing the Gut below John Wests meadow, continuing on a strait course leaving dutch Caty's house to the South about thirty feet passing near the upper corner of Captain T. Wests Field to an oak on the side of the Road at the mouth of Colonel Gilpins lane then with the road to a branch near the hop-yard, thence along the side of the hill to Holmes's run opposite the mouth of Messrs. Bird and Hawkins's lane- We have also viewed the road leading from Alexandria to Difficult bridge as far as a white oak standing on the Road about a quarter of a mile above Capt. W. H. Terrett's old field and are of opinion that the straitest and nearest way is from Princess Street passing near an old Tobacco house formerly Baldwin Dades to the lower corner of Ramsays old field near some Peach trees, thence up the old field to a pine standing on the top of a hill above Ramsay's quarter, thence to the oak above mentioned, and are of opinion that the way for each road as described above may be made good. George Gilpin, Charles Little, James Wren, Thomas Lindsay, William Payne." Whereupon it is Ordered that the said Roads be opened agreeable to the said report, and that the present road from Cameron into Alexandria also remain open, to which opinion of the Court Jesse Taylor William Hepburn, Robert McCrea, Robert Allison, Dennis Ramsay, Peter Wise, James Hendricks, Thomas Conn, William M^c.Knight, John Lomax, and William Duvall objected so far as it relates to the entrance into Alexandria to the South of King Street, and pray an appeal for the following reasons, First, that the persons who viewed the road sat and determined the question. Secondly that the present entrance in Town is into King Street, and all the Public buildings such as the Market house, Courthouse the Tobacco warehouse and the Publick Ferry are all to the north of the present entrance into Town, and that the said entrance has been used & occupied ever since the said Town was by law established. Ordered that they give security in the Clerks Office for prosecuting the said appeal in the General Court with effect.

19 July 1786, page 252
Ordered that the male labouring hands do work on the several Roads in Fairfax district agreeable to a report returned by the persons appointed to allot said hands.

19 July 1786, page 252
John Harper is appointed surveyor of the road in the room of Charles Little Gent.

19 July 1786, page 252
William Scott is appointed Surveyor of the Road from the old Courthouse to the road leading to the falls.

COB 1783-1788

19 July 1786, page 253
William Fallin is appointed Surveyor of the road from the forks of Avery's road to the road leading from Alexandria to George-town.

19 July 1786, page 253
Charles Craik is appointed Surveyor of the road from Avorys Road to the Falls.

19 July 1786, page 253
Joseph Birch is appointed Surveyor of the road from the George-town Road at R. Adams's to Avery's road.

18 September 1786, page 286
Elihu Harden is appointed Surveyor of the road from Avery's road to the George Town Road in the room of Joseph Birch.

19 September 1786, page 287
Licence is granted McKinsey Talbutt to keep an Ordinary in this County who with Thomas West his security acknowledged a bond for the same.

16 October 1786, page 290
On the Petition of Amos Fox setting forth that he owns land on one side of Difficult run and intends to build a water Grist Mill on the said run and prays to have an Acre of land belonging to [blank] on the opposite side viewed and valued according to Law - Whereupon it is Ordered that the Sheriff summons a Jury of Twelve Freeholders of his Bailiwic to meet upon the said land, who being met and duly sworn, are diligently to view and examine the said Land adjacent thereto on both sides the said run which may be affected or laid under water by building the said mill together with the timber and other conveniences thereon and report the same with the true value of the acre petitioned for, & of the damages to the said [blank] or to any other person, under their hands and Seals, and that the Sheriff return the report to the next Court And it is Ordered that the Surveyor of this County survey and lay off the said Acre of land.

19 December 1786, page 314
Ordered that Charles Broadwater, George Gilpin, William Payne and George Minor or any three of them view the best & most convenient way for a road leading from the great falls of Potowmack beginning at difficult run to the Town of Alexandria and report the same to the next Court.

19 December 1786, page 314
Ordered that Alexander Henderson & William Deneale Gent. do allot the hands to work on the road leading from Piney Branch into the Ox road near the Church & report to the next Court.

COB 1783-1788

19 December 1786, page 314
Licence is granted William Turner to keep an Ordinary who with Gilbert Simpson & John Evans acknowledged a Bond for the same.

16 January 1787, page 316
James Deneale is appointed Surveyor of the road from Piney branch into the Ox road by the Church, and it is Ordered that the hands on the Plantations of John Gibson, Peter Mauzy, and John Buckley work on the said road.

16 January 1787, page 318
Marmaduke Beckwith is appointed Surveyor of the road in the room of Edward Ford.

19 February 1787, page 319
Licence is granted John Brown to keep an Ordinary at Colchester - who with Samuel Bayly his security acknowledged a bond for the same

20 February 1787, page 321
Ordered that John Wren, Simon Summers, Thomas Gunnell and Benjamin Talbut, or any three of them view the road leading from Wolf trap run to the old Courthouse and report to the next Court the conveniences and inconveniences attending the said Roads.

19 March 1787, page 329
Licence is granted James Collins to keep an ordinary in Fairfax County who together with Robert Allison acknowledged a bond for the same.

19 March 1787, page 329
Licence is granted William Simpson to keep an Ordinary in this County, who together with Michael Gretter his security acknowledged a bond for the same.

18 June 1787, page 417
Licence is granted Thomas Songster to keep an Ordinary in this County who together with Edward Ford his security acknowledged a bond for the same.

18 June 1787, page 417
Amos Fox tendered in Court Eight pounds five shillings being the value of an acre of Land belonging to Nehemiah Davis and petitioned for by the said Amos for the purpose of erecting a water Grist mill, whereupon it is ordered that the said Acre of Land be vested in the said Amos, & on his motion the Jury's & surveyors report respecting the said land is Ordered to be Recorded.

COB 1783-1788

18 June 1787, page 417
Ordered that Samuel Bayly, William Donaldson, William Reardon and Henry Reardon or any three of them view the way from Vernons Mill on Pohic to the stand or meeting house, and report on oath to the next Court whether a Road is necessary.

17 July 1787, page 427
George Gilpin, Charles Little, James Wren, William Payne, Roger West, John Moss, Charles Broadwater, George Minor, Thomas Gunnell, Charles Alexander, & David Stuart Gent. are appointed to lay off the several Roads within Fairfax district into convenient precincts and allot the hands to work on each precinct.

17 September 1787, page 446
Giles Cook, Alexander McDonald, John Jones & Charles Dodson or any three of them are appointed to view a way for a Road across by John Wards Mill to the road that leads to Alexandria and report to the next Court on Oath, the most convenient way to open the same.

17 September 1787, page 447
The persons appointed to lay off the several Roads in Fairfax district into precints and allot the hands to work on the same returned a report. Whereupon it is Ordered that the hands work on the several Roads agreeable to the said Allotment.

17 September 1787, page 447
Ordered that Martin Cockburn, Thomas Pollard, James Deneale & James Waugh lay off the several Roads in Truro district into convenient precincts, and allot the hands to work on each precinct, and report to the next Court.

15 October 1787, page 452
On the Petition of Thomas Pollard Gent. setting forth that he Owns land on one side of Giants castle branch, and is desirous of erecting a water Grist mill thereon, and prays to have an acre of land on the opposite side the property of John Gibson vewed & valued according to law: Whereupon it is Ordered that the Sheriff summon a Jury of twelve freeholders of his Bailiwic to meet upon the said Land who being met & duly sworn are to view and examine the land adjacent thereto on both sides the said run which may be affected or laid under water by building the said Mill together with the timber and other conveniences thereon and report the same with the true value of the Acre Petitioned for, and of the damages to the said John Gibson or to any other person, under their hands and seals, and that the Sheriff return their report to the next Court; and it is Ordered that the Surveyor of this County survey and lay off the said Acre of Land.

16 October 1787, page 454
Charles Beaty together with Samuel Bayly & Peter Wise his securities entered into and acknowledged a bond for keeping the Ferry at George Town.

COB 1783-1788

16 October 1787, page 455
Ordered that the hands belonging to Henry Gunnell at his home house & Quarter work on the road from the old courthouse to Doctor Dick's mill.

20 November 1787, page 470
Ordered that the Ferry rates from Col°. Masons land over to George Town be established as they were by an act of Assembly passed in the Year 1748 and that the person keeping said ferry do keep one scew, one foot boat, one horse boat and four hands for the dispatch of Passengers.

17 December 1787, page 485
The persons appointed to view a way for a Road from the Colchester road to the road that leads to Alexandria returned the following report, Viz^t. "Pursuant to the Order of Court hereunto annexed we the subscribers being first sworn have viewed the road thereinmentioned, and find the most convenient way for a Road is beginning at the Colcheser road near John Wards Mill, from thence following the bridle way thro' William Rogers's plantation, and continuing the said bridle way to the corner of Richard Sanfords fence, from thence thro' the woods down to the road leading to Alexandria opposite to Capt. Harpers fence. Given under our hands this 17° day of December 1787. Giles cooke, Alex^r. McDonald, Charles Dodson." Whereupon it is Ordered that William Rogers & Richard Sanford be summoned to appear at the next court to shew cause if any, why a road should not be opened agreeable to the said Report.

18 December 1787, page 486
On the motion of James Halley, Ordered that John Wren, James Deneale, George Winn, & William Deneale, or any three of them View the line between James Hally & John Halley, and report to the next Court whether they think it will be a convenient way to remove the road.

18 December 1787, page 486
Simpson Halley is appointed Surveyor of the road from difficult to Songsters.

18 February 1788, page 489
Charles Broadwater, George Gilpin, William Payne, Charles Little, John Jackson & John Allison Gent. or any three or more of them are appointed to view the most convenient way for a Road to lead from the Great Falls of Potowmac, and to continue within the County of Fairfax to the Town of Alexandria.

18 February 1788, page 490
James Hurst is appointed a Commissioner of the roads in addition to George Gilpin & Charles Little the present Commissioners.

COB 1783-1788

20 February 1788, page 491
William Darne, George Thrift, William Wren, Robert Powell and John West or any three or more of them are appointed to view a way for a road from Samuel Adams's Mill, to the place where Benjamin Sebastian formerly lived, and report on Oath to the Court.

20 February 1788, page 491
On the motion of Thomas Herbert leave is given him to erect a Water Grist Mill on Holmes run, on his own land.

16 June 1788, page 532
Robert T. Hooe, Richard Conway, George Minor and William Payne Gent. or any two of them are appointed to settle the Commissioners account of the road tax & report the same to the Court.

FAIRFAX COUNTY COURT ORDER BOOK 1788-92, PART 1

21 July 1788, page 2
George Mason Ju^r. is appointed Surveyor of the road in the room of Martin Cockburn Gent.

21 July 1788, page 3
John Summers Junr. is appointed Surveyor of the Road in the room of William Payne Gent.

22 July 1788, page 5
Robert Adam, David Stuart & Charles Alexander Gent. ore any two of them are appointed to allott the hands to work on the turn pike road from Alexandria to George Town ferry.

22 July 1788, page 5
William H. Terrett, George Minor, Thomas Gunnell & Robert Powell are appointed to allot the hands to work on the turn-pike road from Alexandria to Difficult

22 July 1788, page 5
Thomas Pollard, William Deneale, Benjamin Dulany and Roger West are appointed to allot the hands to work on the turn-pike road from Alexandria to Newgate.

16 September 1788, page 47
William Peake is appointed surveyor of the road from Cameron run to the Gum spring in the room of William Dorrell.

22 October 1788, page 51
On the Petition of Messrs. Hepburn & Dundas setting forth that they hold lands on one side of back lick run and also on one side of Indian run, and are desirous of building a water Grist Mill thereon, and pray that an Acre of land on each of the said runs opposite to the Petitioners may be viewed and valued according to law. Thereupon it is Ordered that the Sheriff summon a Jury of twelve freeholders of his Bailiwic to meet upon the lands Petitioned for who being met and duly sworn are diligently to view and examine the said land, and the Land adjacent thereto on both sides the said Runs which may be affected or laid under Water by building the said Mill together with the timber and other conveniencies thereon and report the same with the true value of the two acres Petitioned for and of the damages to the Proprietors or to any other person or persons under their hands and seals and that the Sheriff return the said report to the next Court And it is further Ordered that the Surveyor of this County survey and lay off the said two acres of land.

22 October 1788, page 52
Giles Cook, John Ratcliff, Samuel Arell, and William Henry Terrett or any three of them are appointed to view a way for a road, to lead from the main road at or near Prices to Hepburn & Dundas's Mill on back lick run and from thence into the main road leading to Alexandria and report to the next Court.

COB 1788-92, PART 1

15 December 1788, page 75
James Wren, William Darne, and John Darne, or any two of them, being first sworn for that purpose before some Justice of the peace for this County, are appointed to view a way for a road leading from the North run of Holmes to the Falls Church, and report to Court truly and impartially the conveniences and inconveniences which will result as well to individuals as to the publick, if such Road shall be opened.

15 December 1788, page 75
William Lyles, Thomas Herbert, and William Ward, or any two of them being first sworn for that purpose before some Justice of the Peace for this County are appointed to view the most convenient way for turning the road leading from Cameron run into the turn pike road, and report to Court of the conveniences and inconveniences which will result as well to individuals as to the publick if such road shall be turned.

16 December 1788, page 77
On the motion of George Washington esq[r]. leave is granted him to alter and open roads on his own Lands, agreeable to a survey and plat laid before the Court.

16 December 1788, page 80
The Persons appointed to view a way for a road leading from the main road at or near Prices to Hepburn's & Dundas's Mill on back lick run and from the said Mill to the Main road leading to Alexandria, and report to Court, now returned a report in these words "We the Subscribers being appointed by the Worshipful Court of Fairfax to view a way for a road, leading from the main road near Price's to Hepburn and Dundas's Mill and from thence to Alexandria, do report That we conceive the nearest and best road to be as follows (Viz[t].) Beginning at the dividing line of William Fitzhugh & the Heirs of Henry Fitzhugh where they intersect the road leading to Newgate, thence with a straight line as near as may be to the Colchester road and to the dividing line of John and Barbara Ratcliff, thence along the said dividing line of John and Barbara Ratcliff as far as may be convenient, and thence with as straight a course as may be through a point of the said Barbara's land and through the land of the Heirs of James Murray to Hepburn & Dundas's Mill as aforesaid - and from the said Mill to the Corner of Hepburn & Dundas's land and David & Sam[l]. Arells land near Minor's gate, thence along the lane that divides the said Arell's land & the land of Charles Lee Esq[r]. as far as the said Lee's fence along the said lane from thence with as straight a line as may be to the Corner of William Bird's fence on the Newgate road, and thence along the said road to Alexandria - John Ratcliff, WH[y]Terrett, Sam[l]. Arell December 8th 1788." Whereupon it is ordered that the road be opened agreeable to said Report.

COB 1788-92, PART 1

21 April 1789, page 137
The Persons appointed to view a way for a Road from the Great falls of Potomack to the Town of Alexandria returned their Report in these words, to wit, "Fairfax County ss In obedience to an order of the Worshipful Court of this County, directing us to view the Ground between the Town of Alexandria and the great Falls of Potowmack, after viewing the ground and considering every circumstance, we are of opinion that the nearest and best road may be made in the following direction, to wit, from the lower Ford over Difficult Creek by pursuing the present great Road as it now runs to the top of a Hill near the Corner of Edward Lanhams field to the line of William Swink then with the line of Swink along his lane past his House to the Land of Joseph Simpson, then with his line to an Oak standing by his Fence, then across his Field by a Locoast to an old field of Gustavus Scott Esqr. then through Scott's Field to the land of John Jackson then through his land to the Land of Charles Thrift, then through a small corner of Thrift's and John Jackson's Land to the place on Scott's run where W^m. Shortridge's old Mill stood, then by a strait line thro' Turbervilles land to an oak corner to Turberville, Scott and George Hunter, then with G. Hunter's line a small distance to a Branch, then across a small corner of G. Hunter's and John Jackson's Land to the line between Richard Conway & Jno. Jackson thence with said line to the line between Philip Darrell and W^m. Gunnell thence with their line to the edge of Gunnell's high woods, thence across a small part of his woods in a direction which shall strike the fence of Gunnell's at the place where foot people crosses to the house, thence by a strait line to the great Road leading from Leesburg to Alexandria at or near that part where James Wrenn's Store stands, thence by the great Road as laid out by the Commissioners of the Turnpike road, and from said Store to the Town of Alexandria. George Gilpin, Charles Little, WPayne, Geo. Minor, John Jackson" and the Court being of opinion that the said road will be necessary and useful, It is therefore ordered that the same be opened agreeable to the said Report, and that the Sheriff Summon Edward Lanham, William Swink, Joseph Simpson, George Smith agent for Gustavus Scott Esqr. John Jackson, Charles Thrift, [blank] agent for John Turberville, George Hunter, Richard Conway, Philip Darrell and William Gunnell to appear at the next Court, to shew cause, if any they can, why the said road should not be opened, and Simon Adams, John Errenshaw, and Joseph Simpson, or either of them, are appointed to superintend the opening and clearing the same.

21 April 1789, page 138
Richard Chichester, Thomas Pollard, John Moss, David Stuart, George Minor, Thomas Gunnell, and William Payne, or any four of them, are appointed to allot the Hands to work upon the several publick Roads in this County, and make report thereof to the Court

18 May 1789, page 139
Ordered that the Sheriff Summon the several Overseers of the Roads through this County, to appear at Price's Tavern on Friday the 29th Instant, to assist the Persons appointed to allot the several hands to work on the Publick roads in making their Report.

COB 1788-92, PART 1

18 May 1789, page 140
It appearing to the Court, that the Persons appointed to view a way for a road leading from the Great falls of Potomack to Alexandria, returned an improper Report. Thereupon Jeremiah Moore, John Moss, John C. Hunter, Rezin Offutt, Lewis Hipkins, William Adams, and Simon Adams, or any three of them, being first sworn before some Justice of the Peace for this County, are appointed to view the most convenient way for a Road leading from the Great falls of Potomack to Alexandria, and report to the Court truly and impartially, the conveniences and inconveniences which will result as well to individuals as to the publick, if such road shall be opened.

15 June 1789, page 160
Ordered that the Sheriff Summon the several Surveyors of roads in this County to appear at Price's Tavern on the Wednesday preceeding July Court next, to assist the Gentlemen appointed to allot the several hands to work upon the several publick roads in this County.

15 June 1789, page 161
On the motion of John Dowdall for leave to erect a Water Grist Mill on Four mile run, It is ordered that a Writ of ad quod damnum do issue according to law.

15 June 1789, page 161
On the motion of John Dowdall, Ordered that George Minor, William Darne, Thomas Crawford Junr. and Thomas Taylor, or any three of them, being first sworn for that purpose before some Justice of the Peace, do view the road from George Town to the Falls church and the ground along which the said John Dowdall proposes turning the same, and report to Court the comparative conveniences and inconveniences which will result as well to individuals as to the publick, if the said road shall be turned.

COB 1788-92, PART 1

15 June 1789, page 161
The persons appointed to view the road from the Great Falls of Potomack to Alexandria returned their Report in these words, to wit, "In obedience to the annexed order of Court, we the subscribers being first duly Qualified before James Wren Gent. Have carefully view'd the most convenient way for a road leading from the Great falls of Potomack to Alexandria and are of opinion no way can be had so good and convenient, nor to do, so little damage to the lands thro' which it must pass, as the way pointed out in the report return'd by the former viewers. Excepting the lands of M^r. George Hunter, Thro' which they have directed the Road to pass we think it may be well, for the said Road to continue on the line, between M^r. Hunter and M^r. Scott nearer said Hunter's Corner. In doing so we conceave it will not make the distance fifty yards further, spare more of said Hunter's wood land and only take off from M^r. Jackson a little more old field not under a fence. We farther give it, as our oppinion, That a good Road may be had. That it is highly necessary it should be open'd as soon as possible. That it will not only be of publick utility, But realy usefull to that thick settled neighbourhood, all the way from Difficult Run to Col^o. Wren's Store. As to the inconveniences. They are as follows, Viz^t. (1st) Intended Road will pass thro a part of Edward Lannum's field, who has no objections (2^d) thro' part of a field of W^m. Swink's. No Objections. (3^d) Thro part of a field of Joseph Simpson. No objections. (4th) Thro a small part of a worn out, and much Gullied old field, not under a fence, which is rented by M^r. Daniel Jenkins from M^r. Scott, said Jenkins objects - for what, we cannot conceive. (5th.) Thro' a part of M^r. John Jackson's lands where he lives. (no objections) (6th.) Thro' a very small part, of a wourn out Gullied old field (under a fence) Rented by M^r. Charles Thrift of Robert Carter Esqr. objections - For what, we can't see, - But are convinced it may be an advantage (7th.) Cross Scot's Run, which must be causey'd seventy or Eighty yards. (8th.) Thro' a part of M^r. John Turbervilles lands. no objections. (9th.) Thro M^r. Hunter's and M^r. Jackson's lands discribed before. (10th.) Cross Pimmits Run, which must be causey'd more than one hundred yards (11th.) Thro a piece of woodland, about one Quarter of a mile, belonging to M^r. William Gunnell. (Objections) (12) Thro' a part of M^r. John West's lands, no objections. (13^h) Thro' M^r. Sampson Trammill's lands. no objections (14^h.) Thro' a part of Col^o. Wren's lands to the Turnpike Road at his Store. Given under our hands this 11th. day June 1789. W^m. Adams, Simon Adams John Moss." Whereupon it is Ordered that s^d. Daniel Jenkins, Charles Thrift and William Gunnell be Summoned to appear at the next Court to show cause, if any they can, why the said road should not be opened through their Lands.

16 June 1789, page 164
On the motion of James Hailey a Licence is granted him to keep an Ordinary at his House in the Town of Alexandria for one year from this day and from thence till the next Court to be held for this County Whereupon the said James, with Baldwin Dade his security entered into and acknowledged a Bond according to Law.

16 June 1789, page 166
Thomazin Ellzey is appointed by the Court, a Commissioner of the Turnpike roads.

COB 1788-92, PART 1

20 July 1789, page 170
On the motion of Cumberland Furguson a Licence is granted him to keep an Ordinary at his house in this County for one year from this day, and from thence till the next Court to be held for the said County, he having with John Harper his security entered into and acknowledged a Bond according to Law.

21 July 1789, page 171
Charles Thrift, William Gunnell and Daniel Jenkins having been summoned to appear at this Court to shew cause, if any they cou'd why a road should not be opened from the great falls of Potomack to Alexandria, which would lead through their Lands, and the said Charles and William not appearing, and the said Daniel waving his objections It is Ordered that the road be opened agreeable to the report of the Persons appointed to view a way for opening the said road.

21 July 1789, page 171
Simon Adams is appointed Overseer of that part of the Road leading from the Great falls of Potomack to Alexandria, which leads from Col. Wren's Store to the falls road.

21 July 1789, page 171
Gilson Whaley is appointed Overseer of that part of the Road leading from the Great falls of Potomack to Alexandria, which leads from the falls road to Difficult.

22 August 1789, page 196
Thomas Gunnell, John Jackson, and Rezin Offutt, are appointed to value the Damages sustained by Daniel Jenkins in consequence of the road leading from the Great falls of Potomack to Alexandria passing thro' his Land, which said Damages are not to exceed the sum of Ten pounds. and ordered that they return a Report thereof to the Court.

21 September 1789, page 204
On the motion of Charles Thrift, John Jackson, George Smith, John Shortridge, and Joseph Simpson, or any two of them, being first sworn before some Justice of the Peace for this County, are appointed to view a way for a road leading from Charles Thrift's near the Little fall's road, to the new cut road leading from the Great fall's to Alexandria, and report to Court truly and impartially, the conveniences and inconveniences which will result as well to individuals, as the Publick, if such road shall be opened.

FAIRFAX COUNTY COURT ORDER BOOK 1789-91

22 September 1789, page 25
The Persons appointed to lay off the several Publick roads in this County into Districts, and to allot the hands to work on the said Roads, returned their Report, Whereupon it is ordered that the male labouring Tithables that now or hereafter may be within the District and on the Plantations of Sebrek Scott, and all the hands above the Run and between the Road from the ferry opposite George town to four mile run, George Chapman's home house excepted, work on and keep the said Road from the ferry opposite George town to Four mile run, in repair according to Law.
=

The road from four mile run to the Turnpike gate - all the hands between the road and river within the District and the male labouring tithables of George Chapman's home plantation as last
=

On the road from the entrance of Avory's road near Marle's house to the Turnpike road leading to the Fall's Church - the tithables of W^m. Fallin, John Bowling, John Kirby Thos. Taylor, W^m. Piper [blank] Hedrick, Jacob Payne, Caleb Brady, Caleb Richards, Archibald Swain, Benj^a. Moulds, John Lister, John Goldop, Joseph Nicholson, Thos. Pendoll, Benj^a. Swain, Thomas Shearwood, Elijah Wood, Walter Hill, John Scott, Annanias Payne, Thomas Payne, John Ball, as last. W^m. Follin Overseer
=

On the road from Robert Adam's Mill to Avory's road, the Tithables within the District &^{c}. of Nace Farthing, John Avory, George Chapman's Quarter, Jere: Thrift, Benj^a. Watson, Moses Ball Junr., as last, Moses Ball J^r. Overseer
=

On the road from Avory's road to the Mountain Road that leads to the Falls landing - the tithables within the District, &^{c}. of Charles Craig, John Massey, Matthew Earp Junr. Edward Earp, W^m. Earp, W^m. Nelson, Rich^d. Nelson, John Robertson, W^m. Baker, James Baker, John Mattinly, James Mattinly, James Sewell, Robert Sewell, as last, Thos. Muse Over.
=

On the road from the Falls Church to the road leading from the falls warehouse to Alexandria, the tithables within this District, &^{c}. of Thos. Muse, John Wilson, Nelson Reed, James Reed, Matthew Earp, Jere. Spurlin, James Mattinly, James Thomas, W^m. Darne, John Butcher, as last, W^m. Darne Over.

22 September 1789, page 26
On the road from the Falls landing to the Red house, the tithables that now are, &^{c}. of Thos. Bladen, Lewis Hipkins's Store, Ludwell Lee's quarter, Geo. Minor's quarter, Ann Minor, Thomas Moody, George Blake, as last, Thos. Moody, Over.
=

On the road from the Red house to the Turnpike Road, the tithables that now are, &^{c}. of Henry Lee's quarter, John Sattridge, Joel Earp, John Williams, John Anderson, Jacob Ruxby, as last, George Smith.
=

COB 1789-91, PART 1

22 September 1789, page 26

On the road from the falls road to the old Courthouse, the Tithables that now are, &ᶜ. of Benjamin Harris, as last Benjamin Harris O.

=

On the roads from the forks of the road near the old Farm to Difficult, the tithables that now are, &ᶜ., of Towlson Mills, Ayler Williams, Jacob Moxley, Thomas Moxley, Thoˢ. Owens, John Moxley, John Turberville's Quarter, Milkey Pearson, Joseph Simpson, Wᵐ. Swink, as last. Jnᵒ. Henderson O

=

On the road from the Turnpike Gate at the intersection of the George Town Road to the White Oak at the Widow Tucker's - the tithables which now are, &ᶜ. of Joseph Carroll, Joseph Gowing, Carlyle's Farm, Ramsay's Quarter, John Ellis, Robert Sanford, Fielden Isaac, Guy Evans, John Violett, Wᵐ. Hʸ Terrett, Mʳˢ. West, Daniel Mᶜ.Alister, John Hawkins, Wᵐ. Herbert's Quarter, Thoˢ. Daniel, Widow Tucker, James Webb, Dade's old Quarter, Chaˢ. Curtis's place, Geo. Solomon, Samuel Bowling, Wᵐ. Bowling, Elisha Harden, Wᵐ. Lyles, James Kenny, Patty Moxly, Fitzgerald's Quarter, and all those living on the land formerly Townshend Dade's and Charles Alexander's, on the west of the west of the George Town Road, and also all those living on the Land of Thoˢ. & John West, between the turn pike road and the road leading to Colchester, Joel Cooper, Robert Adam's plantation. John Geesling, as last.

=

On the road from the white oak at the Widow Tucker's to the falls Church - the tithables which now are &. of James Daniel, John Dowdall, Hall Harding, James Ball, Wᵐ. Donaldson, Andrew Donaldson, Wᵐ. Piper, Thoˢ. Taylor, Thoˢ. Donaldson, Wᵐ. Carlin, Moses Ball, Charles Little's Quarter, Wᵐ. Adams, Wᵐ. Bladen, Wᵐ. Merryday, Vincent Taylor, Edward Skidmore, John Thomas, Simon Pearson, John Pearson, Whiting's Church Quarter, John Adams, John Darne, Robert Powell, John Wren, Thomas Grafford's place, James Wren's Quarter, George and Mʳˢ. Minor's, Ann Bowling, Wᵐ. Thomas, Moses Harden, Mary Frizzel's, Joseph Thompson, John Mills, Wᵐ. Crump Junʳ., Joseph Powell's plantation, Caleb Bladen, Henry Hussey, Benjᵃ. Dulany's two Quarter's, Jonathan Ward, Thomas Pearson's, George Thrift, Charles Charters, James Robertson, Samuel Macatie, as last

=

COB 1789-91, PART 1

22 September 1789, page 27
On the road from the Fall's Church to Deakin's near the Court house Run, - The tithables which now are, &^{c}. of Thos. Lindsay James Wren, M^{rs}. Lindsay, Col. Broadwater's quarter, John West, Benjamin Ballenger, John Hurst's Quarter, W^m. Adam's quarter, W^m. Waters, Henry Darne, Mary Peak's place, W^m. Barry, Thos. Trammell, Charles Simms, Thos. Beach, George Hunter, Philip Dorrell, Sampson Trammell Jun^{rs}. plantation, W^m. Middleton, Widow Lester's, Henry Layne, Luke Fields, John Peptisco, Charles Little at Danby, Watson's place, where Joseph and Negroe Tom lived, Widow Gordon's, Gustavus Scott's farm, John Hunter, William Henderson, Edward Bates, Fenix Duffin, James Collins, and all the Tenants at his place, Nicholas Garrett, Timothy Carrington, W^m. Sims, Francis Rose, David Allen, W^m. Davis, Robert Finley, John Wood, Andrew Beatle, Simon Jenkins, W^m. Wren, Gerrard Trammell, Charles Jenkins, Ralph Hoskins, Joseph Cockerille, Thomas Baker, Col. Broadwater's, Samuel Adams Junr., Simon Adams, W^m. Walker, W^m. Williams, W^m. Gunnell, Cravin Simms, Lewis Hipkins, Allen Davis, as last.
=

The road from Deakins's near the old Courthouse Run, to Difficult, The tithables which now are, &^{c}. of John Shepperd, Jere: Moore, Jesse Moore, George Smith, Charles Thrift, Samuel Jenkins, Bryan Fairfax and Quarter, Reisin Offield, John Lester, James M^c.Carty, Cornelius M^c.Carty, Thomas Fairfax, Thomas Gunnell, Henry Gunnell, and Quarter, Edward Adams, West's old Quarter, Daniel Jenkins, John Jackson, Widow Gullatt's, Franklin Perry's old place, Henry Garrett, Alexander Beach, Benjamin Ballenger as last.

22 September 1789, page 28
The road from the Falls Church by Ricketts' to the Ox Road - the tithables which now are, &^{c}. of Sampson Cockerille, Thos. Dayly, Sarah Dulin's, Jere: Williams, Richard Chichester's quarter, Thos. Burgess, John Ballenger, John Horseman, W^m. Scisson, Nicholas Sebastian, W^m. Lewis, Charles Lewis Broadwater, Simon Summers, Widow Summers, James Dove, Widow Coxon, James Hurst's, John O.Daniel, W^m. Presgrave, John Arskin's, Daniel Storer, Matthew Bosswell, W^m. Johnson's plantation, Summers's plantation, John Pasley, Lewis Saunders, Benj^a. White, Thos. Curtis, John Skinner, as last. Sampson Cockrill O
=

The road from the corner of Richard Scott Blackburn's fence to where it intersects with the old Courthouse Road, The tithables that now are, &^{c}. at Richard Scott Blackburn's Quarter, as last, Stephen ODaniel O.
=

The road from the old Courthouse to W^m. Turner's Tavern, - The tithables which now are, &^{c}. - of Edward Blackburn, Joshua Kidwell, W^m. Gunnell, Michael Keller, Strutton Johnson, Jesse Kidwell, John Mourland, George Williams, James Carroll, John Edmonds, Sarah Monroe's, John Robertson, Nicholas Mooney, John Fenley, John Hurst, Samuel and Daniel Thompson, Aaron Nally, Young Rhoby, Spencer Monroe, Andrew Monroe, Francis Poston, Mary M. Talbott's, Nehemiah Davis, Henry Gunnell's Quarter, Joseph Bennett, Ben: Talbutt, John Wren, John Eels John Chapman Hunter, as last. Edwd. Blackburn O.
=

COB 1789-91, PART 1

22 September 1789, page 28
The road from the old Courthouse Road opposite Joseph Bennett's Plantation, to where it intersects the Ox road - the tithables which now are, &c. of Oliver Burch, Thomas Palmer, Sampson Martin, Miles Rhoby, as last. Thos. Palmer.
=

The road from Difficult bridge where the Ox road crosses to Tho. Songster's - the tithables which now are &c. of James Halley, Thomas Simpson Shoemaker, Edward Taylor, John Dawson, Ch^s. Beach, W^m. Harrison, Simpson Halley, Francis Taylor, Edward Taylor Junr. W^m. Lewis's quarter, John Mansell, Hopkins Rice, Thomas Pollard, James Deneale, James Deneale j^r. W^m. Deneale, as last, Simpson Hally O.

22 September 1789, page 29
The road from the County line to the white oak spring, being part of the turnpike road leading to Alex^a. - The tithables which now are, &c. of James Gill, Joseph Wood, Merryman Harrover, James Gilpin, Thomazin Ellzey, David Stuart's quarter, Marmaduke Beckwith, Newman Beckwith, Jonathan Jackson, Orr's Tenant's, W^m. Turner, Barnaby Woolbright, James Wren j^r. John Buckley, John Gibson's quarter's, Philip Pritchard, W^m. Buckley, Thos. Johnson, Rich^d. Ratcliff, George Summers, Thos. Black, as last,
=

The Road from the white oak Spring to Price's Tavern, the tithables which are, &c. of Samuel Smith, Henry Fitzhugh's Estate, Eli Stone, John Dulin, Tho. Lewis, Opie Lindsay, Syvelster Hall, Samuel Weedon, Isaac Davis, Thos. Johnson, W^m. Stiel, John Peacock, John Mackintosh, John Ashford, M^{rs}. Hepburn's, W^m. Skinner j^r., James Morely, Charles Knowland, Lancelot Johnson, W^m. Fitzhugh's Estate, Jacob Woolbright, W^m. Johnson, W^m. Gilpin, John Jones, John Allen, Sarah Pilcher's, Robert Sisson, W^m. Skinner, Walter Johnson, John Johnson, Lewis Tustar
=

The road from Price's Tavern to Holmes's run, - The tithables which now are, &c. of Charles Burgess, Peter Dow's Plantation, W^m. M^c.Carty, Jeremiah Jefferson, Samuel Arell, W^m. Talbutt, M^c.Lain's place, [blank] Browning, Jefferson's old place, W^m. Hepburn's Mill, W^m. Youst, Lawrence Monroe, Francis Summers, Doct^r. Hawk, W^m. Frazer, John Moss, Thos. Dove, Zachariah Morriss, John Ratcliff, Barbary Ratcliff, Thomas Beach, W^m. Bushby's plantation, M^{rs}. West's quarter, M^c.Kenzy Talbutt, M^{rs}. Prices, Doctor W^m. Brown's quarter, Chandler Spinks, Joseph Hamilton, John Davis, W^m. Davis j^r., Ignatius Lucket, Thos. Lambert, John Payne, Henry Davis, John Swallow, Thos. Baggatt, Jeremiah Jefferson, Joseph Smith, Ann Popejoy's place, John Atkinson, David Thomas, Thomas Kirby, Charles Burgess, as last.
=

COB 1789-91, PART 1

22 September 1789, page 30
The road from Holmes's run, to the turnpike Gate, where W^m. Simpson lives, The tithables which now are, &^{c}. of Chas. Lee's farm, Col. Hendricks, W^m. Bird, Benj^a. Dulany, John Hawkins, W^m. Lightfoot, W^m. Hartshorne, and all his Tenants, Elizabeth Cox, Richard Sanford, Richard Sanford j^r. Thomas Herbert, John Hulls, W^m. Ward, Thos. West's plantation and all his tenants under the Hill, as last.
=

The road from Pinny branch to the New Church. - The tithables which now are &^{c}. of Peter Mauzy, Rich^d. Smith, W^m. Carico, as last, James Deneale O.
=

The road from the Leesburgh road, to Rickett's - The tithables which now are &^{c}. of W^m. Simms Junr., Alexander Simms, as last, Jos. Cockrill O.
=

22 September 1789, page 30
The road from Ricketts to Frazer's The tithables which now are &^{c}. of Col. W^m. Payne, John Summers 3^d, James Richards, Andrew Allen, W^m. Gooding, W^m. Richards, Thos. Johnson, Daniel Mills, W^m. Crump, Ch^s. Cavery, Edward Davis J^r. Samuel Johnson, as last. Nicholas Fitzhugh O.
=

The road from Nicholas Fitzhugh's Gate, to the back lick run, - the tithables, &^{c}., of Gerrard Spinks, Barton Martin, Edwd. Potter, Thos. Kirby, Giles Cooke, John Murry, Chas. Dodson, John Kent, Aler Williams, as last. Gerr^d. Spinks O.
=

The road from Backlick run to where it intersects the road from Colchester to Alex^a. at the end of John Hereford Jun^{rs}. lane, - The tithables which now are &^{c}. of W^m. Ogden, Thos. Ogden, Alex^r. McDonald, W^m. Payne Baylis, Joseph Ballenger, Jacob Hall, John Devaughn, George Mills, Charles Smith, as last. Chas. Smith
=

The road from the turn pike Gate down the Colchester road, to Dogues run at M^{rs}. Moon's - The tithables which now are &^{c}. of Clish plantation, Bryan Fairfax's plantation adjoining Clish, W^m. Hunter, Archibald Johnston, Wilfred Johnston, Col. Mason's two Quarter's at Hunting Creek and Dogues run, M^{rs}. French's, John Compton, Charles Hagon, Francis Hagon, W^m. Keaton, Jere: Adams, W^m. Simpson's, John Harper, Edward Lewis, Walter Brooke, Widow Dorrell, W^m. Dorrell, Thos. Hatford, M^{rs}. Slaughter's, W^m. Boyd, [blank] Jones, (Taylor), W^m. Freeman, Samuel Baggett, Widow Moon, as before. John Compton Os.

COB 1789-91, PART 1

22 September 1789, page 31
The road from Dogue's run at M^{rs}. Moon's to Accotinck - the tithables which now are, &^{c}. of Rich^d. Chichester's home house and Quarter on Accotinck, John Hereford j^r., Thomas Bayliss, John Johnson j^r., Joseph Warner, Rich^d. Scrivener, Zepheniah Swallow, Burgess M^c.Pherson, W^m. Bladen j^r., Jacob Hall j^r. Ann Wiggs, W^m. Williams, Thos. Davis, W^m. Johnson j^r. John Alliston, Thomas Alliston j^r. William Rogers Senr., John Ballenger, John Hill, Jonah Hill, William Cash, W^m. Violett, Francis Ashford, W^m. Triplett Senr., Michael Ashford, Henry Tyler, W^m. Tyler, Henry Worthin, Jo^s. Javins, Edward Worthin, Lund Washington, Zachariah Ward, Ben: Burton, Ebsworth Payne, Casandra Stone's, George Wyley, Samuel M^c.Lean, Samuel Stone, John Fowler, as last. Rich^d. Chichester O

The road from Accotinck to Colchester - the Tithables which now are &^{c}. of M^{rs}. Sarah Triplett's, Robert Boggess, Gilbert Simpson, Joseph Potter, George Mason of Pohick and all his Tenants, W^m. Weston's old place, John Atcheson, Leonard Atcheson, Widow Ragan, Robert Speake, W^m. Weston, Ben: Rigg, Doctor Coningham, Saml. Bayly, Lee Massey, W^m. Thompson, W^m. Lindsay, Jesse Williamson, Lund Washington Junr., Robert Cochran, Lawrence Washington, John Sexton, Walter Findley's old place, Samuel Alexander's old place, W^m. Essex, Terrence Conner j^r. W^m. Harris, Col. Wagener's hands, Ludwell Lee's tenants, Robert Mollihon, Jacob Moreland Junr., Robert Church, Thomas Church, W^m. Triplett j^r. W^m. Reardon, Henry Reardon, W^m. Donaldson, Peter Coulter, W^m. Simpson, (Buck), Widow Boggess, as last. Peter Wagener O.

22 September 1789, page 32
The Road from Robert Boggess's to Genl. Washington's Mill, also from Pohick new Church to intersect the above mentioned road - the tithables which now are, &^{c}. of Daniel M^c.Carty Jun^{ra}. hands, W^m. Potter, Syrus Payne, John Meara, W^m. Pool, John Javins, Lawson Parker, Thos. Pool, Dennis Baptist Hambleton, Thos. Alliston, Daniel Stone, Samuel Gates, M^{rs}. Marton's hands, Col. McCarty's hands, Henry Loyd, W^m. Gray, Francis Spencer, as last. Daniel McCarty Junr. O.

The road from General Washington's ferry to his Mill, from thence to his tumbling dam, thence along his new road to intersect the river side old road above the Gum spring, - The tithables that now are, &^{c}. of General Washington, as last Geo. A. Washington O.

The Road from Cameron Run or Creek by Clish to where General Washington's new road intersects the river side old road - The tithables which now are, &^{c}. of Roger West, Smith's quarter, Buckner Stith, Thompson Mason, Abednego Adams, Mrs Peakes', W^m. Peake, John Frazer, Lawrence M^c.Ginnis, Robert Alexander, Zachariah Ferguson, M^{rs}. Wilson, Alexander Brinston, as last. Thompson Mason O

The road from Pohick old Church to Holland point ferry, The tithables that now are &^{c}. of Col. Mason's home house, and Quarter's, George Mason j^{rs}. home house, and Quarter's - as last Geo. Mason J^r. O.

COB 1789-91, PART 1

22 September 1789, page 32

=

The road from Hallace's Old field to where the road intersects the Ox Road near Thos. Songster's - The tithables which now are, &^{c}. of W^m. Mills, Thos. Woodward, Ja^s. Martin, Saml. Woolbright, Bennett Johnson, Bozwell Williams, W^m. Hally, Dennis Mackeltamare, W^m. Hamilton, Henry Gorin, Abel Carrico, W^m. Carrico, W^m. Dewitt, Solomon Jackson, as last. Will. Mills O

22 September 1789, page 33

The road from the fork of Charles Beache's fence on the Ox road to the Ponds on the Pohick road. - The tithables that now are &^{c}. of W^m. Thompson's place, Nehemiah Morris, Francis Coffer, Barnaby Logan, George Simpson, M^{rs}. Hartshorne's, James Simpson's quarter, Thomas Windsor, James Simpson, John Rigg, Ben: Suddith, Susanna Simpson, Leonard Harper, Aaron Simpson, John Pameroy, James Tascox, Nicholas Hammond, W^m. King, James Hunt, Peter Riley, Nehemiah Morris as last Aaron Simpson O

=

The road from the Ponds to Pohick Church - The tithables that now are &^{c}. of W^m. Keene, W^m. Barker, Widow Williamson, Francis Keene, Widow Ward, Widow Rogers, James Rattle, Jonathan Denty, John Harley, George Harley, W^m. Hicks, as last. Will. Kean O.

=

The road from Hallace's old field to where it intersects the Pohick road by John Fowler's Plantation - The tithables that now are, &^{c}. of Thos. Halley, Moses Barker, W^m. Barker's Plantation, Grafton Kirk, Daniel Kent, Zebedee Compton, John Powell, Gilbert Deavers, James Grimsley, W^m. Rogers Junr., Josiah Clark, Samuel Smitherman, John Ward, Benjamin Riley, Charles Holden, W^m. Hall, Widow Rigg's place, Henry Wingate, John Wiggs, Sanford Potter, as last. Josiah Clarke

=

The road from Bull run at Col: John Hooe's old Mill to the Mountain Road near Thos. Songster's - The tithables which now are, &^{c}. of W^m. Whaley, W^m. Coe, Robert Wickliff, Ann West's, Daniel Douglass, Joseph Jackson, Hanson Croson, Henry Davis, John Tillett j^r., John Harraver, as last. Robt. Wickliff

=

The road from the County line on the Newgate Road to David Loofbourrow's - The tithables which now are &^{c}. of John Brumback, Sanford Payne, W^m. Simpson, Cornelius Thompson, James Waugh, Alexander Henderson's Plantation, Tylor Waugh, Ed^d. Ford, Sarah West, John Chappell, John Marshall, Cornelius Kincheloe, Thos. Songster, Henry Roussaw, James Hickie, [blank] Prichard, (Schoolmaster) Richard Simpson, George Tillett, James Tillett, as last Will Simpson Son of W^m.

=

COB 1789-91, PART 1

22 September 1789, page 34
The road from David Loofbourrow's to Woolfpitt hill - The tithables which now are, &^{c}. of Caleb Stone, Joseph Blansit, Jesse Bayley, John Blansit, Jere: Woodyard, Gerrard Barnett, John Tillett, Rich^d. Wheeler, Drummond Wheeler, John Reed, John Harrison, Thos. Gossom, John Powell, W^m. Eaton, W^m. Holoman, Henry Hunt, W^m. Simpson, John Spragg, Joseph Blansit, John Row's Plantation, as last. Thos. Gossom, O

=

The road from Woolfpit hill to Colchester, - The tithables which now are, &^{c}. on the Plantation of Francis Stone, Ed^d. Washington, John Coffer, John Scott, John H. Manley, Robert Lawson, John King, M^{rs}. Moseley's Quarter, W^m. Mills, John Philips, Joseph Hampton, John Hampton j^r. John Hampton Senr. Gilbert Rowland, Mordecai Jacobs, W^m. Simmonds, Charles M^c.Nabb, W^m. Bryan, John Johnson, (Shoemaker) as last. W^m. Thompson O.

=

The road from the Wolfrun Shoals on Ocaquan to Thos. Songster's - the tithables which now are &^{c}. of Thos. Lucas Chas. Thrift, Ben: King, John Yeamon, Hargess King, Rich^d. Clarke, Yelverton Reardon, W^m. Moon, Thos. Williams, Robert Geesling, Lawrence Suddith, as last, Chas. Thrift

=

The road from the Ox road above W^m. Simpson's to where it intersects the road from the Woolfrun Shoals to Thos. Songsters by Sampson Turley's Plantation. The tithables which now are, &^{c}. of Sampson Turley, Marmaduke Beckwith Quarter, Joseph Jacobs, John Simpson, Sarah Reed, Baxter Simpson, Benj^a. Suddith as last. Sampson Turley O.

=

The road from the forks of the road near Martin Cockburn's, to where it intersects the road from Alex^a. to Colchester, near Bayly's Mill run, - The tithables which now are &^{c}. of Martin Cockburn as last Martin Cockburn O.

COB 1788-92, PART 1

22 September 1789, page 209
On the motion of William Gunnell, Ordered that a Writ of ad quod damnum be issued, to enquire of the Damages which he may sustain by reason of the road leading from the Great falls of Potomack to Alexandria's passing through his Land.

19 October 1789, page 211
On the motion of William ^{Mc}.Knight, a Licence is granted him to keep an Ordinary in Alexandria, for one year from this day and from thence till the next Court to be held for the said County, he having with John Stewart his security entered into and acknowledged a Bond according to Law.

FAIRFAX COUNTY COURT ORDER BOOK 1788-92, PART 2

15 February 1790, page 255
The persons appointed to view a way for a road leading from Charles Thrifts near the little Falls road to the new cut road leading from the great Falls to Alexandria returned the following report Viz^t. "In obedience to an order of Fairfax Court dated September Court 1789 we the underwritten subscribers being first legally sworn have viewed the Road from the little Falls Road to the great Falls Road as follows Viz^t. Beginning at or near M^r. Thrifts gate opposite his house then a direct course thro' the said Thrifts Turnip patch to the top of a hil in his old field thence along that Ridge til it intersects the great Falls road at or near the Corner of John Jackson's fence the distance being about half a mile thro' the lands of Robert Carter esq^r. under rent to the said Thrift which we conceive is worn out - 1st. By so doing we conceive it will save a line of fence through the said Thrifts Plantation about three quarters of a mile part of which is good tendable land, and secondly we conceive it will be of great advantage to the land of John Turberville thro' which the Road as it now stands is about half a mile in length and the two Roads not more than a quarter of a mile apart. We think this road will not be quite as near as the other but equally as good and the turning of it will be of a great advantage to the above lands and no disadvantage to the publick Given under our hands this 2nd. day of February 1790 John Jackson, George Smith, John Shortridge" Whereupon it is ordered that the road be opened agreeable to the said Report.

16 February 1790, page 258
On the petition of Thomas Herbert setting forth that he holds Land on both sides of Holmes run and is desirous of erecting thereon a water Grist mill on his motion it is ordered that a writ of Ad quod Damnum issue according to Law

COB 1789-91

16 March 1790, page 80
Licence is granted William Lindsay to keep an Ordinary - John Stone Sec^y

COB 1788-92, PART 2

16 March 1790, page 267
The Commonwealth against William Young. On an Information for retailing spiritous Liquors without Licence This day came Bushrod Washington Gent. Prosecutor for the Commonwealth of Virginia in the County of Fairfax and the said William having been duly served with the process against him on this Information, and being now solemnly called & failing to appear, It is considered by the Justices here that the said William Young do make a fine and pay to the said Commonwealth of Twenty two pounds ten shillings Current money and that he pay Costs

COB 1788-92, PART 2

17 May 1790, page 291
On the motion of William Gunnell &al. It is ordered that a writ of Ad quod Damnum issue to asscertain the damages sustained by them by a Road being opened leading from the great Falls of Potowmack to Alexandria thro' their Lands.

18 May 1790, page 296
...Whereupon the said Thomas Herbert has leave granted him to build and erect a water Grist Mill and Dam agreeable to his Petition To which opinion and Judgment of the Court Henry Astley Bennett esquire excepts and prays an appeal to the first day of the next Dumfries District court and Charles Little and Robert Townshend Hooe entered into and acknowledged a Bond for prosecuting the same with effect.

19 July 1790, page 324
John Jackson, Thomas Gunnell, John C. Hunter & James Hurst or any two of them are appointed to meet the Commissioners from Loudoun and to agree with some workman to make the necessary repairs at Difficult bridge.

19 July 1790, page 324
William Adams, Samuel Adams, George Minor and William Fallin or any three of them are appointed to view a way for a Road to lead from Georgetown ferry beginning at Lubber branch and extending to Dowdals Corner on the hill by four mile Run and report on oath the conveniences and inconveniences attending the same

20 September 1790, page 348
On the motion of Philip Richard Fendall & Lewis Hipkins setting forth that they hold Lands on both sides of Pimmits Run and are desirous of building a water Grist Mill on the said Run whereupon it is ordered that a writ of Ad quod Damnum be issued as the Law directs

20 September 1790, page 350
Upon the application of Charles Shoemaker & Joseph Kirkbride that a new Road may be opened from the landing opposite their mill upon Occoquan River to the town of Alexandria - George Gilpin, Thomas Herbert, Edward Washington, Isaac McPherson, Martin Cockburn, John Fowler & Lawrence Washington or any three of them are appointed to view the ground along which the said Road is proposed to be conducted and report to the Court truly & Impartialey the conveniences & inconveniences which will result as well to Individuals as to the Publick if such way shall be opened

21 September 1790, page 354
John Moss, William Payne, Thomas Pollard & Richard Simpson or any three of them are appointed to view the Road as turned in Fitzhugh's Lane and report the conveniences & inconveniences attending the turning of the said Road from its former course

COB 1788-92, PART 2

17 October 1790, page 356
Simon Pearson, James Robertson and Thomas Darne or any two of them are appointed to view a way for a Road leading from the Falls Church to the little Falls of Potomack & report on oath the conveniences and inconveniences attending the same

17 October 1790, page 356
The viewers appointed to view a way for a Road to lead from George Town ferry beginning at Lubber branch and extending to Dowdalls corner on the hill by four mile Run, returned the following Report, Viz^t. "In obedience to an Order of Fairfax Court we the Subscribers viewed the said Road leading from Dowdall's Corner/mentioned in the said Order of Court/ to Lubber branch, we think it reasonable that the said Road should be turned on the line between Thomas Darne & Samuel Shreve, the way being nearer and equally as good from the said Corner to Lubber branch, and so on a straight line til intersects the Georgetown Road below the Alexandria Road It is also our Opinion that the Road should be opened by the parties that made application Given under our hands this 12°. day of October 1790 W^m. Adams, Geo: Minor, William Fallin" Whereupon it is ordered that the said Petitioners have leave to oppen the said Road at their own expence agreeable to the said Report

17 October 1790, page 357
Licence is granted James Wiley to keep an Ordinary who with William Summers his Security acknowledged a Bond for the same

21 February 1791, page 394
On the motion of John Hughs, a Licence is granted him to keep an Ordinary at his house in the Town of Colchester for one year from this day, and from thence til the next Court to be held for this County. Whereupon the said John with Terrence Conner his Security entered into and acknowledged a bond for the same -

COB 1789-91

22 February 1791, page 206
John Jackson & Thomas Gunnell are appointed to meet Com: from Loudoun to receive the work done on Difficult bridge.

22 February 1791, page 207
To a proportion of money Levied for build^g. a bridge over Diff. run, for Thos. Barker . . . . £20..0..0..

COB 1788-92, PART 2

22 February 1791, page 398
On the motion of Charles Jones a Licence is granted him to keep an Ordinary at his house in this County for one year from this day, and from thence til the next Court to be held for this County. Whereupon the said Charles with Thomas West his Security entered into, and acknowledged a bond for the same -

COB 1789-91

18 April 1791, page 233
A Writ of Ad Quod Damnum & Inquest for valuing an acre of Land to erect a water Grist Mill, Petitioned for by Philip Richard Fendall & Lewis Hipkins is ret[d]. & Rec[d].

COB 1788-92, PART 2

18 April 1791, page 426
The persons appointed to view the road as turned in Fitzhugh's Lane returned a report in these words (Viz.) "In obedience to the annexed Order of Court, we the Subscribers have viewed the road as turned in Fitzhughs Lane and are of opinion that the present road with the repairs now made on it by M[r]. Benjamin Guinn is nearer and better than any other way that can be had. Given under our hands this 11th. of April 1791. John Moss, Richard Simpson, Tho[s]. Pollard Whereupon it is Ordered that the road be continued agreeable to the said report

19 April 1791, page 429
On the motion of Charles Lewis Broadwater It is Ordered that Jeremiah Moore, Richard Ratcliff, John Wren and John C. Hunter or any three of them being first sworn for that purpose, view a way for a road from Mathew Boswells Shop on Elzeys Church road to the Mouth of the widow Summer's Lane and report to the next Court of the conveniences and inconveniences that will result as well to Individuals as the Publick if such road be opened.

19 April 1791, page 430
On the motion of Samuel Stone and others It is Ordered that Francis Stone, Zebede Compton, Charles Smith and William Keen or any three of them being first sworn according to Law, view a way for a road leading from the neighbourhood of Occoquan Forrest crossing Accotink at John Wards Mill and down to Alexandria and report to the next Court of the conveniences and inconveniences that will result as well to Individuals as the Publick if such road be opened. -

COB 1789-91

18 July 1791, page 247
The persons appointed to lay off the several publick roads in the parish of Truro in this County into Districts, and to allott the hands to work on the said roads, returned their report, Whereupon It is ordered that the male labouring Tithables, belonging to, residing with or that hereafter may reside on the plantations, of the persons following Viz^t. James Gill, Jos: Wood Merryman Harrower, James Gilpin, Thomazin Ellzey, Doct^r. David Stuart, Marmaduke Beckwith, Newman Beckwith, Thos. Pollard, Will. Turner, Vincent Turner, John Barnes, Barnaby Woolbright, Jno. F. Buckley, William Buckley, John Gibson's quarter, Philip Pritchart, J^s. Deneale J^r. quarter, James Deneale, Ch^s. Reed, Thomas Winsor Junr. Richard Ratcliff, James Halley, Henry S. Halley, Jno. Wren, Benj^a. Talbott, George Summers, Jonathan Jackson, Jno. Jackson, Jno. Mansfield, Thos. Simpson (Shoemaker) Tho. Mackey - do work on and keep, in repair the road from the County line on the Turnpike road, lead^g. to Alexandria & to the ox road.

18 July 1791, page 247
The road, from the Ox road on the Turnpike road to Hollis old fields, and opposite to where another road Intersects said Turnpike road leading from the Ox road. the Tithables which now are &c. of, James Conner, John Orr's Tennants, Benjamin Saunders, Ann Farr, Abednego White, Saml. Smith (Shoemaker) W^m. Skinner, William Skinner Junr. Samuel Woolbright, Bazil Williams, Bennett Johnston, William Johnston, Walter Johnston, Lancelott Johnston, John Johnston, (Son of W^m.) Thomazin Ellzey's quar^{tr}. Thomas Woodward, James Martin, Noah Martin, Gerr^d. T. Conns plantation, James Moxley, John McIntosh, Charles Noland, Sarah Pilcher, [blank] Voilett, William Ratcliffs, M^{rs}. Hepburn, John Ashford, Opie Lindsays Quarter, Isaac Davis, William Steel, Elie Stone, Saml. Weeden, John Weeden, Jno. Harle, Sylvester Hall, Matthew Bozwell, David Davis, William Hepburns Plantation (on Pohick) John Skinner

=

18 July 1791, page 248
The road from Hollis's old field - to Accotink, on the Turnpike road, The tithables which now are &c. - of Richard Lane, John Duling, Thomas Lewis, Andrew Allen, John R. Allen, Robert Scisson, Mary Smith, the quarter belonging to the Estate of Henry Fitzhugh dec^d. which is now on the South side of Accotink and one of William Fitzhughs quarters, on the South side of Accotink -

=

The road from Accotink on the Turnpike road to Prices Tavern the Tithables which now are &c - of John Jones, Charles Dodson, Alier Williams, Thomas Kirby, Edward Potter, Giles Cook, Gerrard Spinks, John Murray, the quarters belonging to the Estate of Henry Fitzhugh dec^d. which are on the East side of Accotink, and the quarters, belonging to William Fitzhugh, which are on the East side of Accotink

COB 1789-91

18 July 1791, page 248
The road from the Turnpike road to the ox road, near the brick Church, The tithables which now are &c. - of Peter Mauzey, William Carrico, Uriah Morris, & William Mittchels quarter, James Deneale, Overseer

18 July 1791, page 249
The road from the County line on the mountain road, to Thomas Sangsters, The Tithables which now are &c - of William Simpson J^r. John Brumback, Tyler Waugh, Sanford Payne, Cornelius Thompson, James Waugh, Alexander Henderson's Plantation /on Bull run/ Edwd. Ford, Sarah West, John Butlers, quarter, Lawrence Butler, M^{rs}. Stuarts, John Chappel, John Marshall, Cornelius Kincheloe, Henry Russau, Benjamin Wests, Spencer Colvert, & David Loofborough, - William Simpson Junr. Overseer
=

The road, from the mountain road above Sangsters to Bull run, at Hooes old Mill, The Tithables, which now are &^{c}. - of William Whaley, William Coe, Robt. Wickliff, Henry Davis, Spence Simpson, John Harrower, Yelverton Reardon, William Powell, Sarah Reed, William West, Laurence Suddith King, William Moone, John Moone, M^{rs}. Sarah Wardens Plantation & Joseph Williams - Cornelius Kincheloe, Overseer
=

The road from Thomas Sangsters, to Bull run, at the Wolf Shoals The tith^s. which now are &c. - of Thomas Lucas, Benj^a. King, John Yeoman, Ch^s. Thrift, Hargess King, Richard Clark, Richard Simpson, Moses Simpson, Robert Gozeling, John Yeoman, Junr. George Tillett, Junr. James Tillett, Elijah Kent, Benjamin King of /Loudouns, Tennants/ - Charles Thrift, Overseer.
=

The road, from the Ox road above William Simpson's to where it Intersects the road from Sangsters to Bull run, The Tith^s. &c. - of Sampson Turley, Marmaduke Beckwiths, quarter /on Wolf run/ Jos. Jacobs J^r. John Simpson/son of W^m./Baxter Simpson Benjamin Suddeth, Junr. & Henry Pell, - Sampson Turley Overseer

18 July 1791, page 250
The road from Difficult on the ox road, to Thomas Sangsters Shop, The Tithables which now are &c. - of Nehemiah Davis, Edward Taylor, Edward Taylor Junr. James Taylor, John Dorson, Ch^s. W. Beach, Jacob Valentine, George Harrison, William Lewis's quarter, & William Harrison - Henry S. Halley, Overseer

COB 1789-91

18 July 1791, page 250

=

The road, from Sangsters Shop, on the ox road, to the white oak Harbor, the Tithables, which now are &c. - of Caleb Stone, Joseph Blancitt, Jno. Blancitt, Jos. Blancitt Junr. Jesse Bayly Jere: Woodyard, Gerr^d. Barnett, John Powell /son of W^m./ Nehe: Morris, Rich^d. Wheeler, Drummond Wheeler, John Reed, John Harrison, Thomas Gossom, William Eaton, William Holliman, H^y. Hunt, Pricilla Hunt, William Simpson /son of Geo./ George Nichols, Thomas Athey, John Rows Plantation, John Sprag, Edward Washington, John H. Manley, Edwd. Pearson, the Frenchman on the plantation formerly Jno. Tillett - Thomas Gossom Overseer

=

The road, from the old white oak Harbor, on the ox road - to Colchester at the old sign post the Tithables, which now are &c., - of John Coffer, Francis Stone, Robert Lawsons quarter, John King, John Hampton, John Hampton Junr. Joseph Hampton, William Mills, John Philips, Gilbert Rawland, Mordicai Jacobs, William Simmons, Ch^s. McNabb, William Bryan, John Johnston, /Shoemaker/. William Triplett Junr. William Simpson /brick/. Thomas Church, Robt. Church, Josiah Boggess, William C. Bayly, John Alton, Daniel Linch, Rich^d. Hall, Jno. Scott, /B.Smith/. Zachariah Bond, [blank] Edwards, Zachariah Burns, William Church, Danl. Dorsey & J^s. Spilman - Francis Stone, Overseer

18 July 1791, page 251

The road from Colchester ferry, along the road leading to Alexandria to Accotink run, the Tithables which now are &c. - of Peter Wagener, William Lindsay, William Thompson, Robert Cochran, Samuel Bayly, [blank] Carson, Lund Washington Junr. Jesse Williamson, Terrence Conner J^r. William Harris, Peirce Bayly, John Meara, Jno. Hewes, Cornelius Wells, Lee Massey, William Weston, Will. Reardon, H^y. Reardon, W^m. Donaldson, Ludwell Lee's tenants, Lawrence Washington, Geo. Mason/Pohick/. and all his tenants, or tithables on his lands, Robt. Boggess, Sarah Triplett, John Vernon, Tho. Bayliss /on P.k. creek/. Gilbert Simpson, Joseph Potter, John Atcheson, William Essex, Jere: Hutcheson, Ann Reagan, Robt. Speak, John Short, John Sexton, Walter Findlys old place, Samuel Alexanders D.^{o}. rob^t. Mollerhorne, Jacob Noland, Jacob Noland Junr. John Cranford, Walter Hutcheson, Ch^s. Wright & his mother - Peter Wagener Overseer

=

The road from Accotink, on the Alexandria road, to the Parish Line at Dougues run, the tithables which now are &c. - of Rich^d. Chichesters home house, & quarter, on Accotink, Ann Wiggs, William Williams, Tho. Davis, William Johnston, Junr. John Allison, Tho. Allison Junr. John Hill, Josiah Hill, Will. Voilett, William Cash, W^m. Triplett, Fran.Ashford, Mich^l. Ashford, H^y. Tyler, W^m. Tyler, H^y. Worthing, Edwd. Worthing, Jos: Javins, Zach. Ward, Benj^a. Burton, Alesworth Payne, Casandra Stone, Geo. Wiley, Lund Washington, William Johnston, /Cap^n. Archibald Johnstons plantation, Col^o. Ch^s. Simms quarter, W^m. Hunter quar., Col^o. Geo. Mason's /quar^r. on Dougues run/ [blank] Benson, & Sarah Darrell. - Rich^d. Chichester, Overseer.

=

COB 1789-91

18 July 1791, page 251
The road from the Ox road near Sangsters, to Hollis's to the Turnpike road, The Tithables, which now are &c. - of William Mills, Tho. Sangster, Francis Coffer, Joshua Coffer, W^m. Gilpin, Dennis McNamarr, Will. Hamilton, Henry Gorun, Abel Carrico, William Carrico Junr., Solomon Jackson, William Dewer & William Green - William Mills, Overseer

18 July 1791, page 252
From the Alexandria road at Jno. Herefords lane, to Prices tavern, on the old Courthouse road, the Tith^s. which now are &c. - of Ch^s. Smith, William Ogdon, Tho. Ogdon, Alexander McDaniel, Will. Payne Bayliss, Tho. Bayliss, Joseph Ballenger, Jno. Devaughn, Geo. Mills, Jno. Johnston, /son of James/. Zach. Swallow, Jos. Warner, Burgess McPherson, John Hereford, Junr. W^m. Bladen Junr. [blank] Speake, Peter Pool, Rich^d. Pool, W^m. Rogers,/Blacksmith/. Edwd. Williams, John Kent, Jane Kent, James Glauster, Rich^d. Conway's Plantation or Tenant. - Charles Smith, Overseer
=

The road from the ox road at Ch^s. Beach's, to the Ponds on the Pohick road, The Tith^s. which now are &c. - of Aaron Simpson, Geo. Simpson, M^{rs}. Marg. Hartshorne, Bazzel Hunt, James Simpson's quar^r. Jno. Rigg, Benjn. Suddeth, Susanna Simpson, Leo^d. Harper, Rich^d. Simpson's quar^r. Jno. Pumroy, W^m. King, J^s. Hunt, Thos. Winsors, Zach: Dove & Hancock Lees quar^r. - Aaron Simpson Overseer
=

The road from the Ponds, on the Pohick road, to Pohick Church The tith^s. which now are &c. - of Francis Keen, W^m. Keen, John Fowler, Peter Coulter, J^s. Halley, Junr. Jacob Hall, Jonathan Denty, Mary Rogers, William Hicks, Mary Ward, John Scott, Sarah Athey, George Harley, Tho. Simpson, & Will. Barker Junr. - Francis Keen, Overseer
=

The road, from the Turnpike road, at Hollis's to where it Intersects the Pohick road at Jno. Fowlers fence. The tith^s. which now are &c - of Josiah Clark, Tho. Halley, Moses Barker, William Barker, John Barker, Grafton Kirk, Danl. Kent, Zebedee Compton, Jno. Powell /Pohick/. J^s. Grimsley, William Rogers Junr. Samuel Smitherman, & Son's. John Ward, Benjn. Riley, Ch^s. Riley, Ch^s. G. Holden, W^m. Hall, John Proctor, Aaron Johnston, Benjamin Rogers, Samuel Stone, John Allen /Sawyer/ Cheurie & Durand /F/. Gilbert Deavers, Sandford Potter & Jane Williamson - Josiah Clark, Overseer

18 July 1791, page 253
The road from William Turners tavern, to the parish line on the old Courthouse road, the Tith^s. which now are &c - of John B. Findlay, Hezekiah Kidwell, Elijah Kidwell, Jesse Kidwell, [blank] Dayly, Jno. Edmonds, Sarah Monroe, Spence Monroe, Andrew Monroe, William Monroe, Sybel Hurst, Samuel Thompson, Aaron Nalley, Saml. Thompson, Junr., Danl. Thompson, Elijah Robey, Francis Poston, John Ballenger, Mary M. Talbutt, John Eals, William Stephens & Edwd. Davis Junr. - John B. Findlay, Overseer
=

COB 1789-91

18 July 1791, page 253
The road from the ox road to the Parish line, on Ellzey's Church road, The Tith^s. which now are &c. - of James Dove, J^s. Hurst, Marg^t. Wilcoxen, Jno. O'Daniel, Jno. Askins, Courtney Askins, Jno. C. Robertson, Danl. Storey, Price Skinner, Lewis Saunders, Sampson Martin, Francis Taylor, & Patrick McCarty, - Stephen O'Daniel, Overseer

=

The road from Ellzey's Church road, to the old Courthouse road at Bennetts, The tith^s. which now are &c. of Tho. Palmer, Oliver Burch, Jos: Bennett, Mich^l. H. Robey, & John Thompson, - Tho. Palmer, Overseer

=

The road from old Pohick Church, to Holland Point Ferry, the tith^s. which now are &c. - of Col^o. Geo. Mason's home house & Quarter's in the neck, Geo. Mason Jun^{rs}. home house & quarter, George Mason Junr. Overseer

18 July 1791, page 254
The road from the forks near M^r. Cockburn's, to the Alexandria road, near Bayley's old Mill, The tith^s. &c. of Martin Cockburn - Martin Cockburn Overseer

18 July 1791, page 254

=

The road, from Robt. Boggess's to the General's Mill, & from said road to Pohick Church, The tith^s. which now are &c. - of Daniel McCarty Junr. William Potter, John Javins, Lawson Parker, Daniel Stone, Tho. Allison, Tho. Pool, W^m. Pool, Denison B. Hamilton, Saml. Sam^{l}Gates, Jane Morton, John Ballenger /at Belvoir/. Col^o. Daniel McCarty, Henry Loyd, Will. Gray, Fran: Spencer, W^m. Spencer, Simms /a free negro formerly M^r. Barnes's/. Daniel McCarty Junr. Overseer

=

The road from the Gen^{ls}. Mill by his Mill dam & along his new road, to the parish line - The tith^s. & - of Genl. George Washington's home house & quarters, in Truro Parish - Geo. A. Washington Overseer.

=

The Persons appointed to lay off the several Publick roads in the Parish of Fairfax in this County, into Districts & to allott the hands to work on the said road, returned a report, Whereupon It is ordered, that the male labouring Tith^s. belonging to residing with or that may hereafter reside on the plantations of the Persons following, Viz^t. Jno. Shepard Doct^r. Dick's Plantation, & Mill, & Danl. & Cornelius McCarty, Henry Gunnell Senr. & his quar^r. Tho. Gunnell W^m. Gunnell, Jno. Lester, Edward Adams, Tho Fairfax, Rezin Offord, Alex. Beach, Jacob Ballenger, Saml. Wheeler, Jno. & Geo. Anderson, Toulstons quar^r. Geo. Smith, Hamilton Thrift, Henson Vermillion, M^{rs}. Hassett, Simon & Marg^t. Jenkins, James Edwards & Tho. Bates, do work on & keep in repair according to Law, the road from Difficult bridge to the Courthouse run.

COB 1789-91

18 July 1791, page 255
The road from the Courthouse run to the Falls church, the Tith^s. which now are &c. - of Tho. Lindsay Ch^s. Little at Danby, Ch^s. Broadwaters quarters, Ch^s. Simms, Ch. Jenkins, Tho. West, Craven Simms, Edwd. Bates, Gerr^d. Trammell, Henry Kean, Henry Darnes, James Wren, Jno. West, Jno. Hunter, Js. Collins, Jno. Wood, Jos. Cockrill, Jno. Hipsico, Gustavus Scotts,farm, Luke fields, Saml. Howard, Nicho. Garrett, Robt. Finlay, Susanna Lindsay, Tho. Trammell, Timothy Carrington, Tho. Beach, W^m. Wren, W^m. Gunnell, Edwd. Conner, Francis Rose, Elijah Blundal, W^m. Davis, D^d. Allen, Gedion Shoemaker, John Wilson, & Joseph Fagan.

18 July 1791, page 255
The road from Prices Tavern to the forks of the road above Summers's, The tith^s. which now are &c. - of Saml. Smith, W^m. Yost, McKinzey Talbutt, M^{rs}. Wests quar^r. Benoni Price, Jno. Ratcliff, Barbara Ratcliff, Tho. Dove, John Moss & Nehemiah Morris -
=
The road from the forks of the road above Summers's to Brides bridge, The tith^s. which now are &c. - of Leonard Smith, Jesse Loudoun, Jno. Potts, Jno. Gibson, Ch^s. Burgess, Jno. Moore, Saml. Arell, Ch^s. Tucker, W^m. Talbutt, M^{rs}. Cox's quarter, Robt. McCrea, Peter Dow, Jere. Jefferson, Hendricks place, Ben. Dulany, Ch^s. Lee's place, W^m. Bird, Lawrence Monroe, Jos. Bushby, Tho. Lambert, Tho. Baggott, Fran^s. Summers, Jno. Mullichan, Jno. Payne, Lewis Tristler, Terretts school house, [blank] Pagett, W^m. Gladin, Terretts quar^r. Doct^r. Brown's quarters, Saml. Johnston & Ch^s. Currie.
=
The road from Birds bridge to the Turnpike, the tith^s. &c - of John Wises place, Henry Davis, W^m. Hartshorne, J^s. Abercrombie, Gray Evans a Waggoner, Saml. Roberts, George Kelly, Robt. Sandford, Tho. West & his tenants, Tho. Herbert, Ludwell Lee, the tenants, at Clish, Rev^d. Bryan Fairfax, Roger West, Samuel Lightfoot, Boyds quar^r. Rich^d. Sandford Senr. Rich^d. Sandford Junr., M^{rs}. Coxes' home house, Ch^s. Harper & Jno. Bright.

18 July 1791, page 256
The road from Cameron run to Douges run, The Tiths. which now are &c. - of Will. Dorrell, Cap^n. Harper, Walter Brooks, Mary Darrell, W^m. Keating, Edwd. Lewis, M^{rs}. French, Fran^s. & Ch^s. Hagan, John Compton, Jerry Adams' place & Col^o. Masons upper quarter - William Darrell, Overseer.
=

18 July 1791, page 256
The road from Cameron run to the Gum Spring. the Tith^s. which now are &c. of Bennett & W^m. Freeman, Geo. Wilson, Will. Jones, W^m. Boyd, Hepburns place, Henry Bozwell, M^{rs}. Slaughter, Buckner Stith, Thompson Mason, Mary Peake, Abednego Adam's, Major Washingtons quarter, James Pettit, Robt. Alexanders Tenants, Laurence McGinnis, W^m. Robinsons, Jno. Armstrong, Henry Brent, & Thomas Flattford - William Peake, Overseer.
=

COB 1789-91

18 July 1791, page 256
The road from the Turnpike near Alexandria to Four Mile run, the Tith^s. which now are &c. - of Baldwin Dade, Robert Vincents, Alexander Gordon's, Phil. Fendalls, Ch^s. Alexanders & his tenants between the road & the river, Widow Adams's, M^r. Doughlass, Talbutts place, Elihu Harding & Hugh McNail -
=

The road from Four Mile to Geo. Town Ferry, Robt. Alexander & his tenants, Will. Washington, David Stuarts quarter, Robt. Hedrick & Geo. Chapman.
=

The road from Marle's to the Turnpike road leading to Leesburgh, the Tith^s. which now are &c. - of Sandfords quar^r. Jacob Paynes, Caleb Brady, Caleb Richards, Archibald Swain, Benjamin Moulds, John Lister, Jno. Galdup, Joseph Nicholson, Tho. Pendall, Benjamin Swain, Tho. Sherwood, Tho. Pearson, Walter Hill, Jno. Scott, Zach. Scott, Annanias Payne, Tho. Payne, Jno. Ball, Francis Payne, Jeremiah Thrift, Rich^d. Leonard, Tho. Darne's quart^r. & Sabrett Scott - Samuel Shreeve Overseer.

COB 1788-92, PART 2

19 September 1791, page 465
On the motion of Thomas Sangster, a Licence is granted him to keep an Ordinary at his house in this County for one year from this day and from thence til the next Court to be held for this County. Whereupon the said Thomas with William Deneale his Security entered into and acknowledged a bond for the same. -

19 September 1791, page 465
On the motion of Cornelius Wells a Licence is granted him to keep an Ordinary at his house in the Town of Colchester for one year from this day and from thence til the next Court to be held for this County. Whereupon the said Cornelius with William Thompson his Security entered into and acknowledged a bond for the same.

19 September 1791, page 466
On the motion of William Lindsay a Licence is granted him to keep an Ordinary at his house in Colchester for one year from this day, and from thence til the next Court to be held for this County. Whereupon the said William with John Stone his Security entered into and acknowledged a bond for the same. -

19 September 1791, page 467
Edward Blackburn, John Jackson, George Williams and Edward Adams or any three of them being first sworn according to Law are appointed to view a way for a Road from Colonel Broadwaters Mill to the old Court house and report to the Court truly and impartially of the conveniences and Inconveniences which will result as well to individuals as the Public if such road shall be opened. -

COB 1788-92, PART 2

20 September 1791, page 470
Ordered that Benjamin Dulany, Thomas Herbert and William Henry Terrett or any two of them being first sworn according to Law view a way for a Road from William Hartshornes Mill to the Turnpike road and report to the Court truly and impartially of the conveniences and inconveniences which will result as well to Individuals as the Public if such road shall be opened. -

19 December 1791, page 504
On the motion of James Wren, a Licence is granted him to keep an Ordinary at his house in this County for one year from this day and from thence til the next Court to be held for this County whereupon the said James Wren with Jesse Moore his Security entered into and acknowledged a Bond for the same. -

19 December 1791, page 504
Francis Coffer, James Waugh, Peter Mauzey and Thomas Woodard or any three of them, are appointed to view a way for a Road leading from William Deneale's Mill to the Road leading from Hollis's old Field to Thomas Sangsters, and report to the Court truly and impartially of the Conveniences and Inconveniences which will result as well to Individuals as the Publick if such Road shall be opened.

20 December 1791, page 510
Ordered that the Prosecutor for the Commonwealth Commence Suit on a Bond given by Charles Beatys and his Securities for the faithful keeping of the Ferry at George Town. -

20 February 1792, page 514
Ordered that Samuel Smith, Chandler Spinks, Benoni Price John Summers and Simon Summers, or any three of them being first sworn according to Law, view the back lick Road from Prices Ordinary to the fork of the road above Nicholas Fitzhughs Gate and report to the Court truly and impartially of the Conveniences and Inconveniences which will result as well to Individuals as the Publick if the said Road shall be turned. -

21 February 1792, page 516
Ordered that the allotment of hands to work on the several public Roads in this County made the last year and the several Overseers appointed be continued to work the respective roads for the present year.

FAIRFAX COUNTY MINUTE BOOK, 1791-93

17 April 1792, page 95
Licence is granted James Wiley, to keep an Ordinary Jesse Moore Sec^y.

17 April 1792, page 96
The Viewers appointed to view a way for a road leading from William Deneale's Mill to the road leading from Hollis's old field to Sangsters, made a report, Whereupon It is ordered that the said road be opened accordingly.

17 April 1792, page 96
On the Motion of John Hooe Gent. ordered that, Fran^s. Coffer, John Hampton, Edward Washington, William Simpson, James Halley, Lawrence Washington, Martin Cockburn, William Thompson, & Robert Boggess, or any five of them being first sworn view a way for a road leading from Alexandria to the said Hooe's ferry, on Occoquan, near the old Warehouse Landing, & report the Conveniances & Inconveniances

21 May 1792, page 102
Licence is granted William Turner, to keep an ordinary, W^m. Payne Bayliss & Mich^l. Keller Se^{cys}.

16 July 1792, page 119
Francis Coffer, John Hampton, Edward Washington, W^m. Simpson, & James Halley, viewers appointed, upon the petition of John Hooe, for a road, from Alexandria to Hooes Ferry on Occoquan, made their report, which being considered It is the Opinion of the Court that the road therein mentioned will be convenient to the publick. Whereupon It is ordered, that the several proprietors & Tenants through whose Lands the said road is to run, be summoned to appear at the next Court, to shew cause why the said road shall not be opened agreeable to the said report.

21 August 1792, page 129
The Commissioners of the Turnpike roads exhibited their accounts of the receipt and Expenditures of the money, received on said roads, who made oath thereto, and the same being Examined by the Court are allowed and ordered to be lodged in the Clerks Office for Inspection of the Publick.

17 September 1792, page 147
William Thrift is appointed Overseer of the road from Marl's to the Leesburgh road, in room of Samuel Shreiv

FCMB 1791-93

18 September 1792, page 151
On the Peitition of George Mason /Pohick/ . setting forth that he own's Land on one side of Pohick run, and is desirous of Erecting a Water Grist Mill thereon. Whereupon It is ordered that a writ of ad quod damnum issue, directed to the Sheriff of the said County, to be Executed on the second Monday in October next, And it is further Ordered that the Sheriff, do give Robert Boggess, owner of the adjoining Lands, ten days notice of the time and place of Executing the said writ.

18 September 1792, page 152
On the motion of John Hooe; William Simpson, for Robt. Lawson, John H. Manley, Samuel Bayly, Hancock Lee, Robert Boggess, Benjamin Gwinn for W^m. Fitzhugh, and John Ward having been severally summoned to appear at this Court and shew cause, why a road should not be opened, from the aforementioned Hooe's, ferry on Occoquan river to the Town of Alexandria, agreeable to a report returned therein, and they being solemnly called and failing to appear, It is ordered that the road be opened agreeable to the aforementioned report.

18 September 1792, page 152
John Hampton is appointed Overseer of the road, from Hooe's ferry on Occoquan river, to Samuel Bayly's Lands.

18 September 1792, page 152
James Halley is appointed Overseer of the road from Samuel Bayly's Land, to Ward's Mill.

18 September 1792, page 152
Benjamin Dulany, William Payne, John Moss, James Wren, G. Minor, William Deneale Gent. or any two of them are appointed to allott. the Hands to work, on the road, leading from Hooe's ferry on Occoquan river, to the Town of Alexandria, and make report to the next Court.

15 October 1792, page 156
On the motion of James Coleman setting forth that he owns Land on one side of Difficult Run and is desirous of building a Water Grist Mill thereon Whereupon it is ordered that a Writ of Ad Quod Damnum issue directed to the Sheriff, to value an acre of Land opposite agreeable to Law, giving legal Notice to the Proprietor thereof

15 October 1792, page 157
On the motion of James Richards Ordered that Simon Summers, William Payne Ed. Dulin & John Summers or any three of them view a way for a Road thro' his plantation to Grymes Mill & Report on oath the conveniences & Inconveniences attending the same

FCMB 1791-93

17 December 1792, page 167
A Writ of ad quod Damnum, for Condeming an acre of ground and valuing the same, adjoining the Lands of James Coleman, on Difficult run, who is desirous of Erecting a water Grist Mill thereon, was presented in Court, by said Coleman, & It appearing to the Court that Rezin Offutt, whose Lands was condem^d. had legal notice of this motion, the same is admitted to record. Damages p^d.

18 December 1792, page 171
Charles Smith, is Appointed, Overseer of the road from Wards Mill to the road leading from Fitzhughs quarters near M^{rs}. French's.

21 January 1793, page 174
Benjamin Gwinn, Thomas Lindsay, Alexander McDonald Simon Summers or any two of them are appointed to view the road from Prices Ordinary, to the forks of the road above Nicholas Fitzhughs, plantation & report the Conveniances & inconveniances, of continuing the same.

21 January 1793, page 174
Licence is granted James Wren to keep an Ordinary Thomas Lindsay Sec^y.

21 January 1793, page 175
Licence is granted John Abert to keep an Ordinary John Riddle Sec^y.

21 January 1793, page 176
On the Motion of George Mason, /Pohick/ . setting forth that he holds Lands on one side Pohick run and is desirous of Erecting a Water Grist Mill thereon Whereupon It is Ordered &c. to be Executed on the second Monday in February next.

22 January 1793, page 179
Licence is granted William Ward to keep an Ordinary William Summers Sec^y.

18 February 1793, page 180
William Bird, Francis Peyton, James Patton, and Walter Brooke or any three of them are appointed to view way for a road, leading from Roger West house into the main road leading to the Town of Alexandria, at the foard of Cameron run and report the conveniances and inconveniances attending the same.

18 February 1793, page 182
Courtney Askins is Appointed Overseer of the road from the Ox road to the Parish line on Ellzey's Church road, in room of Stephen ODaniel.

FCMB 1791-93

19 F   1793   1 7
The viewers Appointed to view a way for a road leading from Roger West's house into the main road leading to the Town of Alexandria at the foard of Cameron run made a report, Whereupon It is ordered that John Richster and Bryan Fairfax, through whose Lands the said road is to be run, be summon'd to appear at ap^l. Court, next to shew cause why the said road shall not be opened agreeable to the said report [following crossed out: on this motion motion, Jno. Richster by his Council, move'd that the viewers shou'd be ordered to report Specially the Conveniances & Inconveniances attending the report of the said opening road to the next Court, which was overruled.]

19 March 1793, page 191
The report of a Jury for Condemning an Acre of ground on Pohick run, whereon George Mason is desirous of erecting a Water grist Mill, was returned and on Motion of the said Mason is ordered to be recorded. Damages paid in Court.

16 April 1793, page 211
On the motion of William Simpson; James Halley & Drummond Wheeler are appointed to view a way for turning the ox road, near William Simpsons house & make report to the next Court.

16 April 1793, page 214
Presly Gunnell is appointed Overseer of the road from Col^o. Wrens Store to the falls road in room W^m. Middleton.

20 May 1793, page 215
The viewers appointed to view a way for turning the ox road near, William Simpsons house, through his Land, made a report, Whereupon It is ordered that the said road be Opened accordingly.

20 May 1793, page 218
John Moss, Benjamin Gwinn, Gerrard Spinks & McKinzie Talbutt, or any two are Appointed to view the Colchester road, from Samuel Smiths lane, through Benoni Price's Plantation, and report the Conveniences & Inconveniences attending the Continuance of said road.

19 June 1793, page 234
Ordered that the Allottment of hands to work on the several roads in this County, made for the year 1792, be continued for the present year, together with the several Overseers Appointed to the same.

15 July 1793, page 248
William Powell, John Jones, Charles Smith, Gerr^d. Spinks or any three of them are appointed to view a way for a road from John Summer's house to the Colchester road & make report to the next Court.

FCMB 1791-93

15 July 1793, page 249
On the motion of Amos Fox & others, for a convenient road from the Traff hill, to the said Amos Fox's Mill, on Difficult run, It is ordered that W^m. Payne, Richard Ratcliff, John Wrenn, Thomas Palmer, or any three of them, do view the Lands through which a Publick road may be made from the said Hill to Amos Fox's Mill most conveniently, and report to the Court the Conveniences & Inconveniences as well to Individuals as the Publick, in making s^d. road.

15 July 1793, page 251
John Butcher is Appointed Overseer of the road from the falls Church, to the forks of the road near the falls in room Geo. Thrift.

16 July 1793, page 258
William Peake, William Triplett William Darrell & Walter Brooke, are appointed to view a way for a road from Genl. Washingtons new road, into the old road & make report to the next Court.

16 July 1793, page 258
Weathers Smith, J^r. is appointed overseer of the road leading from the old Courthouse to the red house.

16 July 1793, page 258
Caleb Richards, is appointed Overseer of the road leading from Avery's road to where it Intersects the road leading from Fendall & Hipkins's Mill.

16 July 1793, page 259
Solomon Casity is Appointed Overseer of the road in the room James Hunter resigned.

16 July 1793, page 259
George Gilpin, Ch. Little & Saml. Love exhibited to the Court, an a/c. of the money received & Expended by them as Commissioners of the Turnpike roads, on Oath, and the same being ex^d. by the Court, is allowed & Ordered that the same be filed with the Clerk of this Court for the Inspection of the Publick.

16 July 1793, page 260
Geo. Gilpin, & Charles Little, came into Court and resigned their Office of Commissioners for the Turnpike roads.

16 July 1793, page 261
Geo. Minor and John Moss, are appointed Commissioners of the Turnpike roads in the room of Geo. Gilpin & Charles Little Gent. resigned.

FCMB 1797-1798

20 March 1797, page 3
John Sangster is appointed Overseer of the road from the ox road near Sangsters & to Hollies at the Turnpike road in room of William Mills resigned.

17 April 1797, page 16
The viewers app^d. to view [illegible] Ox Road from the corner between Simpson Halley & Rich^d. Ratcliffe and along [illegible] to the Alex^a. road [illegible] that the said Rich^d. Ratcliffe have leave to open the said road agreeable to the said report [illegible] produced [illegible] Surveyor of the County Certificate that the Survey [illegible] thereof is accurate [illegible] obtaining leave from Simpson Halley to let the road run on his land [illegible] and also that the said Ratcliffe do keep the same in repair for three years

17 April 1797, page 17
Licence is granted William Pelling to keep an Ordy. Jno. Dundas Secy

17 April 1797, page 17
Rich^d. Clarke appearing upon his Sum^s. to shew cause why a road should not be opened from the Wolf Run Shoals to the shallow ford on Bull run Whereupon it is ordered that a Writ of Ad Quod Damnum issue according to law to be executed the 27th of the month.

15 May 1797, page 21
Licence is granted Danl. Dougherty to keep an Ordinary John [illegible] his Security

15 May 1797, page 21
Robert Ratcliff agreeable to a former Order of the Court, produced a Cert. from William Payne Surveyor of Fairfax County that the measurement of a road for altering the Ox road is accurate, & also produced from Simpson Halley leave to alter the same, Whereupon leave is granted him to open the said road agreeable to the said report.

15 May 1797, page 22
The persons appointed to view a way for altering the road leading from Jesse Moores black Smith Shop to Alexandria on the motion of Ch^s. Broadwater & Corbin Washington returned a report, Whereupon leave is granted them to open the said road at their own expence.

15 May 1797, page 22
Ord^d. that Richard Clark be summoned to appear at the next Court to shew cause why a road should not be opened from the Wolf run Shoals to the Shallow ford on Bull run agreeable to the Report of Viewers.

FCMB 1797-1798

15 May 1797, page 23
James Reed is appointed Surveyor of the road from the Sycamore near the little falls Warehouse to the river and the following hands are allotted to work on said road towit W^m. Macknear, Rich^d. Lloyd, Edward Earp, George Williams, Alexander Butcher, John Butcher, Benj^a. Goodrich, Benj^a. Goodrich Junr., James Baker Junr. and Jeremiah Simmons

17 July 1797, page 52
Licence is granted William Turner to keep an Ordinary - Andrew Munroe Sec^y.

17 July 1797, page 53
Licence is granted John Towers to keep an Ordinary Andrew Munroe Sec^y.

17 July 1797, page 54
John B. Finley is appointed Overseer of the road from Turners tavern to the Parish line on the old Courthouse road in the room of Hezekiah Kidwell resigned

17 July 1797, page 54
Ordered that it be Certified to the Auditor of Public Accounts that Daniel Curtin was overcharged by the Commissioner in 1796 for one ordinary licence and for one black tithable

17 July 1797, page 60
George Minor Gentleman came into Court & resigned his Office as a Commissioner of the Turnpike roads.

17 July 1797, page 60
Presley Gunnell is appointed a Commissioner of the Turnpike road in the room of George Minor resigned

17 July 1797, page 61
Licence is granted William Patterson to keep an Ordinary, Thomas West Sec^y.

18 July 1797, page 64
Licence is granted Edward Jacobs to keep an Ordinary John Vowell Security

FCMB 1797-1798

18 July 1797, page 65
Ordered that Daniel McCarty, Edward Washington, William Deneale, Charles Little, Presley Gunnell, Nicholas Fitzhugh, Roger West, Samuel Boling, George Minor & John Jackson or any six of them do allot the several hands to work upon the public roads in this County and recommend Overseers for the same. And ordered that they meet on the 4th day of August next at Col^o. Wrens tavern and that they be at liberty to adjourn from day to day until the business be compleated. And it is further ordered that the Sheriff of this County do attend the said Commissioners with the books of the Commissioners of the Revenue and it is further ordered that the hands belonging to the estate of George Mason & others of Masons Neck be taken from the ferry road and put to the Public road.

21 August 1797, page 69
Licence is granted Hannah Griffith to keep an Ordinary George Deneale Security

22 August 1797, page 77
Nicholas Fitzhugh came into Court and resigned his Office as a Commissioner of the Turnpike road

23 August 1797, page 80
William Henry Terrett is appointed Commissioner of the Turnpike roads in the room of Nicholas Fitzhugh resigned.

18 September 1797, page 93
Licence is granted John Skinner to keep an Ordinary Thos. Pollerd Jr & Richd Wheler Secy^s.

19 September 1797, page 95
Ord^d that Daniel McCarty, Thompson Mason, William Thompson & John Fowler or any three be app^d. to view a way for opening a road from the head of Pohick Creek to the road leading from Alex^a. to the Colchester road & report &^c

19 September 1797, page 97
Ord^d. that George Summers, Richard Ratcliffe, W^m. Henry Terrett, Jesse Whery, & W^m. Payne be added to the number formerly app^d. to allot the several hands to work upon the public Roads in this County & to recommend Overseers of the roads. And the Sheriff is directed to summon them to attend at Col^o. Wrens Tavern on Monday the first day of October.

16 October 1797, page 104
The persons appointed to view a way for a road from the head of Pohick Creek to the road leading from Alexandria to the Colchester road, made a report, whereupon it is ordered that Robert Smock and Geo Mason the persons thro' whose lands the said road will run, be summoned to appear at the next Court to shew cause why the said road should not be opened

FCMB 1797-1798

16 October 1797, page 109
The Commissioners appointed to recommend Overseers of Roads and to allot the hands on the Roads in this County returned a report in thire Words to wit, Pursuant to an order of Court, dated September Court 1797, We the subscribers met at Winter Hill the 2^d of October 1797 and proceeded to allot the hands and recommend Overseers to work the public Roads in the following manner to wit, From the County line on the Turnpike road leading from Newgate to Alex^a. as far as the Ox Road we allot the Male tithables belonging and residing on, or that may hereafter reside on the plantations of the persons following to wit, Thomazin Ellzey, James Gill Junr. Joseph Wood, Meryman Harrower, those living on the plantation where James Gilpin formerly lived, David Stuart, Marmaduke Beckwith, Thomas Pollard, William Turner, Thomas Turner, Vincent Turner, Thomas E. Turner, John F. Buckley, William Buckley, Philip Prichard, John Gibsons Quarters, James Deneale, George Deneale & Edward Deneales Quarters, William Deneale, Thomas Windsor, Richard Ratcliffe, Henry S. Halley, John Wren, George Summers, [blank] Dawson, John Mansfield, Bennett Hill, William Young, Thomas Pollard Junr., John Harrowers, William West, Nehemiah Davis, Samuel Farr, James Moxley and Jacob Valentine.
=

From the Ox Road, along said Turnpike road to the Fork of the road at Hollices old field near Price Skinners we allot &^{c}. James Conner, Francis Adams's Tenants, Benjamin Landers, James Bradley, William Skinner, Samuel Woolbright, Bazil Williams, Bennett Johnson, Walter Johnson, Launchley Johnson, William King, John Barnes, Thomas Woodard, James Martin, Noah Martin, Henry Martin, the place where Sarah Taylor lives, Benjamin Talbut, Laughlin McIntosh, John McIntosh, John McAtee, Charles Nolan, John Harrell, William Green, John Ashford, William Davis, Benjamin Davis, William Steel, Samuel Weedon, Thomas Finley, Martha Finley, Matthew Bozwell, David Davis, John Wilkinson, John Harrison, Jesse Daileys Plantation, Edward Taylor, John Scisson, Geo. Harrison, John Dawson, William Harrison, John Daniel, Price Skinner, John Asken, Michael H. Robey, Coatney Askin, Elias Roby, Levy Roby, Marshall P. Kidwells place, James Love, Saml. Martin, Noah Martin, John Wren, James Bradley and Daniel Bradley
=

From Hollices old field in the fork of the Road Price Skinners to Prices Tavern on said Road we allot &^{c}. Richard Lane, Atwell Dulin, Thomas Lewis, John R. Allen, Robert Scisson, Richard Fitzhugh, William Fitzhughs quarter, Opie Lindsays Quarters, Oliver Williams, Thomas Kirby, Nicholas Fitzhugh and William Gooding Junr.
=

From Prices Tavern to the forks of the road above Summers we allot &^{c}. Mordecai Cooke Fitzhugh, Edwd. Potter, Giles Cooke, Gerrard Spinks, John Murray, Saml. Smith, Augustine J. Smith, Henry Rose, M^{rs}. Wests Quarter, Benoni Price, John Ratcliffs place, George Fielder, John Moss, Zach^h. Morris, Thomas Dove, John Thompson, Stephen Crump, Benjamin Padget, John Payne, Thomas Bozwell, William Rodgers and and William Crump.

FCMB 1797-1798

16 October 1797, page 111
From the forks of the road above Summers to Holmes's Run, we allot &^{c}. Elisha Beech, Ch^s. Burges, Josiah Ward, Joel Ellis, William Talbut, Theophilus Harris, Alexander McConnell, William Harper, Negro Cambridge, William Morgan, Stephen Lomax, Alexander Simms, Francis Summers Junr., William Gates, Thomas Lambert, Jere Jefferson, William Hepburns Farm, [blank] Norris, Jeremiah Jefferson Junr., Bush Hill Farm & ^{c}., Charles Lees place, Michael Madden, William Harley, John Harley, Richard French, Zach^h. Dove, Saml. Johnson, Francis Summers, Alexander McClain, Joseph Bushby, William Bushby, Terretts Quarter, Negro Mingo, John Dove and Peter Hawk

16 October 1797, page 111
=
From Holmes's Run to the Stone bridge below the Turnpike, we allot &^{c}. Josiah Watson, Benjamin Dulany, Thomas Baggot, Law^r. Munroe, Michael Fielding, Andrew Munroe, W^m. Hartshorne, Daniel McDaniel, Guy Evans, Geo. Kelley, Thos. West, McKinzie Talbut, Richard Sanford, Jun^{rs}. place, Charles Harper, John Bright, John T. Ricketts & his Workman, Mandevills Mill, William Simpson, John Korn, Eliz^a. Simmons, Michael Omeara, John Geesland, Saml. Hulls, Mary Bloxham, P. Callahans lot, Doc^t. Potts, Joseph Fagan, John West and his Tenants, Charles Allen, John Gooding, Daniel Magie, Jacob Hynaman, Richard Weightman, John Longdon, Thomas Conn, John Balinger, John Simpson and Abram Lukehouse.
=
From Difficult bridge on the Turnpike road to the old Courthouse Run, We allot &^{c}. John Anderson Junr., John Merchant, John Shepherd, Jesse Wherry and his hirelings, Daniel McCarty, Thomas Dyer, John Listers place, Edward Adams, Thomas Fairfaxs place, Reazin Offutt, Alex^a. Beech, John Femisters, Samuel Wheatons place, David Griffiths place, George Thrift, Hamilton Thrift, Hanson Vermillion, John Trammell, Margaret Jenkins, James Edwards's place, Thomas Bates, Solomen Cassady, George Hunter, James Wiley, William Deneals quarter, John Hunter, James Collins, Gustavus Scots farm, Timothy Carrinton, William Henderson, Daniel Bradley, William Gilpin, Edmund Cudick and Henry Gunnell

16 October 1797, page 112
From the Courthouse Run to the Falls Church we allot &^{c}. Thomas Lindsay, Charles Little /at Danby/ Charles Broadwaters Quarter, Charles Simms, Charles Jenkins, Edward Bates, Gerrard Tramell, James Wren, John Wests Plantation, John Wood, Joseph Cockran, Luke Fields, John Pearson, Robert Finley, Opie Lindsay, Thomas Trammell, Thomas Beech, William Wren, Allen Gunnell, Edwd. Conner, Francis Rose, Elisha Blunden, David Allen, John Wilson, David Hayes, Presley Gunnell, Hanson Jenkins, Thomas Darnes Quarter, William Simms, John Cartwright, James Harris, Widow Jenkins, John Smith, Philip Darrell, William Donaldson, Henry Cain, James Thomas Junr., David Ammontree, Simon Pearsons old place, Henry Allen, Negro Tobey, Negro Adam.
=

FCMB 1797-1798

16 October 1797, page 112
From the Falls Church, to the Widow Tuckers we Allot &^{c}. John Rollins, James Ball, Geo. Williams, Thomas Donaldson, William Carlin, Charles Littles Quarter, William Adams, Thomas Bladen, Edward Skidmore, John Thomas, Carlyle Whitings Quarter, John Darne, Samuel Shreve, Thomas Darne, James Wren, Jemima and George Minor, Kenol Bladen, George Hussey, John Mills, Joseph Powell, Henry Hussey, William Darne, Benjamin Dulaneys Quarter, Jonathan Ward, Levy Talbut, Bazil Ball, Richard Leonard, Richard Glover, Simon Summers, Charles Murray, Richard Thompson, William Skidmore, John Kirby Junr.& father living on Carlin's land, George Minor Junr., negro Searor, William Follin, Charles Thompson, Jacob Bontz, John Luke, negro Daniel, William Glover, John Davis, Geo. Gerdon.

16 October 1797, page 113
From the Widow Tuckers to the Turnpike Alexandria we allot &^{c}. William Boling, Samuel Boling, Carlyle Whitings Quarter, Alexander's, Conways, FitzGeralds and Adams Tenants Westward of the George Town Road, W^m. H Terrett and his Tenants, Ludwell Lee, Jesse Taylors Quarter, Thomas Glover, W^m. Agullenil near Ramsays old Field.
=
From the Turnpike to Four Mile Run, we allot &^{c}. Baldwin Dades place, John Fitzgerald, Robert Vincent, Alexander Gordon, Philip R. Fendall, Charles Alexander and his Tenants between the Road and the River, Moses Coats and all living at Widow Adam's Mill, William M. Green, Elisha Harden, Samuel Howard, Ja^s. Frazier & Wm Russell.
=
From four Mile Run to George Town ferry, We allot &^{c}. Robert Alexander and all his Tenants, W^m H Washington and his Tenants, David Stuarts Quarter, Geo. Chapman, John Masons hands and Nicholas Hengston.
=
From Marles to the Turnpike road leading to Leesburg, we allot &^{c}. Jacob Payne, Caleb Brady, Caleb Richards, Archibald Swain, Benjamin Moalds, John Lister, John Goldrope, Joseph Nicholson, Thos. Pindals place, Benjamin Swain, Thomas Sherwood, John Ball, William Thrift, Walter Hills place, John Scott, Zachariah Scott, Annanias Paynes place, Francis Payne, Jeremiah Thrift, Sabret Scott and Craven Simms. Caleb Richards OSeer

16 October 1797, page 114
From four mile run below the old Mill, to Averys Road we allot &^{c}. William Herbert Quarter and his Tenants, M^{rs}. Robinson, Ignatius McFarling, Geo. Chapmans Quarter, Jacob Bent, Thomas Harden, Jesse Robinson, Geo. Tucker and Anthony Harden. / John Shreve Overseer/.
=
From Averys Road to the little falls Road We allot &^{c}. Rev^d. Thomas Davis, John Massey, Edward Earp, John Robinsons Widow, Edward Lanhams place, Benjamin Goodrick, Thomas Wrights place, Widow Baker, James Baker, William Baker, Thomas Elliott and Charles Craing. / John Massey Overseer/.

FCMB 1797-1798

16 October 1797, page 114
From the falls Church to the little Falls We allot &c. William McNear, Geo. Thrift, Matthew Earp, John & James Matney, James free negro, James Balinger, John Boucher, Henry Jones, Nelson Reid, James Reid, Matthew Earp, James Walker, Joseph Sewell, [blank] Jones, living at old Sewells place, William Philips, Richard Loyd, John Reid, Samuel McAtee, and all the Male tythes at the little falls. /William McNair, Overseer/
=
From the falls Church to the Cross Road at Cockrells old place we allot &c. Edward Dulin, William Richards, Sarah Dulin, Canel [Kenol?] Bladen, Sarah Gordon, John Horsman and Wm. Johnson. /William Johnson Overseer/.
=
From the Cross Roads at Cockerills old place along Ellzeys Church Road to the Parish line, We allot &c. Charles L Broadwater, Mary Summers, Nicholas Sebastian, Daniel M Chichester and James Fergason. / James Fergason Overseer./

16 October 1797, page 115
From the cross roads near Cockrills to the forks of the road below Fitzhughs We allot &c. John Summers 3d Joseph W. Harrison, Thomas Johnson, William Gooding, Danl Mills, James Richards, William Payne, Andrew Allen, William Scisson, John Lewis. / William Scisson, Overseer /.
=
From the forks of the Road below Fitzhughs, to the intersection of the Turnpike road at Fraziers old place we allot &c. Giles Fitzhughs Quarter, John Powell, William Frazier, Nathan Smith, Joseph Smith, Mrs. Browns, Joseph Powell Junr. /Joseph Powell Overseer/.

16 October 1797, page 115
=
From the forks of the road below Fitzhughs along the backlick road to back lick run we allot &c. Chandler Spinks, John Jones, Miss McDonalds, Charles Dodson, John Summers of Francis. /Chandler Spinks Overseer/.
=
From the back lick run to the intersection of the Colchester road at Herefords we allot &c. Charles Smith, John Devaughn, William Rogers, Widow Ogdens, Barton Davis old place, Wm P. Bailis, Thomas Bailis of Wm., George Mill, William Bailis of Wm. & Thos. Ogden. /Wm. P. Bailis Overseer/.
=
From Colo Wrens Tavern up the Great Falls Road to Maddens Shop We allot &c. William Waters, Saml. Adams, John Adams, William Payne, Henry Darne, George Hunter, John Honesty, Samson Trammell, Danl. Hodgkins, Wm. Adams Quarter, Widow Hipkins. / John Adams Overseer/.
=

FCMB 1797-1798

16 October 1797, page 115
From Maddens Shop to Difficult Run, We allot &^{c}. Elisha Lanham, John Turbervilles Quarter, and Corbin Turbervilles Plantation, John Jackson, Hamilton Thrift, Daniel Jenkins, Swinks Plantation, Hugh Ogden, John Jefferson, Thomas Moxly, Thomas Oxens, John Shortridge, Thomas Simpson, Job Moxly, Spencer Moxly, John Moxley, James Douglas millurs and Coopers, William Moxley, Thomas Moxly Junr. Ruben Rooksberry, Alier Williams, [blank] Mills living at the foot of the red Hill. /James Douglas Overseer/

16 October 1797, page 116
From Maddens Shop to the little falls Warehouse, We allot &^{c}. Ralph Hodgkins, Henry Lees Quarter, Sarah Ann Harris, Alexander Kidwell, John Williams, Joel Earp, Thomas Asberry, John Walker, Paul Tanner, Geo. Blake, Danl. Holbert, Jeremiah Spirlin, Hanson Spirlin, George Minors Quarter and Ann Minor. /George Blake Overseer./
=
From the old Courthouse to the Parish line we allot &^{c}. Edward Blackburn, Lewis Blackburn, John C Hunter, Jeremiah Moore, George Williams, Grayhams Plantation, Col^o. Broadwaters, Joshua Kidwell, Edward Davis, James Callahan, Saml. Adams, Michael Keller. Jeremiah Moore Overseer.

16 October 1797, page 116
=
From the fork of the Road at Blackburns fence to Difficult Run / Col^o. Broadwaters Mill road/ we allot &^{c}. Robert Gunnells Quarter, William William Davis, William Garner, Robert Gunnell Junr. and Rhodam Talbert. / William Davis Overseer/.

16 October 1797, page 117
From the Mountain road above Songster to Hooes old Mill on Bull run We allot &^{c}. Hector Kincheloe, Widow Coe, Robert Wickliff, John Reed Junr., Nathaniel Wardens Tenants, Yelverton Reardons Plantation, William Powell, Joseph Powell of W^m., Sarah Reed, Laurence S. King, William Moore, John Moore, Henry Pell, Daniel Douglas, Roger McCason, Zach^a. Leatherwoods Quarter, Cornelius Kincheloe. / Cornelius Kincheloe Overseer/.
=
From Thomas Songsters to the Wolf run Shoals on Bull Run, We allot &^{c}. Thomas Lucas, Benjamin King, John Yearman Junr., Charles Thrift, Hargus King, Richard Clark, Richard Simpson, James Foleys Tenants, Elisha Kent, Benjamin King, Moses Simpson. /Moses Simpson Overseer/.
=

FCMB 1797-1798

16 October 1797, page 117
From the Turnpike Gate, along the Colchester Road, to Dogue Run we allot &^{c}. William Simms, John Richsters place and Tenants, Bryan Fairfax, Thomas Dyer, William Williams, Richard Freeman, Bennett Freeman, William Freeman, negro Jasper, William Dorrell, George Wilson, Samuel Baggot, Walter Brook, Saml. Lightfoot, Thompson Masons Quarter, on the West side of the Gum Spring road, James Irvin, John Cherm, Wilkerson Grigsby, William Jones, William Caten, James Saven, Huslers place, Solomon Cole, M^{rs}. Frenchs place, Francis Hagan, John Compton, William Peakes widow, Michael Ashfords place, Thomas Flatford, William Boyd, Henry Brent, Charles Tucker and Michael Ashfords Sons place. / James Irvin Overseer/

16 October 1797, page 118
From Dogue Run to Accotink on the Colchester road we allot &^{c}. Robert Benson, Wm Johnson, Baldwin Johnsons place, William Hunters place, Charles Simms place, M^{rs}. Washington, William Triplett, M^{rs}. Chichester, D. Pitt Chichester, Thomas Bailis, William Johnson /Fowlers place/John Allison, Thomas Allison Son of John, John Hill, Josiah Hill, William Violett, William Cash, Henry Tylor, W^m. Tylor, Henry Warthin, Edward Warthin, Joseph Javins, Zachariah Wards place, Benjamin Barton, William Tripletts Tenants, Thomas Triplett, George Wiley, John McFarlins place, John Groves, William Baggot, William Bladen Junr., Levin Maubley, John Moore, Manley Moore, Miss Molly M^{C}Cartees. John Hereford, Hugh Violett, and Joseph Warner. /Robt. Benson Overseer/.

16 October 1797, page 119
From Accotink along the Colchester road to Pohick Run, We allot &^{c}. Daniel McCartys /white marsh Quarter/, [blank] Preston, John Harley, M^{rs}. Tripletts place, Robert Boggess and Gilbert Simpson / Robt. Boggess Overseer/.
=
From Pohick run to Colchester We allot &^{c}. Geo. Mason /Pohick/ and all his Tenants, John Atchison, William Essex, Jeremiah Hutchison, Ann Reagan, Robert Speak, Joseph Potter, William Mills /Bricklayer/ Richard Pool, Michael Thorn, John Cranford, Walter Hutchison, William Weston, William Church, Dudley Dennison, Samuel Green, William Jordan, Luke Rollins, William Speak, John Speak, Laurence Washington, Ludwells Lees Tenants, M^{rs}. Rerdons old place, Robert Lawson, John Cofer, Lee Massey, Cornelius Wells, Terrence Conner, Jesse Williamson, Samuel Baily, Wm Thompson, Peter Wagener, Martin Cockburn, M^{rs}. Masons home house and Quarters, Thomas Mason and all the Inhabitants of Colchester. /Thomas Mason Overseer./

FCMB 1797-1798

16 October 1797, page 120
From Cameron Run to the Gum Spring We allot &^{c}. Rogers West, William Craigs Quarter, M^{rs}. Slaughter, James B. Nichols, Thomson Masons home house, Mary Peake, Abednego Adams place, Tobias Lears Quarter, James Petitt, Fielding Lewis and Tenants on Alexanders land, Laurence McGinnis, Nathaniel Hines, Charles Hines, Geo. Cole, Zach^e Ferguson, Genl. Washingtons Quarters in Fairfax Parish, John Fergason, John Boyd Aquilla Emmerson and Henry Bozwell. /Laurence McGinnis Overseer/.
=

From the Gum Spring to Genl. Washingtons Mill We allot &^{c}. George Washingtons home house and all his Quarters in Truro parish and /John Anderson Overseer/
=

From General Washingtons Mill to Robert Boggess we allot &^{c}. Daniel McCartys home house, W^m. Potter, John Javins, Lawson Parker, Danl Stone, Thomas Allison, Widow Pool, Dennison B. Hambleton, Widow Gates, Henry Loyd, William Gray, Francis Spencer, Negro Simms & Tenants on on Belvoir Estate.
=

From Difficult Run on the Ox Road to Songsters Shop we allot &^{c}. Ch^s. Watson Beech, Sol^o. Beech, Th^o. Songsters place, Geo. Sangsters place, W^m. Gilpins place, Jno. Harrison, George Tillet, Ja^s. Tillet, Edwd. Ford, Geo. Summers & Priscilla Hunt. / George Summers Overseer./

16 October 1797, page 121
From the County line at the Turnpike road to Thos Songsters we allot &^{c}. John Brumbacks place, Abigail Pain, James Waugh and all his Tenants, Alexander Hendersons Quarter at Moorehill, Sarah West, John Butlers Quarter, John Chappel, Geo. Pain, Jno. Marshall, Henry Rassau's place, Ben^j. Wests place, M^{rs}. Stuart and David Luffburrow. /William Simpson Junr. Overseer./
=

From Thomas Songsters Shop on the Ox Road, to the White Oak harbour, We allot &^{c}. Caleb Stone, Gerrard Barnett, Jesse Baily, Jeremiah Woodard, Jno. Powell of W^m., John Skinner, John Reid, Thos Gossum, Henry Easton, Edward Washingtons Quarter, Henry Hunt, William Simpson of Geo., Geo Nicholas, Jno. Sprag, Jos. Johnsons place, John S. Stone, Francis Peytons Tenants, William King, John Pomery, Uriah Morris, James Eldridge, William Bryan, Catherine Brown, Samuel King, Elisha Kent, Jacob Tull, Ch^s. McNab, Edwd. McPherson, Henry Woodyard, and /Thomas Gossum Overseer /.
=

From the white Oak Harbor, on the Ox Road to the Town of Colchester, we allot &^{c}. Francis Stone, Robert Lawsons Upper Quarters, John King, W^m. Miller, John Philips, Gilbert Rowland, Mordecai Jacobs, W^m. Simms, W^m. Bryan, John Johnson, William Triplett, William Simpson, /Beech/ Th^o. Church, Robert Church, Josiah Boggess, Zach^e. Bond, William Church and Jesse Church. /Fran^s. Stone Overseer./

FCMB 1797-1798

16 October 1797, page 122
From the Ox Road at the Wood Cutting to Occoquan and County line leading to the Mill We allot &c. Joseph Hampton, Edwd. Washington home house hands and black Smiths, Jno. Hampton & John Hampton Junr. /Joseph Hampton Overseer/

=

From the Ox Road near Tho Songsters to Hollices old field at the Turnpike road we allot &c. William Mills, Francis Cofer, Joshua Cofer, Dennis McNamara, Wm Hambleton, Wm Greens place, Tho. Beech, Geo. Triplett & Bazil Hunt / Wm Mills Overseer./

=

From the Ox Road at Ch^s. Beeches along the Pohick road and to the Ponds we allot &c. Richard Windsor, Aaron Simpson, Geo Simpson, Mrs Hartshorne, Bazil Hunt, Ja^s Simpson of Geo., Silent Suddith and her Son James Suddith, Susanna Simpson, Presley Simpson, James Hunt, M^{rs}. Windsor and Hancock Lee./ Richard Windsor Overseer./

=

From the ponds on the Pohick road, crossing the Stage road to the intersection of the road leading from Boggess to Genl. Washingtons Mill, We allot &c. Peter Coulter, Fran^s. Keene, W^m. Keene, John Halley Junr., Barton Hall, Jonathan Dainty, Mary Rogers, William Hix, Mary Ward, William Rodgers Tenants, Thomas Davis, Thos. Simpson, W^m. Baker, W^m. Gray, Geo. Harley, Francis Hix, John Rigg & Geo. Halley. /Peter Coulter Overseer./

16 October 1797, page 123
From the Turnpike road at Hollices old field to where it intersects the Pohick road at John Fowlers plantation We allot &c. Josiah Clark, John Barker, John Simpson of Thomas, W^m. Barker, Leonard Barker, Mary Barker, W^m. McPherson, Sanford Potter, Zebedee Compton, [blank] Powell on Pohick, James Grimsley, Saml. Smitherman and Sons, John Ward, Benj^a. Riley, Ch^s. G Holden, Wm Hall, John Proctor, Saml. Stone, John Allen, Jane Williamson, Gilbert Deavers, Thos. Bailis Son of Hannah, Jno. Clark, Danl Kent, and James Thompson. /Josiah Clark Overseer/

=

From William Turners Tavern to the Parish line on the old Courthouse Road we allot &c. Hezekiah Kidwell, John B. Finley, Sarah Munroes, Sybil Hurst, Spencer Munroe, Wm Munroe, Saml. Thompson, Danl Thompson, Saml. Thompson, Junr. Aaron Alley, Francis Posten & Benjamin Thrifton /Hezekiah Kidwell Overseer./

=

From the Ox Road to the Parish Church line on Ellzeys church road we allot &c. James Dove, Thomas Flood, Margaret Wilcoxin, John C. Robinson, Peter Hursts place, Oliver Burch, Joseph Bennett, Jesse Moore and Charles Low James Dove Overseer.

=

From Ellzeys Church road, to the old Courthouse road we allot &c. Thomas Palmer, Elisha Robey, John Thompson, William Whaley, Archibald Noland Thomas Palmer Overseer.

20 November 1797, page 125
Licence is granted James Wren to keep an Ordinary. John Duling Security.

FCMB 1797-1798

20 November 1797, page 127
Licence is granted Amos Fisher to keep an Ord^y Gilbert Simpson Security.

19 December 1797, page 153
Licence is granted Elizabeth Simmons to keep an ordinary Charles Allen Sec^y..

19 December 1797, page 153
Licence is granted Thos. Lindsay to keep an Ord^y John Wood Sec^y

15 January 1798, page 160
Licence is granted Andrew Munroe to keep an Ordinary. Gilbert Simpson Sec^y

15 January 1798, page 161
Licence is granted Ja^s. Harris to keep an Ordy. Andrew Munroe Sec^y.

15 January 1798, page 163
Licence is granted James Wiley to keep an Ord^y. Thomas Darnes J^r Sec^y.

19 February 1798, page 177
Lewis Pritchard is appointed Overseer of the old Courthouse road from the mouth of Popehead run to the corner of John Gibsons fence at the road leading to the Newgate Road and Ord^d. that W^m. Deneale, David Stuart and Richd Ratcliffe do allot the hands to work on the same.

19 February 1798, page 177
Thomas Sinclair is appointed Overseer of the Road in the room of Jeremiah Moore

19 February 1798, page 180
Licence is granted Samuel Farr to keep an ordinary. John Skinner Sec^y

19 February 1798, page 182
Licence is granted Joseph Smith to keep an ordinary. Augustine J Smith Security.

19 February 1798, page 183
On the motion of William Payne, It is ordered that Charles Noland, Laughlin McIntosh, John McIntosh and Rezin Wilcoxin or any three of them do view a way for a Road from Ellzeys Church Road opposite Richard Ratcliffes land passing by William Paynes Mill and from thence to the Alexandria Road and make Report &^{c}

FCMB 1797-1798

20 February 1798, page 188
On the motion of Isaac McPherson & Nathaniel Ellicot, It is ordered that Edward Washington, Hancock Lee, Thomas Gossum, William Simpson and Aaron Simpson or any three, do view a way for a road from Ellicot & C°. bridge on Occoquan, into the road leading from Washingtons Shop to Robt. Boggess's Mill and from thence to the intersection of the said Road to the Alexandria Post Road and report according to Law.

19 March 1798, page 194
On the motion of James Waugh, It is ordered that the Overseer of the Road appointed at the last Court, from Popeshead run to the Corner of John Gibsons fence at the road leading to the Newgate road, do not proceed to open the same until the further order of this Court

16 April 1798, page 215
Licence is granted Charles Jones to keep an Ord^y Richard Weightman and John White Securities.

16 April 1798, page 215
Licence is granted Raphael Hogskins to keep an Ord^y. Perry Green Hodgkins Sec^y

16 April 1798, page 217
Ordered that William Lane Thomas Gunnell and Amos Fox be Appointed to view the most convenient and practable way for a Road from Foxes Mill to Alexandria taking into view all Circumstances resulting as well to the Publick as individuals and make a report to the next Court -

17 April 1798, page 217
Ordered that the Parish Line be extended from its present Termination at Col. Broadwaters Mill Dam, along the Lawyers road, til it Intersects the ox road, and thence along the ox road to frying pan run & thence down the said run to the present County Line.

17 April 1798, page 219
The Persons appointed to view a way for a Road from Ellzeys Church Road opposite Richard Ratcliffes land passing by William Paynes Mill and from thence to the Alexandria Road made report, whereupon It is ordered that the persons through whose land the said Road will run, be Summoned to appear at the next to shew cause why the same should not be opend agreeably to the said report.

17 April 1798, page 220
Richard Bland Lee, George R.L. Turberville, Hugh Stuart, John Keene and Newton Keene, or any three of them are appointed to veiw a way for a road from Samuel Loves Mill on rocky run, into the Turnpike road and make report to the next Court.

FCMB 1797-1798

17 April 1798, page 222
Licence is granted Thomazin Ellzey to keep an Ord^y. Richard Ratcliffe Sec^y.

17 April 1798, page 223
On the motion of Hepburn & Dundas it is ordered that Nicholas Fitzhugh, W^m. Payne and Geo. Summers or any two do veiw the most convenient way for a road from their Mill to the Colchester road near Cameron run, and make report &^{c}.

17 April 1798, page 224
On the Petition of Thomazin Ellzey setting forth that he holds lands on both sides of Popeshead run and is desirous of erecting a Water Grist Mill thereon, Whereupon it is ordered that a Writ of Ad Quod Damnum issue as the law directs, to be executed on the 25th day of May next.

21 May 1798, page 234
Licence is granted John Hubbal to keep an ord^y W^m. Mitchell Sec^y

21 May 1798, page 234
Licence is granted Hanson Vermillion to keep an ord^y. Henry Gunnell Sec^y

21 May 1798, page 235
Thomas Glover is appointed Overseer of the road in the room of John Massey resigned.

21 May 1798, page 236
Thomas Braden is appointed Overseer of the road in the room of James Reed, resigned.

22 May 1798, page 243
On the motion of William Carlin, setting forth that he owns lands on one side of Four Mile run the bed whereof belongs to the CommonWealth and is desirous of erecting a Water Grist Mill thereon, It is ordered that the Clerk issue a Writ of Ad Quod Damnum as the law directs to be executed on the 12th day of the next month.

22 May 1798, page 243
The persons appointed to veiw a way for a Road from Hepburn & Dundas's Mill to the Colchester road near Cameron run, made a report, whereupon it is ordered that the several persons thro' whose land the said road is proposed to be conducted be summoned to appear at the next Court to shew cause why the same should not be opened.

FAIRFAX COUNTY COURT ORDER BOOK 1799-1800

20 May 1799, page 1
License to keep a Tavern, is granted Mary Bloxam, who thereupon together with Alexander Smith as her Security executed & acknowledged a Bond, conditioned, as is by Law in such Case required.

20 May 1799, page 3
License to keep an Ordinary at his house in this County is granted Edward Jacobs who thereupon together with John Korn, as his Security, executed & acknowledged a Bond, conditioned as is by Law in such Case required.

20 May 1799, page 6
License to keep a Tavern at his house in this County is granted John Workman, who thereupon together with Joseph Cary and Joshua Riddle, as his Securities executed & acknowledged a Bond conditioned as is by Law in such Case required.

20 May 1799, page 6
License to keep a Tavern, at his house in this County is granted to Amos Fisher, who thereupon together with Henry Selecman, as his Security, executed and acknowledged a Bond conditioned as is by Law in such Case directed and required.

20 May 1799, page 8
The Persons through whose Lands a road is proposed to be opened from Daniel Lewis's Mill, to the Turnpike road, at the Mine Branch, having been summoned to appear, and failing to do so - it is ordered that said road be opened agreeably to the report to the Court made.

19 August 1799, page 153
License is granted William Thompson, to keep an Ordinary - who thereupon together with George Deneale as his Security, entered into, & acknowledged a Bond, conditioned as is by Law in such Case directed & required.

19 August 1799, page 163
License to keep a Tavern, at his House in this County, is granted Benjamin Posey, who thereupon together with Wm H. Washington as his Security entered into and acknowledged a Bond conditioned as is by Law in such Case directed & required.

19 August 1799, page 163
License to keep a Tavern is granted Thomasin Ellzey, who thereupon together with Thomas Millan as his Security entered into and acknowledged a Bond conditioned, as is by Law in such Case directed & required.

COB 1799-1800

21 August 1799, page 178
License to keep a Tavern at her house in this County is granted Anne Meeks, who thereupon together with James Coleman as her Security entered into and acknowledged a Bond for the same, conditioned as is by Law in such Case directed & required.

(Pages 225 through 420 missing from book. The index of the book lists the following for the period between 16 September 1799 and 18 March 1800.)

Mills: Courtney Askins petitioned to build a mill

Ordinary Licenses granted: John Towers, Ralph Hodgkins, John Stuart, James Sawkins and William Simpson

Roads. Overseers: William Phillips, John Ball, Moses Cotes, James Gill, Thomas Baxter

Roads: Proposed by Chichester; at Boggess's; Proposed by William Lane, Jr.)

20 March 1800, page 445
The CommonWealth agst. Moses Coats - Deft. On a Presentment of the Grd Jury The Defendant appearing upon summons to shew cause why an Information should not be filed against him, as an Overseer of the Road, for not keeping a Bridge in repair upon hearing the reasons assigned in his behalf, or excuse it is ordered that said Presentment be discontinued.

21 April 1800, page 492
License is granted to William Tuel to keep a Tavern at his house in this County; said Tuel together with Francis Adams as his Security entered into and acknowledged a Bond conditioned as is by Law in such Case directed & required

21 April 1800, page 492
Ignatius Wheeler is appointed Overseer of the Road in the room of William Noddy - and it is ordered that William Gunnel do allot a sufficient number of hands to work on the same. -

21 April 1800, page 494
License to keep a Tavern is granted John Ratcliffe, who thereupon together with William Payne, Gent, as his Security entered into and acknowledged a Bond conditioned as is by Law in such Case directed & required. -

COB 1799-1800

21 April 1800, page 496
On the Motion of John Jackson, Junior - William Middleton, James Marshall, Samuel Marshall, Joseph Gardner, & Samuel Adams are appointed, or any three of them to view a Way for a road from the Widow Brewster's landing to Ballandine's Mill, and make report on Oath to the Court the most convenient way for continuing, or altering the said road - and report the Conveniencies or inconveniencies, resulting as well to the Public as individuals therefrom. -

21 April 1800, page 500
License to keep a Tavern at his House in this County is granted Charles Russell who thereupon, together with Thomas Gunnell, as his Security entered into & acknowledged a Bond, conditioned, as is by Law in such Case directed & required.

22 April 1800, page 504
The persons appointed to view a Way for a road from Yates's ford on Bull Runn, to Songster's made a report whereupon it is ordered that the Heirs of Daniel Kincheloe and Robert Wickliffe, the Persons through whose Lands the said road is proposed to be conducted be summoned to appear at the next Court to shew Cause why the said Road shall not be opened agreeable to said report.

22 April 1800, page 509
License to keep an Ordinary at his house in this County is granted Richard Lane, who thereupon, together with William Millan and John Skinner as his Securities entered into and acknowledged a Bond, conditioned as is by Law in such Case directed & required. -

22 April 1800, page 509
License to keep an Ordinary at his house in this County is granted John Skinner, who thereupon together with Richard Lane, as his Security entered into and acknowledged a Bond, conditioned as is by Law in such Case directed & required. -

22 April 1800, page 509
License to keep an Ordinary at his House in this County is granted Ellzey Turner, who thereupon together with Henry S. Hally as his Security, entered into and acknowledged a Bond conditioned as is by Law in such Case directed & required.

22 April 1800, page 510
On the Motion of Charles Broadwater - it is ordered that John Shortridge, John Wilson, George Beard & George Smith, or any three of them do view a way for a road from the Cross-Roads, at Hodgkins', to the Mouth of Beard's Lane, on Lee's land, and make report of the Conveniencies or inconveniencies, that apparently may result to the Public, or individuals, from opening & establishing a road in the direction, as aforesaid. -

COB 1799-1800

22 April 1800, page 510
On the Motion of William Lane Junr. it is ordered that he be allowed the sum of three hundred and twenty eight dollars, and thirty three cents, in the next County Levy, for his said Lane's building a Bridge over Flat-lick runn $328.33/100

22 April 1800, page 511
License to keep an Ordinary at his house in this County is granted George Newman, who thereupon together with Thomas Wrenn, as his Security, entered into & acknowledged a Bond conditioned as is by Law in such Case directed and required. -

19 May 1800, page 513
License to keep a Tavern at his House in this County is granted to Thomas Lindsay, who thereupon together with William Davis as his Security entered into and acknowledged a Bond, conditioned as by Law in such Case directed and required -

19 May 1800, page 519
License to keep an Ordinary at his House in this County is granted Zachariah Ward, who thereupon, together with Francis Coffer as his Security, entered into and acknowledged a Bond conditioned as is by Law in such Case directed and required. -

19 May 1800, page 522
On the Motion of James Richards, it is ordered that the road, commonly called Grymes's Mill road, leading through the said Richards's plantation be discontinued. -

19 May 1800, page 522
Hector Kincheloe, and Jesse Kincheloe, two of the Persons through whose Lands, a road is proposed to be conducted from Yates's Ford, on Bull Run, to Songster's, appeared upon their Summons to shew cause why the said Road shall not be opened agreable to said report - and prayed that a Writ of Ad quod damnum may issue - Whereupon it is ordered that such Writ do issue according to Law, to be executed on the tenth day of June next, and that the Jury do meet on the lands of said Kincheloe -

19 May 1800, page 523
Ordered that Richard Ratcliffe, William Payne, James Colman, William Stanhope, Francis Adams, Francis Coffer, Simon Summers, James Irvin, John Jackson, Charles Little, Nicholas Fitzhugh, George Beard, Edward Washington, and William Middleton, or any five of them do allott the hands to work on the Public roads in this County; and that they do meet for that purpose, at the Courthouse on the first thursday in June next, or the next fair day -

COB 1799-1800

19 May 1800, page 524
On the Motion of John Jackson. Jr, it is ordered that William Middleton, James Marshall, Samuel Marshall, Joseph Gardner, and Samuel Adams, or any three of them, do view a Way for a road from the Widow Brewster's, to the intersection of Fairfax's Mill Road, on the line of Fairfax - Hugh Conn, and John Jackson, Senr & report to the Court, the Conveniencies, & inconveniencies resulting as well to the Public, as to individuals from opening a road in the direction aforesaid. -

Index

This index is arranged by subject: Bridges and Causeways; Churches, Chapels, Glebes, Meeting Houses, Parishes; County Government; Ferries; Fords; Land Features; Mills and Mill Dams; Miscellaneous Subjects (houses, landmarks, neighborhoods, plantations, quarters, stores, towns, etc.); Ordinaries; Ordinary and Tavern Licenses and Applications; Personal Names; Rivers, Runs, Springs, Creeks, and other Water Features; Roads.

Bridges and Causeways

Bridge at Alexandria, 115
Bridge to be made by Sybell West, 33
Bird's/Bride's bridge, 149(2)
Broad Run/over Broad Run, 14, 19, 22, 24(4), 28(2), 33, 36
Buckhorn bridge, 112
Bridge at Cameron, 46
Bridge above the old Court house, 88
Dirt bridge (over Bridge Branch?) near the old Court house, 94, 95
Bridge near the old Court house (to be made by Sybell West), 33
Cub/over Cub Run, 29, 31
Difficult bridge, 37, 50, 53, 54, 55, 71, 77, 87, 94, 95, 101, 103(2), 108, 110, 119, 120, 135, 141, 142(2), 148, 161
Bridge built by Charles Broadwater over Difficult Run, 53, 54, 55
Timber for a bridge (Difficult Bridge), 71, 110
Ellicot & Co. bridge on Occoquan, 169
Bridge over Flat-Lick Run, 174
Bridge at Grayson's mill, 6
Bridge on the road over the mill race (at George Mason the younger's mill on Pohick Creek), 96, 109
Bridge/bridge and causeway/causeway over Pimmets Run, 26, 84, 130
Causeway over Scot's (Scott's) Run, 130
Bridge to be made by John Seals, 68
Stone bridge below the turnpike, 161
Gerrard Trammell's bridge, 105
Bridge built by John Trammell/John Trammel, Jr. (Broad Run bridge), 22, 24, 28(2), 36
Tobacco levied for building Broad Run bridge to be sold, 23, 24

Churches, Chapels, Glebes, Meeting Houses, Parishes

Chapel on the Beaver Dam, 14
The brick Church on Ox road, 145
Broad Run Chapel, 30
The Church 1, 11(2), 13, 57(2), 116,
Falls Church, 1, 49, 53, 60, 63, 66, 71, 79, 81, 88, 97, 99(2), 100(2), 105, 106, 112, 113(2), 114, 116, 127, 129, 132(2), 133, 134(2), 142, 149, 156, 161, 162, 163(2)
Goose Creek Chapel, 33
The new Chapel, 18
The new Church, 20(2), 93, 117, 136
New church built by Mr. Edward Payne, 87
The old Church, 116
NOTE: see also Church roads; Ellzey's Church road
Pohic/Pohick Church, 9, 46, 60, 83, 84, 87, 106, 109, 114, 138, 147, 148
Pohick new church, 137
Pohick old church, 137, 148
Rocky Run Chapel, 17, 18, 22, 25(2), 26
Rocky Run Church, 8
Upper Church, 93
The stand or meeting house, 123
Glebe of Cameron parish, 33
Cameron Parish, 29(2), 33
Fairfax parish, 85, 86, 90(2), 94, 99, 100(2), 105(2), 106, 110(2), 118, 166

Both parishes (to go to the new church built by Mr. Edward Payne), 87
The parish line, 148, 164
Parish line on the old courthouse road, 147, 158, 167
Parish line on Dogue Run, 146
Parish line on Ellzey's church road, 148, 154, 163, 167
Parish line extended, 169
Truro parish, 29, 58, 60(2), 61, 80, 82, 89(5), 90(4), 96, 99, 100, 105(2), 109(6), 144, 148, 165
Truro parish churchwardens, 21(2)

County Government

Courthouse 1, 6, 8, 9, 11, 14, 134, 174
Old courthouse 32, 33, 37, 39, 47, 48, 49, 54, 55, 56, 57, 62(2), 63, 66, 67(2), 68, 72(2), 80(2), 82, 83, 86, 87, 88, 94(2), 95, 99, 105, 107, 110, 112(3), 114, 115, 117, 120, 122, 124, 133, 134(2), 135, 147(2), 148, 150, 156, 158, 164, 167(2), 168
NOTE: See also: Courthouse road; old courthouse road
Public buildings 120
Public wharf 21, 22
Commissioners of the turnpike road, 128, 152, 156(3), 158, 159
Individual commissioners:
    Thomazin Ellzey, 130
    George Gilpin, 156(2)
    Charles Little, 156(2)
    Samuel Love, 156
    George Minor, 156, 158(2)
    John Moss, 156
    Presley Gunnell, 158
    Nicholas Fitzhugh, 159
    William Henry Terrett, 159
Trustees of the road from Vestal's and Williams's gaps leading to the towns of Alexandria and Colchester, 102(2), 111
Fairfax County, 63(2), 80, 82(2), 85, 86, 89(5), 90(6), 93, 94, 96, 99(2), 100(2), 105, 106(2)
County line, 58(2), 135, 138, 144, 145, 167, 169
    At the Blue Ridge, 20
    At Loudoun line, 92, 98, 100, 101, 103, 112(2), 114, 116, 145
    On the Turnpike road, 160, 166
County Districts
    Fairfax District, 120, 123(2)
    Truro District, 123
Other counties
    Loudoun County/ Loudoun County Court, 50, 77, 87, 101, 141, 142
    Prince William County/Prince William County Court, 1(2), 3, 21, 61, 64
    Westmoreland County, 47, 50(2), 51
Maryland, 91

Ferries

Ferry, 18, 159
Alexandria ferry, 2, 4, 9, 38, 54, 68
Public ferry to Alexandria, 120
Ferry on Josias Clapham's land (on Potomac River), 39
Clifton's/William Clifton's ferry, 17, 52, 53
Colchester ferry, 146
Court's ferry, 84
Georgetown ferry, 87, 97, 113, 119, 123, 126, 141, 142, 150, 151, 162
Ferry opposite Georgetown, 132
Goose Creek ferry/scew, 2, 3, 6, 8, 14, 16, 18, 24, 38, 42, 44
(John) Hereford's/Herryford's ferry, 5, 9, 45(2), 46, 109, 116
Holland Point ferry, 137, 148
Hooe's ferry on Occoquan near the old warehouse landing, 152(2), 153(3)
Hunting Creek ferry, 38, 70, 81, 88, 93, 102
(Samuel) Johnson's ferry, 67
Ferry from Hannah Johnson's plantation

to the (opposite) landing in Maryland, 91
(Ferry) landing in Maryland, 122
George Mason's ferry/landing, 88
Ferry from Col. Mason's land over to Georgetown, 124
Mason's ferry opposite to Rock Creek (see also Rock Creek ferry), 18
Samuel Mead's ferry, 28
Noland's ferry, 27
Occoquan ferry, 2(2), 3, 6, 7, 9, 14, 19(3), 21, 24, 28(2), 31, 36, 38, 42, 44(2), 47, 50(2), 51, 52, 56, 61, 65, 71(2), 84, 117
Posey's/Pozey's (John) ferry, 30, 32, 33, 39, 44, 59, 73, 74, 117
Poultney's ferry, 7
Rock Creek ferry (see also Mason's ferry opposite to Rock Creek), 89, 115
George/Gen. Washington's ferry/landing, 88, 137
Ferry keepers:
   Henry Awbrey, 35
   Samuel Johnson, 67
   Samuel Mead, 28
   Thomas Saunders (at Hunting Creek), 70

Fords

Ford over Accotink, 61
John Ballendine's ford, 50
Shallow ford on Bull Run, 157(2)
Yates's ford on Bull Run, 173, 174
Ford of Cameron Run, 154, 155
Lower ford over Difficult Creek, 128
Dawson's ford, 41
Four Mile Run ford, 4
Goose Creek ford by Samuel King's, 11
Ford over Holmes Run, 120
Hunting Creek ford, 4
Ford over Piney Branch, 76, 77

Land Features

Ashby's gap, 10
The Blue Ridge/the Ridge, 10, 16, 20, 32, 41
Hill below Boggess's, 69, 72
Cameron Hill, 47, 106, 116, 118
Mr. Carlyle's ditch, 96
Hill by Four Mile Run, 141, 142
Rock hole on Four Mile Creek, 68(2)
Hill near Holmes Run, 112, 120
Island on the Potomac River belonging to John Trammel, 37
Keen's Hill, 83, 87
Hill near Edward Lanham's field, 128
Mason's Neck, 159
Hill by John Mason's fence, 74
The mountain, 22, 98
Hill above Ramsay's Quarter, 120
The Red Hill, 164
Gap of the Short Hills, 36
Traff Hill, 156
Vestal's/Vestel's gap, 30, 102(2), 111
Gut below John West's meadow, 158
Hill near Thomas West's plantation, 136
Williams's gap, 6, 8, 14, 26, 102(2), 111
Wolf Pit Hill, 50, 106, 116, 117, 139(2)

Mills and Mill Dams

Adam's mill, 113
Robert Adam's mill on Four Mile Creek, 68(2), 69, 132
Samuel Adams's mill on Pimmets Run, 118, 119
Samuel Adam's mill (same as above??), 125
William Adam's mill 162
William Adams's mill on Pimmets Run, 119
Charles Alexander's mill on his land on Four Mile Run, 92
Courtney Askin's mill, 172
Mr. John Balendine/Ballandine's mill, 49, 173

John Ballendine's mill at the Falls landing, 74
Bayly's mill, 139
Bayley's old mill, 148
Boggess's mill on Pohick, 102
Robert Boggess's mill, 15, 169
Robert Boggess, Jr.'s mill, 109
Col. Broadwater's mill/dam, 150, 164, 169
Mr./Major Charles Broadwater's mill at Difficult Run, 94, 95
William Carlin's mill on Four Mile Run, 101, 170
John Carlyle's mill at Four Mile Run/ Creek, 62, 64, 68(2)
Josias Clapham's mill on his land where he lives, 43
James Coleman's mill on Difficult Run 153, 154
John Colvill, Gent.'s mill on Four Mile Run, 26
William Deneale's mill, 151, 152
Doctor Dick's mill, 124, 148
Difficult mill, 114
John Dowdall's mill on Four Mile Run, 129
Thomazin Ellzey's mill on Popeshead Run, 170
Fairfax mill, 175
William Fairfax, Esq.'s mill on Difficult Run, 11
The Falls mill, 77
Philip Richard Fendall's (and Lewis Hipkins's) mill on Pimmets Run, 141, 143, 156
Old mill on Four Mile Run, 162
Amos Fox's mill on Difficult Run, 112(2), 121, 122, 156, 169
Edward Garret's mill on Beaverdam Branch of Goose Creek, 3, 5, 7
Benjamin Grayson, Gent.'s mill, 4, 6
Col Grayson's mill on Pohick, 23
Grymes's mill, 153, 174
William Hartshorne's mill, 151
Alexander Henderson's mill, 114

Hepburn and Dundas's mill on Back Lick Run (and near Indian Run), 126(2), 127
Hepburn and Dundas's mill, 170(2)
William Hepburn's mill, 135
Thomas Herbert's mill on Holmes Run, 125, 140, 141
(Philip Richard Fendall's and) Lewis Hipkins's mill on Pimmets Run, 141, 143, 156
Hooes'/John Hooes' mill/old mill/on Bull Run, 81, 117, 138, 145, 164
John Hough's mill on leased land in gap of the Short Hills, 36
(Charles Shoemaker and) Joseph Kirkbridge's mill on Occoquan (opposite a landing which is to be the beginning of a new road to Alexandria), 141
George Lampkin's mill on Wolf Run, 92
Daniel Lewis's mill, 171
Peter Lewis's mill, 20(2)
Stephen Lewis Gent.'s mill on Difficult Run (to rebuild old mill), 1, 6, 11
Mr. Thomas Lewis' mill on Difficult Run, 94, 95
Thomas Lewis' mill on Wolf Trap Run, 104
Samuel Love's mill on Rocky Run, 169
Mandevill's mill, 161
George Mason's mill on Pohick, 153, 154, 155
George Mason, Jr.'s/ the younger's mill on Pohick, 96, 109(2)
Edward Masterson's mill on Difficult Run, 12, 17
Thomas Monroe's mill on his land at Quarter Branch, 54
The mill near Occoquan, 167
Edward Payne's (?) mill, 87
William Payne's old mill on Accotink Run, 113(2)
William Payne's mill, 168, 169
Peyton's mill, 35
Henry Peyton's mill on Little River, 20

Thomas Pollard's mill on Popeshead Run, 83
Thomas Pollard's mill on Giants Castle Branch, 123
Charles Shoemaker (and Joseph Kirkbridge's) mill on Occoquan (opposite a landing which is to be the beginning of a new road to Alexandria), 141
William Shortridge's old mill on Scott's Run, 128
Thomas Simon's mill, 51
Moses Simpson's mill on Sandy Run, 110
Steptoe's mill on Accotink, 30
Daniel Thomas mill on Popeshead, 24
Thomas Throgmorton's mill on Bull Run, 95
Tolson's mill/Towlson's mills on Difficult Run, 81(2), 91, 133
John Trammell's mill on Difficult Run, 71, 76
Turley's mill, 47
Vernon's mill on Pohick, 123
Ward's mill, 153, 154
John Ward's mill on Accotink, 100, 123, 124, 143
The General's/Gen.George Washington's/George Washington, Gent.'s mill, 103, 137(2), 148(2), 166(2), 167
General Washington's tumbling dam, 137
Brand for flour of George Washington's mill, 103
John West, Jr.'s mill over Poultney's Run, 69
Joseph West's mill on north branch of Goose Creek, 17

Miscellaneous Subjects (houses, landmarks, neighborhoods, plantations, quarters, stores, towns, etc.)

Robert Adam's land on Four Mile Run, 69
Robert Adam's plantation, 133
(Fitzgerald's and) Adams's tenants west of the Georgetown road 162
Abednego Adams's place, 166
Francis Adams's tenants, 160
Jerry Adams's place, 149
R. Adams's, 121
Widow Adams's at Goose Creek, 32
Samuel Adams's plantation, 119
William Adams's/ Mr. William Adams's/William Adams's, Gent., 66(2), 69, 80(2), 82(2), 83, 84, 92, 96, 97, 100, 105
Mr. William Adams's house/plantation, 4, 19, 63
William Adams's quarter, 134, 163
William Adams's land on Pimmets Run, 119
Tenants on Alexander's land, 166
Charles Alexander's land/tenants on Four Mile Run, 92, 150
Charles Alexander's old land, 133
Charles Alexander's tenants between the road and the river 162
Robert Alexander's tenants, 149, 150, 162
Samuel Alexander's old place, 137
(Town of) Alexandria 2(2), 4, 5, 9(2), 14, 16(2), 17(2), 18, 21(2), 25, 26, 27, 29, 31, 37, 38(3), 41(2), 43, 45, 46(3), 48(2), 49, 51(2), 52, 53, 54, 55, 56, 57, 62, 63(2), 66, 67, 68, 69, 71(2), 72, 74(2), 75, 76, 79(4), 80(2), 82, 83, 84(4), 85, 86(3), 88(2), 91, 93(4), 94, 96(2), 97(3), 98, 99(2), 100(2), 101(2), 102(6), 103(2), 104, 105, 108(2), 109, 111, 112(2), 113, 114(2), 115(2), 116, 118(2), 119(2), 120, 121(2), 123, 124(2), 126(4), 127, 128, 129, 130(2), 131(5), 132, 135, 136, 139(2), 140, 141(2), 142, 143, 144, 146(2), 147, 148, 150, 152(2),

153(2), 154, 155, 157(2), 159(2), 160, 162, 168, 169(3)
Back line of Alexandria 37, 115
Inhabitants of the town of Alexandria 21
Anderson's, 41
David and Samuel Arell's land, 127
Thomas Ashbury's, 49
Michael Ashford's place, 165
Michael Ashford's sons' place, 165
William Ashford's house/land, 35, 36, 37, 49, 54
Widow Atcheson's plantation, 102
Ballendine's Farm, 85
Mr. Barnes's/Barn's/Barns's land/plantation/quarter, 32, 47, 61
Mr. Bayley's land, 102
Samuel Bayly's land, 153(2)
William Bayley's house in Colchester, 45
Charles Beach's/Beache's/Beech's land/fence on the Ox road, 138, 147, 167
Mr. Beckwith's, 73
Marmaduke Beckwith's plantation, 108
Marmaduke Beckwith's quarter on Wolf Run, 145
Beech (see Beach)
Belhaven 32
Belvoir, 148
Tenants on Belvoir estate, 166
Bennett's, 148
Joseph Bennett's plantation, 135
Berryman's quarter, 11
William Bird's fence, 127
Blackburn's fence (at the fork of the road), 164
Mr. Edward Blackburn's, 94, 95(2), 100
Richard Scott Blackburn's fence/quarter, 134
Boggess's, 32, 69, 72, 97, 172
Henry Boggess's house, 9, 18
Boggess's/Robert Boggess's land/house, 4, 12, 45, 46, 59, 137, 148, 166, 167
Robert Boggess's land on Pohick Run, 153

Matthew Boswell's shop on Ellzey's church road 143
Boyd's quarter, 149
Widow Brewster's, 175
Mr. Broadwater's plantation, 26
Col./Mr./Major Charles Broadwater's quarter, 71, 72, 73(2), 134, 149, 161
Mrs. Bronaugh's cornfield, 45
James Brown's house, 58
Richard Brown's house, 3, 13, 15, 37, 38
Dr. Brown's/Dr. William Brown's quarter,135, 149
John Brumback's place, 166
John Buckley's plantation, 122
William Bushby's plantation, 135
Bush Hill Farm, 161
John Butcher's fence, 119
Edmund Butler's fence, 57
John Butler's quarter, 145, 166
P. Callahan's lot, 161
Cameron, 4, 19(2), 22, 27, 29, 33, 41, 43, 45, 46(2), 47(2), 63(2), 66(4), 69(2), 71, 80(3), 82, 83(2), 87, 96, 97(2), 98, 101(2), 103, 105(2), 112, 117(2), 120
NOTE: Some entries for Cameron may refer to Cameron Run or Cameron parish; these sections should be checked as well
Mark Canton's near Little River, 35
Carlin's land, 162
William Carlin's land on Four Mile Run, 101, 170
Carlyle's farm, 133
John Carlyle's plantation, 41
John Carlyle Gent.'s land at Four Mile Run, 62, 69
Mr. Carlyle's ditch 96
Field rented by Mr. Charles Thrift of Robert Carter Esq., 130, 140
Robert Carter's land on Sandy Run, 110
Dutch Caty's house, 120
George Chapman's home house plantation, 132(2)
George Chapman's quarter, 132, 162

Richard Chichester's quarter, 134
Richard Chichester's home house and
    quarter on Accotink, 137
Josias Clapham's land where he lives, 43
Josias Clapham's land (on Potomac
    River), 39
Richard Clark's land on Wolf Run, 92
Mr. William Clifton's, 17
Clish/Clish plantation/tenants at Clish,
    136, 137, 149
Bryan Fairfax's plantation adjoining
    Clish, 136
Mr. (Martin) Cockburn's, 139, 148
Cockerell's/Cockrell's/Cockrill's old
    place, 163(3)
Joseph Cockerill's, 105, 107
Colchester, 42, 45(2), 48(3), 49, 50, 54,
    59(2), 60, 62, 65, 67(3), 68(2), 72,
    74, 75(2), 79, 81, 83, 84, 87, 89, 93,
    98(4), 100, 101(2), 102(3), 103,
    104(2), 107, 109, 111, 116, 117,
    118, 122, 124, 127, 133, 136(2),
    137, 139(2), 142, 146(2), 150(2),
    155(2), 159(2), 163, 165(4), 166,
    170(2)
Inhabitants of Colchester, 165
Robert Colclough's house, 12
James Coleman's land on Difficult Run,
    153, 154
Richard Coleman's house, 39
Collins's fence, 118(2)
James Collins's place (and tenants), 134
James Collins's old field, 112
Col. Colvill's Tuskarora quarter (near
    the Mountain road), 11, 16
John Colvill's land on Four Mile Run,
    26
Thomas Colvill's plantation, 55
The Commonwealth, 170
Line of Hugh Conn, Fairfax, and John
    Jackson, Sr., 175
Gerrard T. Conn's plantation, 144
Richard Conway's plantation/tenant, 147
Line between Richard Conway and John
    Jackson, 128

John Cotton's plantation, 87
The courthouse cornfield, 49
Mrs. Cox's home house, 149
William Craig's quarter, 166
Charles Curtis's place, 133
Dade's old quarter, 133
Baldwin Dade's plantation/place, 41
Baldwin Dade's old tobacco house, 120
Townshend Dade's plantation, 41
Townshend Dade's old land, 133
Jesse Dailey's plantation, 160
Danby, 134, 149, 161
Thomas Darne's quarter, 150, 151
Line between Thomas Darne and
    Samuel Shreeve, 142
Line between Philip Darrell and William
    Gunnell, 128
Allen Davis's lane, 114
Barton Davis's old place, 163
Nehemiah Davis's land on Difficult Run,
    122
Thomas Davis's land/house, 10(2), 19,
    20
Deakins's, 134(2)
Line of Deneale and Fairfax, 118
(James, George and Edward) Deneale's
    quarter, 160
James Deneale, Jr.'s, quarter, 144
William Deneale's corner/land on Wolf
    Trap, 118
William Deneale's quarter, 161
Dr. Dick's plantation, 148
Dickens's, 115(2)
William Dodd's house, 38
James Donaldson's land/house, 51, 53
William Donaldson's, 54(2)
Dowdal's corner on the hill by Four
    Mile Run, 141, 142
Peter Dow's plantation, 135
Dozer's, 59
James Ingo Dozer's/Dozer's house, 39,
    43, 51, 57
James Dozier's, 44
Benjamin Dulany's quarter/two quarters,
    133, 162
Dumfries, 23, 87, 141

Matthew Earp's, 119
James Edward's place, 161
Capt. Lewis Ellzey's, 11, 57(2)
Thomazin Elzey's land on Popeshead Run, 170
Thomazin Ellzey's quarter, 144
William Ellzeys's plantation, 41
Thomas Evans's house, 9
Line of Deneale and Fairfax, 118
Line of Hugh Conn, Fairfax, and John Jackson, Sr., 175
Bryan Fairfax's land, 155
Bryan Fairfax's plantation adjoining Clish, 136
Bryan Fairfax's quarter, 134
Thomas Fairfax's place, 161
The farm/the old farm, 114, 133
Philip Fendall and Lewis Hipkins's lands on Pimmets Run, 141
Mr. Ferguson's cleared ground, 45
Mrs. Mary Ferguson's plantation, 87
Walter Findley's old place, 137, 146
Amos Fisher's house, 171
Fitzgerald's quarter, 133
Fitzhugh's, 163(3)
Fitzhugh's gate, 113
Fitzhugh's lane, 142, 143
Fitzhugh's quarter on Ravensworth, 12, 92
Fitzhugh's quarter, 66, 154
Giles Fitzhugh's quarter, 163
Henry Fitzhugh deceased's quarter on the east side of Accotink, 144
Henry Fitzhugh deceased's quarter on the south side of Accotink, 144
Dividing line of William Fitzhugh and the heirs of Henry Fitzhugh, 127
Nicholas Fitzhugh's gate, 136, 151
Nicholas Fitzhugh's plantation, 154
William Fitzhugh's quarter, 160
William Fitzhugh's quarters on the east side of Accotink, 144
William Fitzhugh's quarters on the south side of Accotink, 144
Dividing line of William Fitzhugh and the heirs of Henry Fitzhugh, 127
James Foley's tenants, 164
John Ford's plantation, 50
The Forge, 98
John Fowler's plantation/fence, 138, 147, 167
Amos Fox's land on Difficult Run, 121
Frazer's, 136
Frazier's old place, 163
Mrs. French's, 154, 165
Mr. French's quarter, 66
Mr. Daniel French's plantation, 63
Cumberland Furguson's house, 131
George Mason's land (plantation where Joseph Garberry lives), 102
William Gardner's/Garner's, 72, 73, 74
William, Gardner's house, 75
Nicholas Garrett's plantation, 65
Walter Garrett's plantation, 95
Gates on roads:
    The two gates on the road through Sampson Turley's plantation, 108
    Gate across the road from Back Lick to Falls Church, 99
    All gates to the removed from the road, 41
    The turnpike gate, 132, 136
    The turnpike gate at the intersection of Georgetown road, 133
    The turnpike gate where William Simpson lives, 136
Georgetown, 87, 94, 96, 99, 113, 119, 121(3), 123, 124, 126, 129, 132, 133, 141, 142, 150, 151, 162
John Gest's house, 35
John Gibson's land on Giants Castle Branch, 123
John Gibson's plantation/fence, 122, 168, 169
John Gibson's quarter, 135, 144, 160
Plantation where James Gilpin formerly lived, 160
William Gilpin's place, 166
John Gladin's house, 41

William Godfrey's, 9
Inhabitants of Little River and Goose Creek, 7
Thomas Grafford's house in Alexandria, 14
Thomas Grafford's place, 133
Michael Grater's house in Alexandria, 51
Grayham's plantation, 164
James Green's plantation, 85
William Green's place, 167
David Griffith's place, 161
Gullat's fence, 115(2)
Henry Gunnell's home house plantation, 124
Henry Gunnell's quarter, 134(2)
Robert Gunnell's quarter, 164
William Gunnell's/William Gunnell's land, 131, 139, 141
William Gunnell's woods/fence where foot people cross to the house, 128
William Gunnell's woodland, 130
Line between Philip Darrell and William Gunnell, 128
Elisha Hall's, 1, 6
Martha Hall's (near the Mountain road), 11
William Hall, Jr.'s house, 1
Hallace (see Hollies/Hollis/Hollice)
James Hailey's house in Alexandria, 130
Halley's, 114
Benoni Halley's/Benoni Hally's plantation, 72, 87
Line between James Halley and John Halley, 124
Simpson Halley's land, 157(2)
James Hamilton's land/house/plantation, 12, 13, 26
Capt. Harper's fence, 124
William Hartshorne's tenants, 136
Hepburn and Dundas's land on Back Lick Run and Indian Run, 126
William Hepburn's plantation on Pohick, 144
William Hepburn's farm, 161
Alexander Henderson's plantation, 138

Alexander Henderson's plantation on Bull Run, 104, 107, 108, 117, 145
Alexander Henderson's quarter at Moorehill, 166
Hendrick's place, 149
Herbert's, 116
Thomas Herbert's land on Holmes Run, 125, 140
William Herbert's quarter and tenants, 162
Hereford's (at the intersection of the Colchester road), 163
(End of) John Hereford, Jr.'s lane, 136, 137
John Herryford's house at the ferry, 5
Ezekiel Hickman's house, 39
John Higgerson's house, 5
Walter Hill's place, 162
Lewis Hipkin's store, 132
Hodgkins's, 173
Hollis's/Hollies's, 85, 93, 147(2), 157
Hollis's/Hallace's/Hollice's old field, 58, 84, 138(2), 144(2), 151, 152, 160(2), 167(2)
John Hollis's, 28, 49, 72
Place where John Hollis formerly lived, 106
The Hop-yard, 120
John Hough's leased land in the gap of the Short Hills, 36
John Hooe's land on Bull Run, 81
John Hugh's house (at Colchester), 142
Hunter's fence, 118(2)
Hunter's old storehouse, 81
George Hunter's land, 128, 130
Line between Mr. Hunter and Mr. Scott, 130
William Hunter's place, 165
William Hunter's quarter, 146
John Hurst's plantation above the Falls Church, 81
John Hurst's quarter, 134
Peter Hurst's place, 167
Husler's place, 165
Benjamin Hutchinson's, 41
Mr. Hutchison's, 7

Edward Jacobs house, 171
John Jackson's land (where he lives), 128, 130, 140
Line of Hugh Conn, Fairfax, and John Jackson, Sr., 175
Line between Richard Conway and John Jackson, 128
Edward Jacob's house, 171
Jefferson's old place, 135
Daniel Jenkins's, 131(2)
Field rented by Mr. Daniel Jenkins from Mr. Scott, 130
Hanah Johnson's plantation, 91
Capt. Archibald Johnson's plantation, 146
Baldwin Johnson's place, 165
Joseph Johnson's place, 166
George Johnston's fence/plantation, 72
William Johnston's land/plantation, 75, 134
Charles Jones' house, 143
James Keen's plantation, 108
James Keith's land on Popeshead Run, 83
Marshall P. Kidwell's place, 160
(Heirs of Daniel Kinchloe's) land of Hector Kinchloe and Jesse Kinchloe, 173, 174
Benjamin King (of Loudoun)'s tenants, 145
Samuel King's, 11
William Kitchen's plantation, 87
George Lampkin's land on Wolf Run, 92
James Lane's plantation, 13
Richard Lane's house, 173
Edward Lanham's/Lannum's field/place, 128, 130, 162
Robert Lawson's upper quarters, 166
Tobias Lear's quarter, 166
Zachariah Leatherwood's quarter, 164
Landings (see entry for Warehouses and Landings at the end of this section)
Leesburg, 115, 128, 136, 150, 152, 162
(Beard's lane on) Lee's land, 173
Charles Lee's/Charles Lee, Esq.'s fence/land/place/farm, 127, 136, 149, 161

Hancock Lee's quarter, 147
Henry Lee's quarter, 132
Ludwell Lee's quarter, 132
Ludwell Lee's tenants, 137, 146, 165
Philip Ludwell Lee's plantation, 49
Col. Thomas Ludwell Lee's land, 95
Hon. Thomas Lee's quarter, 1
Lewis's line, 118
Elizabeth Lewis, infant's, land on Difficult Run, 76
Land of the heirs of Stephen Lewis, dec., 51
William Lewis's quarter, 135, 145
Opie Lindsay's quarter/quarters, 144, 160
Thomas Lindsay's house, 174
William Lindsay's house at Colchester, 150
John Lister's place, 161
Inhabitants of Little River and Goose Creek, 7
Charles Little's quarter, 133, 162
Charles Little's at Danby, 134, 149, 161
Samuel Littlejohn's fence/land, 58, 84, 85
Loofborough/Loofborrow/Loofburrow's, 93, 101, 103(2), 114, 116
David Loofbourrow's, 138, 139
John Lucas's house, 1, 7
Daniel Lyon's house, 11
John McFarlin's place, 165
Daniel McCarty's White Marsh quarter, 165
Daniel McCarty's home house, 166
Madden's shop, 163, 164(2)
Mrs. Manley's plantation 73, 74
Marle's house (near Avery's road) / Marle's/Marl's, 132, 150, 152, 162
John Mercer's land on Little River, 20
Mason's Neck, 159
Charles Mason's house, 40
Col. Mason's land, 124
Col. Mason's upper quarter, 149
Col. (George) Mason's quarter at Hunting Creek, 136
Col. George Mason's quarter at Dogue

Run 136, 146
Col. George Mason's home house and quarter (in the Neck), 137, 148
Estate of George Mason and others of Mason's Neck, 159
George Mason's land, 159
George Mason's land (plantation where Joseph Garberry lives), 102
George Mason, Jr.'s home house and quarters, 137, 148
George Mason (of Pohick)'s land on Pohick Run, 153, 154
George Mason (of Pohick)'s tenants, 137, 146, 165
John Mason's fence, 74
Mrs. Mason's home house and quarters, 165
Thompson Mason's home house, 166
Thompson Mason's quarter (on the west side of Gum Springs road), 165
Peter Mauzy's plantation, 122
Anne Meek's house, 172
Daniel Mills's, 113(2)
Minor's gate, 127
Minor's quarter, 16
George Minor's quarter, 132, 164
Nicholas Minor's, 27, 41
John Monroe's, 62, 63
Thomas Monroe's land at Quarter Branch, 54
Mr. Benjamin Moody's, 100
Mrs. Moon's at Dogue Run, 136, 137
William Moon's plantation, 108
Jesse Moore's blacksmith shop, 157
Moorehill, 166
Mrs. Moseley's quarter, 139
George Mull's house, 40
William Munday's house in Alexandria, 16
Land of the heirs of James Murray, 127
Netherton's, 26
Newgate, 126, 127, 138, 160, 168, 169
George Newman's house, 174
Capt. Newton's quarter on Great Rockey Run, 13
Neighborhood of Occoquan Forrest, 143

Rezin Offutt's land on Difficult Run, 154
John Orr's tenant, 135, 144
Patterson's, 117
Annanias Payne's place, 162
William Payne's plantation, 113
John Peake's house, 12
Mary Peak's place, 134
Simon Pearson's plantation/old field, 57(2)
Simon Pearson's old place, 161
Franklin Perry's old place, 134
Francis Peyton's tenants, 166
Thomas Pindal's place, 162
John Pinkstone's, 88
Joseph Pollard's plantation, 73
Thomas Pollard's plantation, 71
Thomas Pollard's land on Popeshead Run, 83
Thomas Pollard's land on Giants Castle Branch, 123
Ann Popejoy's place, 135
Benjamin Posey's house, 171
The post in the main road, 17
Joseph Powell's plantation, 133
Price's, 99(2), 116(3), 126, 127
Benoni Price's plantation, 155
Ramsay's old field, 120, 162
Ramsay's quarter, 133
Henry Rassau's place, 166
Dividing line of John and Barbara Ratcliff, 127
John Ratcliff's plantation/place, 55, 56, 160
Richard Ratcliffe's/Richard Ratcliffe's land, 157, 168, 169
Ravensworth, 1, 30, 72, 73
Michael Reagan's/Regan's house, 10, 16
William Reardon's plantation, 19
Yelverton Reardon's plantation, 164
Yelverton Reardon's tobacco ground, 108
The Red house, 103, 114, 115, 118, 119, 132(2), 156
Mrs. Rerdon's old place, 165
James Richards' plantation, 153, 174

John Richster's land/place/tract, 155, 165
Place where Ricketts lives at the crossroads, 95
Ricketts's, 115(4), 134, 136(2)
John T. Ricketts's workman, 161
William Rodger's tenants, 167
William Rogers's plantation, 124
John Row's plantation, 146
Charles Russell's house, 173
Peter Ryley's plantation, 108
Edmund Sand's house at Goose Creek ferry, 16
Lewis Sanders's/Saunders's land/plantation, 23, 87, 88, 89, 94
Lewis Sanders, Jr.'s plantation, 19
Sandford's quarter, 150
Robert Sandford's house, 13
Richard Sanford's fence, 124
Richard Sanford, Jr.'s place, 161
Sangster's/Songster's, 84, 114, 116, 124, 145(2), 147, 157, 164, 173
George Sangster's place, 166
Thomas Sangster's/Songster's, 85, 86, 95, 98(4), 100, 105, 107, 135, 138(2), 139, 145(2), 150, 151, 152, 164, 166(2), 167, 174
Sangster's/Thomas Sangster's/Thomas Songster's shop, 114, 145, 146, 166(2)
Saunders's, 117
Lewis Saunders (see Sanders)
School houses:
    The old school house, 33, 34, 42, 60, 66(2)
    School house above Anderson's, 41
    Summer's old school house, 54(2)
    School house near John Summer's, 82, 99
    Old school house near Francis Summers, 99
    Terrett's school house, 149
Mr. Scott's pasture and cornfield, 49
Line between Mr. Hunter and Mr. Scott, 130
Field rented by Mr. Daniel Jenkins from Mr. Scott, 130
The Revd. Mr. Scott's land, 74
Old field of Gustavous Scott Esq., 128
Gustavus Scott's/Scot's farm, 134, 149, 161
Sebrek Scott's plantation, 132
John Seale's plantation, 67
John Seal's, 68
Place where Benjamin Sebastian formerly lived, 125
Mr. Sebastian's lot in Alexandria, 38
Mr. Sebastian's quarter, 71, 72, 73(2)
Old Sewell's place, 163
Shanandore, 27
William Shortridge's, 86
Line between Thomas Darne and Samuel Shreeve, 142
Signposts:
    Old signpost at Colchester, 146
    Post in the Main road, 17
    Post in the Mountain road, 26
Charles Simms's place 165
Col. Charles Simms's quarter 146
James Simpson's quarter, 138, 147
Joseph Simpson's land/field, 128, 130
Moses Simpson's land on Sandy Run, 110
Richard Simpson's quarter, 147
William Simpson's/William Simpson's house/land, 139, 145, 155(2)
Place where William Simpson lives at the Turnpike gate, 136
William Simpson's line, 108
John Skinner's house, 173
Price Skinner's, 160(2)
Nathaniel Smith's house in Alexandria, 9, 16
Samuel Smith's lane, 155
Robert Smock, 159
Songster (see Sangster)
Thomas Songster (see Thomas Sangster)
Thomas Sorrell's house, 12
Balty Stearns's, 62, 63
Steptoe's quarter, 73
James Steptoe's land, 15
The old Storehouse, 110

David Stuart's quarter, 135, 150, 162
Sugar Lands, 18, 52
Sumers's/Summers's land/plantation, 25, 115, 116(2), 134, 149(2), 160, 161
Francis Summer's, 99
Francis Summer's Jr., 106(2)
John Summer's, 22, 46, 82(2), 97, 99
John Summer's house, 155
Widow Sumer's lane, 143
Swink's plantation, 164
William Swink's field/house/line, 128, 130
Benjamin Talbot's/Talbott's/Talbutt's plantation, 60, 67, 86, 87, 88, 90, 94, 95, 112(2)
Samuel Talbott's fence/plantation, 95, 104
Talbut's fence, 113
Talbutt's place, 150
Col. Tayloe's quarter, 3, 13
Jesse Taylor's quarter, 162
Place where Sarah Taylor lives, 160
Templeman's plantation orchard, 26
John Templeman's house, 2
Terrett's quarter, 149, 161
Capt. W. H. Terrett's old field, 120
Daniel Thomas's land on Popeshead Run, 24
Edward Thompson's, 10, 16
Joseph Thompson's plantation, 104
William Thompson's place, 138
Charles Thrift's land near the Little Falls road, 128, 131(2), 140
Field rented by Mr. Charles Thrift of Robert Carter Esq., 130, 140
George Thrift's house, 40
John Tillett's former plantation, 146
Toulston's quarter, 148
Trammell's, 116
Trammell's old field, 54
Gerrard Trammell's old field, 57
Garrard/Gerrard Tramel's /Tramell's/ Trammell's land/plantation, 25, 26, 66, 115
Young Gerrard Trammel's, 81
John Trammel's house, 37

John Trammel's island on Potomac River, 37
John Trammell's land on Difficult Run, 71, 76
Mr. Sampson Trammell's lands, 130
Sampson Trammell, Jr.'s plantation, 134
William Trammell's, 35
Trees:
    The Sycamore near the Little Falls warehouse, 158
    The Chestnut tree, 67
    The Mile tree, 116(2)
    The White Oak harbour, 146(2), 166(2)
    The White Oak at the Widow Tucker's, 133(2)
    The White Oak on the road near Capt. W. H. Terrett's old field, 120
Mrs. Triplett's place, 165
William Triplett's tenants, 165
Widow Tucker's, 133(2), 162(2)
William Tuel's house, 172
Mr. Turberville's plantation, 56
Mr. Turberville's old field, 57
Mr. Turberville's quarter, 106
Corbin Turberville's plantation, 164
Mr. John Turberville's land, 128, 130, 140
John Turberville's quarter, 133, 164
Sampson Turley's/Sampson Turley's plantation, 27(2), 108, 117, 139
Sampson Turley's plantation on Johnimore, 89
Elzey Turner's house, 173
Charles Tyler's house, 4, 8, 15
John Urton's land/plantation, 49, 62
Bond Veale's, 9
Edward Violet's/Violett's house, 12, 15, 43, 53
Valinda/Verlinda/Virlinda Wade's tithables/plantation, 17, 33, 35
John Waller's plantation, 95
Zachariah Ward's house/place, 165, 174
Nathaniel Warden's tenants, 164
Mrs. Sarah Warden's plantation, 145
George Washington, Esq.'s home

plantation, 52, 166
George Washington, Esq.'s land, 127
George Washington, Esq.'s tenants, 51
George Washington's Dogue Run quarter, 59
George Washington's quarter in Fairfax parish, 166
George Washington's quarter in Truro parish, 166
Edward Washington's plantation, 87
Edward Washington's quarter, 166
Edward Washington's home house hands and blacksmiths, 167
Washington's (Edward?) shop, 169
Lund Washington's quarter, 75
Major Washington's quarter, 149
William H. Washington's tenants, 162
Watson's place where Joseph and Negro Tom lived, 134
James Waugh's tenants, 166
Cornelius Wells's house at Colchester, 150
Benjamin West's place, 166
George West's house, 17
Hugh West's house at the ferry at Alexandria, 2
West's old quarter, 134
John West's meadow/land/plantation, 120, 130, 161
John West's tenants, 161
Mrs. West's quarter, 135, 149, 160
Roger West's house, 154, 155
Capt. T. West's field, 120
Thomas West's plantation and his tenants (under the Hill), 136, 149
Thomas and John West's land, 133
William West's house/land, 3, 7, 13, 23, 41
William West's old place, 137
Samuel Wheaton's place, 161
Whiting's Church quarter, 133
Carlyle Whiting's quarter, 162(2)
Robert Wickliffe's land, 173
Mrs. Willis's, 6
Winter Hill, 160
John Wise's place, 149

The Wood Cutting, 167
John Workman's house, 171
Col. Wren's land, 130
James Wren's house, 151
James Wren's plantation, 46
James Wren's quarter, 133
Col. James Wren's store, 128, 130(2), 131,155
Mr. Wren's plantation, 57
Thomas Wright's place, 162
Warehouses and Landings (exclusive of ferry landings):

The (Pohick?) warehouse, 2
Warehouses, 12
Public landings at Alexandria, 21(2), 22, 63
Public tobacco warehouse at Alexandria, 120
Colchester warehouse, 74, 117
Falls landing, 44, 74, 107, 110(2),114, 132(2)
Falls warehouse, 44(2), 48(2), 49(2), 67, 72, 76, 79, 99, 103, 106, 107, 110(2), 114, 132
Hunting Creek warehouse(s) 21, 22
Little Falls warehouse, 158, 164
Pohick warehouse(s), 15, 18, 43, 106
Occoquan warehouse, 48, 49, 59, 152
Old warehouse landing on Occoquan, 152
Widow Brewster's landing 173
Landing opposite Shoemaker and Kirkbride's mill on Occoquan River, 141

Ordinaries

Brown's ordinary, 28, 42, 75(2)
Price's tavern/ordinary, 128, 129, 135(2), 144, 147, 149, 151, 154, 160(2)
William Turner's tavern, 134, 147, 158, 167
Water's ordinary, 58
Col. Wren's tavern, 159(2), 163

Ordinary and Tavern Licenses and Applications

Abert, John, 154
Anderson, John, 20
Anderson, John (presented for keeping a tipling house at Goose Creek), 3
Arrell, Richard, 73,
Arrell, Richard (at Alexandria), 75, 79, 83, 88, 93, 102, 108
Ashford, William, 35
Ashford, William (no license), 44
Awbrey, Henry, ferry keeper, 35
Bayley, Samuel (at Colchester), 67
Bayley, Pierce, 50
Bayley, Pierce//Peirce (at Colchester), 75, 79
Bayley, William (at Colchester) 45
Bayley, William, 55
Beall, Sopia, 108
Bell, James (at Alexandria), 74, 84
Bell, Mary (no license), 106
Bloxam, Mary, 171
Boggess, Henry, 9, 18, 24
Boggess, Robert, 4, 45, 51, 56, 60
Boggess, Robert (attempt to take away his license rejected), 5
Boylston, Thomas (at the Falls), 47
Brawner, John, 105
Brown, John (at Colchester), 122
Brown, James, 58
Brown, James (no license), 90
Brown, James (at Pohick) 92, 104
Brown, Richard, 3, 15, 22, 31, 38
Byrn, Patrick, 63
Burnum, Henry (at the old Court house), 83
Burnum, Henry (no license), 90
Burnum, Henry, 95
Bush, Daniel (at Alexandria), 71
Chew, Joseph, 17, 31
Chew, Joseph, 24
China, Joseph (at Cameron), 87
Colclough, Robert, 12
Coleman, Richard, 35, 39
Collins, James, 122
Conner, John (at Cameron), 71
Courts, Joseph (at Alexandria), 101
Courts, William (at Colchester), 83, 89, 100, 107
Courts, William (no license), 99
Curtin, Daniel, 158
Dalton, John (at Alexandria), 5
Daws, Philip (no license), 99
Demovil, Sampson, 31
Dodd, William, 38
Dougherty, Daniel, 157
Dozer, James Ingo, 39, 43, 46, 51
Dozer, James Ingoe (suspended), 41
Ellzey, Thomazin, 170, 171
Evans, Thomas, 9
Fallon, Elizabeth, 48
Farr, Samuel, 168
Farrow, John, 46
Fisher, Amos, 168, 171
Forbers, Thomas, yeoman (no license), 110
Forbes, Thomas (at the Falls), 96
Furguson, Cumberland, 131
Gardner/Garner, William, 62, 75, 81
Gardner, William (no license), 90
Gates, Isaac (no license), 89
Gates, Samuel (no license), 89
Gladin, John, 33, 41
Gladin, John (no license), 44
Gladin, John (at Cameron), 45
Gord, Joseph (at Alexandria), 86
Grafford, Thomas, 14, 21
Grafford, Thomas (no license), 66
Graham, John (at Alexandria), 104
Grayson, Benjamin, Gent. (at Alexandria), 74
Gretter/Grater/Gratter, Michael (at Alexandria), 51, 57, 64, 79, 85, 93, 102
Griffith, Hannah, 159
Hailey, James (at Alexandria), 130
Hall, Elisha, 1
Hamilton, James, 12
Harden, Edward (at Rock Creek), 93
Harris, James, 168
Harrison, Thomas, Jr., 24

Hawkins, William, 92
Hereford/Herryford, John (at his ferry), 5, 45
Hickman, Ezekial, 39
Higgerson, John, 5
Hill, Bennett, 84
Hill, John, 101
Hodgkins, Ralph, 172
Hogskins, Raphael, 169
Hollis/Holless, John, 47, 51, 56
House, John and Margaret (no license), 94
Hubbal, John, 170
Hubbard, Jacob (contrary to law), 86
Hughes, Nathan, 53, 58
Hughs, John (at Colchester), 142
Irmell, Paul (contrary to law), 60
Irmill, Paul (at Alexandria), 38
Jackson, Eleanor, 45, 48, 55
Jackson, Eleanor (no license), 44
Jacobs, Edward, 158, 171
Jenkins, Samuel, 20, 27, 35
Johnson, Seth, 52, 56
Johnson, William (no license), 90
Johnston, Samuel, ferry keeper, 67
Jones, Charles, 143, 169
Jones, Joseph (at Alexandria), 103
Lane, Richard, 173
Leake, Richard (at Alexandria), 80
Lidyard, Newbury, 22
Lightfoot, Richard, yeoman (no license), 109
Lindsay, Thomas, 168, 174
Lindsay, William, 140
Lindsay, William (at Colchester), 118, 150
Lindsey, Robert, 68
Linton, John, 68
Linton, William (at Colchester), 59, 67, 68
Lomax, John (at Alexandria/Hunting Creek ferry), 84, 88, 93, 102
Lomax, John and Rachel (no license), 82
Loofborrow/Loofburrow, David, 88, 93, 108
Lucas John, 1, 7

Lyon, Daniel (license rejected), 11
Mason, Ann, 45, 55
Mason, Ann (no license), 44
Mason, Charles, 40
McDaniel, James, yeoman (no license), 109
McDaniel, William (at Colchester), 104
McDonald, James (no license?), 93
McKnight, William (at Alexandria), 139
Mead, Samuel, ferry keeper, 28
Meeks, Ann, 172
Mills, Sarah, widow (no license), 89, 109
Minor, John, 61
Moxley, Richard, 32
Mull, George, 40
Mull, George (no license), 44
Munday, William (at Alexandria), 16
Munroe, Andrew, 168
Newman, George, 174
Parrott, George, 65
Parrott, Thomas (John Ward's hireling man; no license), 100
Patterson, William, 158
Pattinson, Susanna, widow, (no license), 109
Peake, John, 12, 20
Pelling, William, 157
Pendleton, James, 37
Perry, Hanah (no license), 100
Plummer, John (at Alexandria), 55
Plummer, John (no license), 66
Posey, Benjamin, 171
Powell, Micajah, 74
Pshaw, Jacob (at Alexandria, at the ferry and Mr. Sebastian's lot), 26, 38(2)
Pshaw, Jacob, 32
Ratcliffe, John, 172
Reagan/Regan, Michael, 10, 16
Rhodes, James (at Hunting Creek ferry), 81
Russell, Charles, 173
Sanders, Thomas (at Alexandria ferry), 68
Sands, Edmund (at Goose Creek ferry), 16
Sanford, Elizabeth (at Cameron), 83, 93

Sanford, Elizabeth (no license), 106
Sanford/Sandford, Robert, 13, 21
Sanford, William (no license), 82
Sangster/Songster, Thomas, 122, 150
Savenus, Nicholas, 23
Sawkins, James, 172
Seal, John, 62
Sebastian, Benjamin, 51, 61
Sebastian, Benjamin (at Alexandria), 56, 67
Seelig, Michael, 58
Sewell, William (at Alexandria), 25
Sexton, Daniel (at Colchester), 65
Shilling, Jacob (at the ferry at Alexandria), 54
Simmons, Elizabeth, 168
Simpson, George, 60
Simpson, William, 122, 172
Skinner, John, 159, 173
Smith, Joseph 168
Smith, Nathaniel (at Alexandria), 2, 9, 16
Smith, Nathaniel, 24
Sorrell, Thomas, 12, 20
Sparrow, Matthew, 68
Steel, John, 92
Stephens/Stevens, Joseph [Revoked/disabled], 47, 48
Stuart, John, 172
Summers, Francis, 88, 91, 104
Summers, Francis (no license), 90
Summers, Francis, Jr., 82
Talbutt, McKinsey, 121
Templeman, John, 27
Templeman, John (at the Falls), 2, 7, 12
Thompson, William, 171
Thrift, George, 32, 40
Thrift, George (no license), 44
Towers, John, 158, 172
Tramel, William (disorderly house), 42
Trammell, William (rejected), 42
Tuel, William, 172
Turner, Ellzey, 173
Turner, Charles (at Alexandria), 82
Turner, William, 122, 152, 158
Tyler, Charles, 4, 8, 15, 42
Tyler, Charles (at Colchester), 45, 54
Vermillion, Hanson, 170
Violet, Edward, 12
Violett, Edward (no license), 90
Wailes/Wales, Andrew, 84, 91
Wailes, Andrew (at Alexandria), 108
Ward, John (at his mill; no license by his hireling man Thomas Parrot), 100
Ward, William, 154
Ward, Zachariah, 174
Waters, Elizabeth (presented for keeping tipling house; no license) 58, 60, 65
Wathen, Edward (no license), 89
Watson, John (at Alexandria), 41
Wells, Cornelius (at Colchester), 150
West, George, 17
West, Hugh (at Alexandria ferry), 2, 9
West, William, 7, 13, 22, 29, 41
Wiley, James, 142, 152, 168
Williams, Elijah (at Brown's ordinary), 75
Willet/Willett, Edward, 45, 48
Willit, Edward (at Alexandria), 43
Workman, John, 171
Wren, James, 151, 154, 167
Young, David, 55, 59, 68
Young, David (at Alexandria) 79
Young, William (no licence), 140
Young, Dorthy/Dortha. (at Alexandria), 86, 93, 102

Personal Names

NOTE: Names may be spelled several different ways; all possible spelling variants should be checked; (_____) indicates a blank space in the record

No surname given:
    Dutch Caty, 120
    Jack (slave), 107
    James (free Negro), 163
    Joseph, 134
    Negro Adam, 161
    Negro Cambridge, 161
    Negro Daniel, 162
    Negro Jasper, 165
    Negro Mingo, 161
    Negro Searor, 162
    Negro Simms/Simms (free Negro), 148, 166
    Negro Tobey, 161
    Negro Tom, 134

Abercrombie, Js., 149
Abert, John, 154
Adam, Robert/Robt. (and Gent.), 39, 44, 48, 68(2), 69(2), 73, 76, 112, 114, 115, 126, 132, 133
Adam, William (and Gent.), 54, 69
Adam, Widow, 162
Adams, 113, 162
Adams, Abednego, 36, 39, 67, 73, 77, 137, 149, 166
Adams, Edward, 134, 148, 150, 161
Adams, Francis, 160, 172, 174
Adams, Jere:/Jerry, 136, 149
Adams, John, 133, 163
Adams, Philip, 107
Adams, R., 121
Adams, Samuel/Saml., 118, 119, 125, 141, 163, 164, 173, 175
Adams, Samuel Adams Jr. 134
Adams, Simon, 128, 129, 130, 131, 134
Adams, Widow, 32, 150
Adams, William/Willm./Wm.(and Mr. and Gent.), 4, 19(2), 35, 53, 58, 59, 63, 66(2), 68, 69, 79, 80(3), 82(2), 83, 84, 92(2), 96, 97, 100, 101, 103, 104, 105, 112, 116, 119(2), 129, 130, 134, 141, 142, 162, 163
Agullenil, Wm., 162
Alexander, 162, 166
Alexander Charles/Chs. (and Gent.), 71, 72, 92, 123, 126, 133(2), 150, 162
Alexander John, 29
Alexander John Jr., 78
Alexander Philip/Phillip, 78, 93, 102, 108, 115
Alexander Robert/Robt., 68, 78, 87, 94, 96, 96, 137, 149, 150, 162
Alexander Samuel, 137, 146
Allen, Andrew, 136, 144, 163
Allen, Charles, 161, 168
Allen, David/Dd., 134, 149, 161
Allen, Henry, 161
Allen, John (and sawyer), 135, 147, 167
Allen, John R., 144, 160
Alley, Aaron, 167
Allison, John (and Gent), 124, 146, 165
Allison, Robert, 120, 122
Allison, Thomas (son of John), 165
Allison, Thomas/Tho., 148, 166
Allison, Thos. Jr., 146
Alliston, John, 137
Alliston, Thos., 137
Alliston, Thomas Jr., 137
Alton, John, 146
Alverson, John, 107
Ammontree, David, 161
Anderson, 41
Anderson, Geo., 148
Anderson, John/Jno., 3, 6, 8, 20, 26, 36, 40, 132, 148, 166
Anderson, John Jr., 161
Anderson, John Baptist, 15
Ansdell, Daniel, 19
Arell, David, 127
Arell, Richard/Rd., 73, 79, 83, 88, 93(2), 102
Arell, Samuel/Saml., 126, 127, 135, 149

Armstrong, Jno., 149
Arrell, Richard, 75, 108
Arskin, John, 134
Asberry, Thomas, 164
Ashby, 10
Ashbury, Thomas, 49
Ashford, Butler Stone Street, 92
Ashford, Francis/Fran., 137, 146
Ashford, John, 2, 15, 21, 35(2), 36, 37, 64, 74, 135, 144, 160
Ashford, Michael/Michl., 137, 146, 165
Ashford, Michael's son, 165
Ashford, William, 44, 49, 54
Asken, John, 160
Askin, Coatney, 160
Askins, Courtney, 148, 154, 172
Askins, Jno., 148
Atcheson, John, 137, 146
Atcheson, Leonard, 137
Atcheson, Widow, 102
Atchison, John, 165
Athell, John, 41
Athey, Sarah, 147
Athey, Thomas, 146
Atholl, John, 26(3), 40
Atkinson, John, 135
Avery, 121(3), 156, 162(2)
Avory, 121, 132(3)
Avory, John, 132
Awbrey, 4
Awbrey, Henry, 35, 38
Baggatt, Thos., 135
Baggett, Samuel, 136
Baggot, Samuel, 165
Baggot, Thomas, 161
Baggot, William, 165
Baggott, Tho., 149
Bailis, Hannah, 167
Bailis, Thomas, 165
Bailis, Thomas (son of William), 163
Bailis, Thos. (son of Hannah), 167
Bailis, William (son of William), 163
Bailis, Wm. P., 163
Baily, Jesse, 166
Baily, Samuel, 165
Baker, James, 132, 162

Baker, James Jr., 158
Baker, Thomas, 134
Baker, Widow, 162
Baker, William/Wm., 132, 162, 167
Balendine, John (and Mr.), 49
Balinger, James, 163
Balinger, John, 161
Ball, Bazil, 162
Ball, James, 133, 162
Ball, John/Jno., 113, 132, 150, 162, 172
Ball, Moses, 83, 92, 105, 133
Ball, Moses Jr., 132
Ballandine, 173
Ballendine, 85
Ballendine, John, 50, 74, 81
Ballenger, Benjamin, 134(2)
Ballenger, Jacob, 148
Ballenger, John, 134, 137, 147, 148
Ballenger, Joseph, 136, 147
Barker, John, 147, 167
Barker, Leonard, 167
Barker, Mary, 167
Barker, Moses, 138, 147
Barker, Thos., 142
Barker, William/Wm., 138(2), 147, 167
Barker, Will. Jr., 147
Barnes, Abraham, 3, 15, 85
Barnes, John, 144, 160
Barnes, Mr., 32, 148
Barnett, Gerrard/Gerrd., 139, 146, 166
Barns, 61
Barns, Mr., 47
Barry, John, 33, 43, 59, 61, 62, 107
Barry, William/Wm., 33, 134
Barton, Benjamin, 165
Bates, Edward/Edwd., 134, 149, 161
Bates, Thomas/Tho., 148, 161
Baxter, Thomas, 172
Baylis, Thomas, 18
Baylis, William/Will./Wm. Payne, 89(5), 90(3), 136, 147
Bayliss, Thomas/Tho., 137, 146, 147
Bayliss, Wm. Payne, 152
Bayley, 148

Bayley, Jesse, 139
Bayley, Pierce, 50
Bayley, William, 45, 48, 51, 55
Bayly, Jesse, 146
Bayly, Mr., 102
Bayly, Pearse/Peirce, 75, 79, 83, 84, 90, 92, 100, 104, 146
Bayly, Samuel/Saml., 67, 91, 104, 122, 123(2), 137, 146, 153(3)
Bayly, William/Wm., 65, 74, 75, 77
Bayly, William C., 146
Beach, Alexander/Alex., 134, 148
Beach, Chs., Beach 135, 147
Beach, Chs. W., 145
Beach, Thomas/Thos./Tho., 60, 83, 134, 135, 149
Beache, Charles, 138
Beall Sophia, 108
Beard, 173
Beard, George, 173, 174
Beatle, Andrew, 134
Beaty, Charles, 123, 151
Beckwith, Marmaduke, 89, 95, 108, 122, 135, 139, 144, 145, 160
Beckwith Newman, 135, 144
Beckwith Mr., 73
Beech, Alex., 161
Beech, Chs. Watson, 166
Beech, Elisha, 161
Beech, Solo., 166
Beech, Thomas/Tho., 161, 167
Bell, James, 74, 84
Bell, Mary, 106
Bennet, Joseph, 61
Bennett, 148
Bennett, Henry Astley, Esq., 141
Bennett, Joseph/Jos:,60, 67, 104, 113, 134, 135, 148, 167
Bennitt, Joseph, 115
Benson, ____, 146
Benson, Robert/Robt., 165
Bent, Jacob, 162
Berkley, Burgess, 4, 10, 26, 28
Berkley, John, 27, 58
Berkley, William, 27

Berry, William, 16
Berryman, 11
Birch, Joseph, 121(2)
Bird, 149
Bird, Mr., 120
Bird, William/Wm., 127, 136, 149, 154
Black, Alex:, 92
Black, Thos., 135
Blackburn, 164
Blackburn, Edward/Edwd. (and Gent.), 51, 53, 54(2), 57, 61, 62, 63, 64, 65, 67, 70, 72, 76, 77, 94, 95, 100, 112(2), 113, 114, 118(2), 134, 150, 164
Blackburn, Lewis, 164
Blackburn, Richard Scott, 134
Blackburn, Thomas, 96
Bladen, Caleb, 133
Bladen, Canel/Kenol, 162, 163
Bladen, Thomas/Thos., 132, 162
Bladen, Wm., 133
Bladen, William/Wm. Jr., 137, 147, 165
Blake George/Geo., 132, 164
Blancitt, Jno., 146
Blancitt, Joseph, Blancitt 146
Blancitt, Jos. Jr., 146
Blansit, John, 139
Blansit, Joseph, 139
Bloxam, Mary, 171
Bloxham, Mary, 161
Blundal, Elijah, 149
Blunden, Elisha, 161
Bo[page torn], Gerard, 71
Boggess, 32, 69, 72, 97, 102, 167, 172
Boggess, Henry, 9, 18, 24, 47
Boggess, Henry Jr., 24, 34
Boggess, Josiah, 146, 166
Boggess, Robert/Robt., 4, 5, 12, 15(2), 18(2), 23, 34, 43, 45, 46, 51, 56, 59, 60, 62, 72, 74, 87, 101, 102, 104, 109, 137(2), 146, 148, 152, 153, 165, 166, 169
Boggess, Robert Jr., 81, 91
Boggess, Vincent, 74
Boggess, Widow, 137
Boling, Garrard/Gerrard, 36, 64

Boling, Samuel, 159, 162
Boling, William, 162
Bolling, Gerrard, 65, 87
Bolling, John, 104
Bond, John, 16
Bond, Zachariah/Zach., 146, 166
Bontz, Jacob, 162
Bosswell, Matthew, 134
Boswell, Benjamin, 85
Boswell, Mathew, 143
Boucher, John, 163
Bowling, Ann, 133
Bowling Gerrard, 93
Bowling, John, 88, 100, 132
Bowling, Samuel, 133
Bowling, William/Wm., 91, 133
Boyd, 149
Boyd, John, 166
Boyd, William/Wm., 136, 149, 165
Boylstone, Benjamin, 103
Boylston, Thomas, 47
Bozwell, Henry, 149, 166
Bozwell, Matthew, 144, 160
Bozwell, Thomas, 160
Braden, Thomas, 170
Bradley, Daniel, 160, 161
Bradley, James, 160(2)
Brady, Caleb, 132, 150, 162
Brawner, John, 105
Brayley, Owen, 78
Brent, Henry, 149, 165
Brewster, Thomas, 20
Brewster, Widow, 173, 175
Bricken, 103
Bride, 149
Bright, John/Jno., 149, 161
Brinston, Alexander, 137
Broadwater, C., 74
Broadwater, Col., 134, 150, 164(2), 169
Broadwater, Charles/Chas./Chs. (and Maj. and Gent.), 1(2), 14(2), 19, 24, 25, 28(2), 32, 33, 35, 47, 48(2), 49, 50, 51, 53(2), 54(2), 55, 56, 57, 58, 60, 61, 63, 67, 70, 71, 72, 73, 77(2), 78, 79, 83, 85, 86, 87(2), 88, 90, 94(2), 95, 100, 104, 110, 113, 118(2), 121, 123, 124, 149, 157, 161, 173
Broadwater, Charles Lewis/Charles L., 134, 143, 163
Broadwater, Maj., 66(2), 72
Broadwater, Mr., 26, 73(2)
Bronaugh, Jeremiah 2
Bronaugh, Mrs., 45
Bronaugh, William (and Gent.), 48(2), 49
Brook, Walter, 165
Brooke, Walter, 136, 154, 156
Brooks, Walter, 149
Brown, 28, 42, 75(2)
Brown, Catherine, 166
Brown, Dr., 149
Brown, James, 15, 58, 90, 92, 93, 104
Brown, James Jr., 15
Brown, John, 122
Brown, Mark, 15
Brown, Mrs., 163
Brown, Richard, 3, 13, 15, 22, 31, 37, 38
Brown, Doctor Wm., 135
Browning, ____, 135
Bruester, Thomas, 14
Brumback, John, 138, 145, 166
Brummitt, William, 91, 102
Bruster, Thomas, 7, 19
Bryan, William/Wm., 139, 146, 166(2)
Buckner, Peyton, 6
Bucklew, William, 42
Buckley, Benjamin, 89
Buckley, John, 122, 135
Buckley, John F., 144, 160
Buckley, William/Wm., 25, 43, 135, 144, 160
Buckman, James/Jams., 68(2), 69, 75, 78
Burch, Oliver, 135, 148, 167
Burges, Chs., 161
Burgess, Charles/Chs., 135, 149
Burgess, Thos., 134
Burns, Zachariah, 146
Burnum, Henry, 81, 83, 90, 95, 110
Burton, Benja./Ben:, 137, 146
Bush, Daniel, 71
Bushby, Joseph/Jos., 149, 161

Bushby, William/Wm., 135, 161
Butcher, Alexander, 158
Butcher, John, 78, 119, 132, 156, 158
Butler, Edmund, 57
Butler, John, 145, 166
Butler, Lawrence, 145
Byrn, Patrick, 63
Cain, Henry, 161
Caldwell, Hugh, 33
Callahan, James, 164
Callahan, P., 161
Calmes, Marquis, 14
Campbell, Oneas (and Gent.), 35, 37, 42
Canaday, John, 72
Canady, John, 30
Canary, Richard, 16
Cannaday, John, 16
Canterbury, Samuel, 58, 60
Canton, Mark, 35, 36[?]
Carico, Wm., 136
Carlin, 162
Carlin, William/Wm., 101, 133, 162, 170
Carlyle, 133
Carlyle, John (and Gent.), 14, 19, 22, 41, 48, 58, 62, 64, 68(2), 69(2), 70, 76(2), 81, 88, 107
Carlyle, Mr., 96
Carpenter, 72(2)
Carrico, Abel, 138, 147
Carrico, Wm., 138, 145
Carrico, William Jr., 147
Carrington, Timothy, 134, 149
Carrinton, Timothy, 161
Carrol, Demsie, 22
Carroll, Dempsie/Demsie, 17, 25
Carroll, James, 134
Carroll, Joseph, 133
Carson, ____, 146
Carter, Peter, 27
Carter, Robert Esq., 110, 130, 140
Cartwright, John, 161
Cary, Joseph, 171
Cash, Joseph, 30, 42, 62, 73, 75, 79, 92,
Cash, William, 102, 109, 137, 146, 165

Casity, Solomon, 156
Cassady, Solomon, 161
Caten, William, 165
Cavener, Henry, 50
Cavery, Chs., 136
Chappell, John, 138
Chapman, George/Geo., 113, 132(3), 150, 162(2)
Chappel, John, 145, 166
Chappell, John, 108
Charters, Charles, 133
Cherm, John, 165
Cheurie, 147
Chew, Joseph, 17, 24, 31
Chew, Mercy, 44
Chichester, 172
Chichester, Daniel M., 163
Chichester, D. Pitt, 165
Chichester, Mrs., 165
Chichester, Richard/Richd., 117, 128, 134, 137, 146
Chilton, George, 19
China, Joseph, 87
Chin, Elijah, 29, 31
Chinn, Elijah 10, 35
Chinn, Elishh 36
Church, Jesse, 166
Church, Robert/Robt., 137, 146, 166
Church, Thomas/Tho., 137, 146, 166
Church, William, 146, 165, 166
Clapham, Josias, 39, 43
Clark, Jno., 167
Clark, Josiah, 138, 147, 167
Clark, Richard, 92, 107, 108, 114, 145, 157, 164
Clarke, Josiah, 138
Clarke, Richd., 139, 157
Clifton, 52
Clifton, Mr., 53
Clifton, William (and Mr.), 17, 79, 80
Coats, John, 113, 118
Coats, Moses, 162, 172
Cochran, Robert, 137, 146
Cockburn, Martin (and Gent.), 84, 87, 94, 116, 123, 126, 139, 141, 148, 152, 165

Cockburn, Mr., 148
Cocke, Catesby/Catesbys (and Maj. and Gent.), 1(2), 6, 18
Cockeril, Joseph, 41
Cockeril, Thomas, 8
Cockerill, 163
Cockerill, Joseph, 50, 52, 53, 105, 107, 115
Cockerill, Sampson, 115
Cockerille, Joseph, 134
Cockerille, Sampson, 134
Cockran, Joseph, 161
Cockrell, 163
Cockrill, 163
Cockrill, Jos., 136, 149
Cockrill, Sampson, 134
Coe, Widow, 164
Coe, William/Wm., 138, 145
Cofer, Francis, 167
Cofer, John, 165
Cofer, Joshua, 167
Coffer, Francis/Frans., 138, 147, 151, 152, 174(2)
Coffer, John, 139, 146
Coffer, Joshua, 147
Coffer, Thomas, 85, 88
Coffer, Thomas Withers, 57, 58, 84(2), 103
Colclough, Robert, 12
Cole, Geo., 166
Cole, Solomon, 165
Coleman, James, 108, 153, 154, 172
Coleman, John, 22
Coleman, Richard, 6, 18, 20(2), 22, 28, 35, 37, 38, 39, 42
Collins, 118(2)
Collins, James/Js., 112(2), 122, 134, 149, 161
Colman, James, 174
Colter, Thomas, 2
Colvert, Spencer, 145
Colvill, Col., 11
Colvill, John (and Gent.), 16, 26
Colvill, Thomas (and Gent.), 21, 23(2), 25, 28, 32, 33, 37, 55

Compton, John, 136(2), 149, 165
Compton, Samuel, 10
Compton, Zebede/Zebedee, 138, 143, 147, 167
Coningham, Dr., 137
Conn, Gerrd. T., 144
Conn, Hugh, 175
Conn, Thomas, 120, 161
Connell, Thomas, 28(2), 35
Conner, Edwd., 149, 161
Conner, James, 160
Conner, John, 71, 144
Conner, Samuel, 15, 48
Conner, Terence/Terrence, 15, 142, 165
Conner, Terrence Jr., 137, 146
Conway, 162
Conway, Richard/Richd., 125, 128, 147
Cook, Giles, 123, 126, 144
Cooke, Giles, 124, 136, 160
Cooper, Joel, 133
Cotes, Moses, 172
Cotton, John, 19, 23, 87, 112(2), 117
Cotton John Jr., 107, 108
Coulter, Peter, 137, 147, 167
Courts, 84
Courts, Joseph, 101
Courts, William, 79, 83, 89, 99, 100, 101, 103, 107
Cox, Elizabeth, 136
Cox, Mrs., 149(2)
Cox, Presley, 25, 47(2), 54, 64(2), 90
Coxon, Widow, 134
Craig, Charles, 103, 132
Craig, William, 166
Craik, Charles, 121
Craing, Charles, 162
Cranford, John, 146, 165
Crawford, Thomas Jr., 129
Croson, Hanson, 138
Crump, Stephen, 160
Crump, William/Wm., 136, 160
Crump, Wm. Jr., 133
Cudick, Edmund, 161
Currie, Chs., 149
Curtin, Daniel, 158

Curtis, Chas., 133
Curtis, Thos., 134
Dade, 133
Dade, Baldwin, 41, 120, 130, 150, 162
Dade, Francis, 47
Dade, Townshend (and Gent.), 1, 37, 41, 46, 48, 94, 96, 133
Dailey, Jesse, 160
Dainty, Jonathan, 167
Dalton, John/Jno. (and Gent.), 5, 21, 22, 31, 38, 53, 55(3), 73, 76, 79, 102, 103, 104
Daniel, James, 133
Daniel, John, 59, 160
Daniel, Thomas, 133
Darne, Henry, 134, 163
Darne, John, 127, 133, 162
Darne, Thomas/Tho., 142(2), 150, 161, 162
Darne, William/Wm., 114, 118, 119, 125, 127, 129, 132, 162
Darnes, Henry, 149
Darnes, Thomas, 75
Darnes, Thomas Jr., 168
Darnes, William/Wm., 72, 79
Darrell, Augustus, 81
Darrell, Mary, 149
Darrell, Philip, 128, 161
Darrell, Sampson (and Gent.), 56
Darrell, Sarah, 146
Darrell, William, 149, 156
Davis, 20
Davis, Allen, 114, 134
Davis, Barton, 163
Davis, Benjamin, 44, 160
Davis, David, 27, 144, 160
Davis, Edward, 115, 164
Davis, Edward/Edwd. Jr., 136, 147
Davis, Henry, 135, 138, 145, 149
Davis, Isaac, 135, 144
Davis, John, 41, 135, 162
Davis, Joseph, 55, 76
Davis, Nehemiah, 122, 134, 145, 160
Davis, Samuel, 16
Davis, Thomas/Thos./Tho. (and Rev.), 10(2), 16, 19, 20(2), 38, 109(3), 137, 146, 162, 167
Davis, William/Wm., 134, 149, 160, 164, 174
Davis, William William, 164
Davis, Wm. Jr., 135
Daws, Philip (coppersmith), 99
Dawson, 41
Dawson, 160
Dawson, John, 135, 160
Dayly, _____, 147
Dayly, Thos., 134
Deakins, 134(2)
Deakins, John, 110
Deavers, Gilbert, 138, 147, 167
Demovil, Sampson, 31
Demoville (D'Moville), Sampson, 60, 61
D'Neal, Jas., 76
Deneal, William, 161
Deneale, Edward, 160
Deneale, George, 159, 160, 171
Deneale (DeNeale, D'Neale), James/Jas.,61, 72, 76, 84, 100, 122, 123, 124, 135, 136, 144, 145, 160
Deneale, James/Js. Jr., 135, 144
Deneale, William/Wm. (and Mr. and Gent.),116, 118(2), 119, 121, 124, 126, 135, 150, 151, 152, 153, 159, 160, 168
Dennison, Dudley, 165
Denty, Jonathan, 138, 147
Devaughn, John/Jno., 136, 147, 163
Devaund, Thomas, 100
Dewer, William, 147
Dewit, John, 16
Dewitt, Wm., 138
Dick, Dr., 124, 148
Dickens, 115(2)
Dodd, William, 38, 43
Dodson, Charles/Chas. 123, 124, 136, 144, 163
Dogan, Henry, 117
Donaldson, Andrew, 133
Donaldson, James (and Gent.), 31, 33, 38, 40, 51, 53, 54, 55(2), 56, 57, 72, 74

Donaldson, Thomas/Thos., 133, 162
Donaldson, William/Wm., 54(2), 123, 133, 137, 146, 161
Dorrell, Philip, 134
Dorrell, Sampson (and Gent.), 51
Dorrell, Widow, 136
Dorrell, William/Wm., 117, 126, 136, 149, 165
Dorsey, Danl., 146
Dorson, John, 145
Dougherty, Danl., 157
Doughlass, Mr., 150
Doughlass, Robert, 109(5)
Douglas, Daniel, 164
Douglas, Eliz. (wife of William), 76
Douglas, James, 164
Douglas, Robert, 62
Douglas, William/Wm., 47, 54, 56, 72, 76
Douglass, Daniel, 138
Douglass, William, 47, 62, 67
Dove, James, 134, 148, 167
Dove, John, 161
Dove, Thomas/Thos./Tho., 135, 149, 160
Dove, Zach./Zachh., 147, 161
Dow, Peter, 135, 149
Dowdal, 141
Dowdall, 142
Dowdall, John, 129(2), 133
Dowdle, John, 113
Dozer, 59
Dozer, James Ingo/James Ingoe, 39, 41, 43, 46, 49, 51, 57
Dozer, Leonard 39, 45, 71
Dozier, James, 44
Dozier, Leonard, 24, 27
Duffin, Fenix, 134
Dulaney, Benjamin, 162
Dulany, Benjamin/Benja./Ben., 114, 126, 133, 136, 149, 151, 153, 161
Dulin, Atwell, 160
Dulin, Edward/Ed., 65, 83, 153, 163
Dulin, John, 135
Dulin, Sarah, 134, 163
Duling, John, 144, 167

Duling, William, 13, 26
Duncan, Blanch Flower, 38, 41
Duncan, Samuel, 40, 41
Dundas, 126, 127, 170(2)
Dundas, Jno., 157
Dundas, Mr., 126
Durand, 147
Duvall, William, 120
Dyer, Thomas, 161, 165
Eals, John, 147
Earp, Edward, 132, 158, 162
Earp, Joel, 132, 164
Earp, Mathew, 119, 132, 163
Earp, Matthew Jr., 132
Earp, Wm., 132
Easton, Henry, 166
Eaton, William/Wm., 139, 146
Edmonds, John/Jno., 134, 147
Edwards, ____, 146
Edwards, Hadon, 19
Edwards, James, 148, 161
Eels, John, 134
Eldridge, James, 166
Elliott, Thomas, 162
Ellis, Joel, 161
Ellis, John, 133
Ellis, Philip, 78
Elsey, Francis, 8
Elzey, 143
Elzey, Francis, 6, 8
Elzey, Lewis, 88
Ellicot, 169
Ellicot, Nathaniel, 169
Ellzey, 148(2), 154, 163, 167(2), 168, 169
Ellzey, Lewis (and Capt. and Gent.), 11, 17, 18, 22, 25(2), 27(2), 36(2), 42, 57(2), 58(3), 61, 72, 74, 76, 119, 120
Ellzey, Thomazin, 130, 135, 144(2), 160, 170(2), 171
Ellzey, William/Wm. (and Gent.), 14, 18, 20, 22(2), 24, 26, 31, 35, 38, 39(2), 41, 45, 46, 47, 75, 89
Emmerson, Aquilla, 166
Emzie, Stephen, 43

Errenshaw, John, 128
Essex, William/Wm., 137, 146, 165
Evans, Gray (waggoner), 149
Evans, Guy, 133, 161
Evans, John, 122
Evans, Thomas, 2, 9
Fagan, Joseph, 149, 161
Fairfax, 118, 175
Fairfax, Bryan (and Gent. and Rev.), 54, 73(2), 87, 95(2), 103(2), 107, 108, 110, 134, 136, 149, 155, 165
Fairfax, George (Gent.), 2
Fairfax, George William/George Wm./Geo. Wm. (and Esq. and Gent.), 8, 14, 17, 21(2), 23(2), 29, 31, 32, 33, 37, 39, 47, 77(2)
Fairfax, Lord, 6
Fairfax, Thomas/Tho., 134, 148, 161
Fairfax, Hon. William Esq., 11
Fallin, William/Wm., 121, 132, 141, 142
Fallon, Elizabeth, 48
Farguson, Joshua, 11
Farr, Ann, 144
Farr, Samuel, 160, 168
Farrow, John, 46
Farthing, Nace, 132
Felangdigam, Thomas, 1
Femisters, John, 161
Fendall, 156
Fendall, Phil., 150
Fendall, Philip Richard/Philip R., 141, 143, 162
Fenley, John, 134
Fergason, James, 163
Fergason, John, 166
Ferguson, Joshua, 13
Ferguson, Mr., 45
Ferguson, Mrs. Mary, 87
Ferguson, Zachariah/Zach., 137, 166
Fielder, George, 160
Fielding, Michael, 161
Fields, Luke, 134, 149, 161
Findlay, John B., 147
Findley, Walter, 137
Findly, Walter, 146

Finlay, Robt., 149
Finley, John B., 158, 167
Finley, Martha, 160
Finley, Robert, 134, 161
Finley, Thomas, 160
Fisher, Amos, 168, 171
Fitzgerald (FitzGerald), 133, 162
Fitzgerald, John, 162
Fitzhugh, 12, 66, 113, 141, 143, 154, 163(3)
Fitzhugh, Giles, 163
Fitzhugh, Henry, 45, 127, 135, 144(2)
Fitzhugh, Mordecai Cooke, 160
Fitzhugh, Nicholas, 136(2), 151, 154, 159(3), 160, 170, 174
Fitzhugh, Richard, 160
Fitzhugh, William/Wm., 127, 135, 144(2), 153, 160
Flatford, Thomas, 165
Flattford, Thomas, 149
Fletcher, James, 30
Flint, John, 78
Flood, Thomas, 167
Foley, James, 164
Follin, William/Wm., 100, 132, 162
Forbers, Thomas, 110
Forbes, Thomas, 96
Ford, Edward/Edd./Edwd., 79, 84, 100, 114, 122(2), 138, 145, 166
Ford, John, 27, 36, 42, 44, 50, 53(2), 62, 67, 69, 74, 75, 77
Ford, Thomas/Thos., 27, 58(2), 71, 73, 103
Fowler, 165
Fowler, John/Jno., 114, 137, 138, 141, 147(2), 159, 167
Fox, Amos, 112(2), 121, 122, 156, 169
Frazer, 136
Frazer, John, 137
Frazer, Wm., 135
Frazier, 163
Frazier, Jas., 162
Frazier, William, 163
Freeman, Bennett, 149, 165
Freeman, Richard, 165
Freeman, William/Wm., 136, 149, 165

French, Daniel/Dl. (and Mr. and Gent.), 21, 23, 33, 52, 63, 74, 77(2), 90
French, Mr., 66
French, Mrs., 136, 149, 154, 165
French, Richard, 161
Frizell, John, 100
Frizell, Nathan, 110(2)
Frizzel, Mary, 133
Frizzell, John, 76
Furguson, Cumberland, 131
Galdup, Jno., 150
Garberry, Joseph, 102
Gardiner, Joseph, 10
Gardner, Joseph, 173, 175
Gardner, William/Wm., 73, 74, 75, 81, 84, 90
Garner, William/Wm., 62, 72, 164
Garret, Edward, 3, 5, 7
Garret, Nicholas 30, 56
Garrett, Henry, 134
Garrett, Nicholas/Nichs./Nicho., 65, 76, 134, 149
Garrett, Walter, 95
Gates, Isaac, 89
Gates, Samuel, 89, 137, 148
Gates, William, 161
Gates, Widow, 166
Geesland, John, 161
Geesling, John, 133
Geesling, Robert 139
The General (see Washington, George)
Gerdon, Geo., 162
Gest, John, 35
Gibson, John/Jno., 122, 123, 135, 144, 149, 160, 168, 169
Giles, 59(2), 102
Gill, James, 135, 144, 172
Gill, James Jr., 160
Gilmore, William, 15
Gilpin, Col., 120
Gilpin, George/Geo., 112, 116, 118, 119, 120, 121, 123, 124(2), 128, 141, 156(3)
Gilpin, James, 135, 144, 160
Gilpin, William/Wm., 135, 147, 161, 166

Gilping, Benjamin, 60
Gladin, John, 33, 41, 44, 45, 47, 50(2), 51
Gladin, William/Wm., 43, 149
Glauster, James, 147
Glover, Richard, 162
Glover, Thomas, 162, 170
Glover, William, 162
Godfrey, William, 9
Goldop, John, 132
Goldrope, John, 162
Gooding, John, 161
Gooding, William/Wm., 136, 163
Gooding, William Jr., 160
Goodrich, Benja., 158
Goodrich, Benja. Jr., 158
Goodrick, Benjamin, 162
Gord, Joseph, 86
Gordon, Alexander, 150, 162
Gordon, James, 78
Gordon, Sarah, 163
Gordon, Widow, 134
Gorin, Henry, 138
Gorun, Henry, 147
Gossom, Thomas/Thos., 139, 146, 166, 169
Gowing, Joseph, 133
Gozeling, Robert, 145
Grafford, Thomas, 14, 21, 66, 133
Graham, John, (and Gent.) 2, 6, 18, 104
Grasty, George, 81, 103
Grasty, Sarshall, 58
Grater, Michael, 51, 64
Gratter, Michael, 57, 63
Gray, William/Will./Wm., 137, 148, 166, 167
Grayham, 164
Grayson, 6
Grayson, Benjamin/Benjn. (and Gent.), 4, 41, 43, 48, 49, 59(2), 74
Grayson, Col., 23
Green, Charles (clerk), 47, 61
Green, James, 85, 91
Green, Samuel, 165
Green, Shadrach, 94, 109
Green, William/Wm., 147, 160, 167

Green, William M., 162
Greenwood, Wm., 78
Gretter, Michael, 79, 84, 85, 86, 89, 93(2), 102, 122
Griffith, David, 161
Griffith, Hannah, 159
Grigsby, Samuel, 59, 67
Grigsby, Wilkerson, 165
Grimes, Philip, 25, 27(2), 28, 30,
Grimes, William, 41
Grimsley, James/Js., 138, 147, 167
Groves, John, 165
Grymes, 153, 174
Grymes, Philip, 87
Guinn, Mr. Benjamin, 143
Gullatt, 115(2)
Gullatt, Peter, 93
Gullatt, Widow, 134
Gunnel, William, 172
Gunnell, Allen, 161
Gunnell, Henry (and Gent.), 2, 5, 32, 46, 47, 48, 49, 50, 51, 53, 54(2), 56, 59, 62, 67, 72(2), 79, 86, 87, 90, 94, 95(2), 100, 101, 103(2), 108, 112, 124, 134(2), 161, 170
Gunnell, Henry Sr., 148
Gunnell, Henry Jr., 92
Gunnell, Presley, 155, 158, 159, 161
Gunnell, Robert, 164
Gunnell, Robert Jr., 164
Gunnell, Thomas/Tho., 112, 114, 122, 123, 126, 128, 131, 134, 141, 142, 148, 169, 173
Gunnell, William/Wm. (and Mr. and Gent.), 48, 49, 56, 60, 71, 76, 77, 95, 128, 130, 131, 134(2), 139, 141, 148, 149
Gunnell, William Jr., 35, 94
Gwin, Benjamin, 116
Gwinn, Benjamin, 153, 154, 155
Hadon, George 30
Hagan, Chs., 149
Hagan, Francis/Frans., 149, 165
Hagon, Charles, 136
Hagon, Francis 136

Hague, Francis 14, 24
Hailey, James, 130
Hall, Barton, 167
Hall, Elisha, 1, 6
Hall, Jacob, 136, 147
Hall, Jacob Jr., 137
Hall, Martha, 11
Hall, Richd., 146
Hall, Sylvester/Syvelster, 135, 144
Hall, Wm., 138, 147, 167
Hall, William Jr., 1
Hallace, 138(2)
Halley, Benoni, 62, 72, 88
Halley, Geo., 167
Halley, Henry S., 144, 145, 160
Halley, James, 17, 18, 19, 22, 23, 25, 61, 72(2), 115, 124, 135, 144, 152(2), 153, 155
Halley, Js. Jr., 147
Halley, John, 124
Halley, John Jr., 167
Halley, Simpson, 124, 135, 157(2)
Halley, Thos./Tho. 138, 147
Halley, William 114
Hally, 114
Hally, Benoni, 87, 94(2), 95
Hally, Henry S., 173
Hally, James, 92
Hally, Simpson, 135
Hally, Wm., 138
Hambleton, Dennis Baptist/Dennison B., 137, 166
Hambleton, Wm., 167
Hamilton, Denison B., 148
Hamilton, James/Jas., (and Gent.), 12, 13(2), 17, 18, 22, 23(2), 25(2), 26, 29, 38, 41, 42
Hamilton, Joseph, 135
Hamilton, Will./Wm., 138, 147
Hammond, Jervis, 88
Hammond, Nicholas, 138
Hampton, John/Jno., 39, 49, 59, 68, 77, 107, 146, 152(2), 153, 167
Hampton, John Sr., 139
Hampton, John Jr, 139, 146, 167

Hampton, Joseph, 139, 146, 167
Hampton, Thomas, 49
Hardage, James, 1
Harden, Anthony, 162
Harden, Edward, 93, 113
Harden, Elihu, 121
Harden, Elisha, 133, 162
Harden, Moses, 133
Harden, Thomas, 86, 162
Hardin, Isaac, 16
Hardin, Thomas, 73
Hardin, William, 85, 91
Harding, Edward, 94
Harding, Elihu, 150
Harding, Hall, 133
Hargis, Abraham, 28
Hargiss, Abraham, 67
Harle, Jno., 144
Harle, William, 1
Harley, George/Geo., 80, 138, 147, 167
Harley, John, 138, 161, 165
Harley, Peter, 95
Harley, William, 161
Harper, Capt., 124, 149
Harper, Charles/Chs., 149, 161
Harper, James, 79, 88
Harper, John, 120, 131, 136
Harper, Leonard/Leod., 138, 147
Harper, William, 161
Harraver, John, 138
Harrell, John, 160
Harris, Benjamin, 133
Harris, James/Jas., 161, 168
Harris, Sarah Ann, 164
Harris, Theophilus, 161
Harris, Thomas, 108
Harris, William/Wm., 137, 146
Harrison, George/Geo., 145, 160
Harrison, John/Jno., 139, 146, 160, 166
Harrison, Joseph, W. 163
Harrison, Robert, 79, 88
Harrison, Thomas, 21, 39
Harrison, Thomas Jr., 24
Harrison, William/Wm., 135, 145, 160
Harrover, Merryman, 135
Harrower, John, 145, 160

Harrower, Merryman/Meryman, 144, 160
Hassett, Mrs., 148
Hartshorne, Mrs., 138, 167
Hartshorne, Mrs. Marg., 147
Hartshorne, William/Wm., 136, 149, 151, 161
Hatford, Thos., 136
Hawk, Dr., 135
Hawk, Peter, 161
Hawkins, John, 133, 136
Hawkins, Mr., 120
Hawkins, William, 85, 92
Hayes, David, 161
Hedrick, ____, 132
Hedrick, Robt., 150
Henderson, Alexander/Alexr./Alex. (and Gent.), 50, 55, 62, 65, 69, 72, 74, 75, 77(2), 81, 92, 102, 103, 104, 107(2), 108, 114, 117, 121, 138, 145, 166
Henderson, Jno., 133
Henderson, William, 134, 161
Hendricks, 149
Hendricks, Col., 136
Hendricks, James, 120
Hengston, Nicholas, 162
Hepburn, 126, 149, 170(2)
Hepburn, Mr., 126, 127
Hepburn, Mrs., 135, 144
Hepburn, William/Wm., 120, 135, 144, 161
Herbert, 116,
Herbert, Thomas/Tho., 118, 119, 125, 127, 136, 140, 141(2), 149, 151
Herbert, William/Wm., 133, 162
Hereford, 45, 46, 109, 116, 163
Hereford, George, 71
Hereford, John/Jno., 45, 147, 165
Hereford, John Jr., 136, 137, 147
Herryford, 9
Herryford, John, 5
Hewes, Jno., 146
Hickie, James, 138
Hickman, Ezekiel, 39
Hicks, William/Wm., 138, 147
Higgerson, John, 3, 5, 13

Hill, Bennett, 77, 80, 84, 87, 89(2), 90, 91, 160
Hill, John, 101, 137, 146, 165
Hill, Jonah, 137
Hill, Josiah, 146, 165
Hill, Walter, 132, 150, 162
Hines, Charles, 166
Hines, Nathaniel, 166
Hipkins, 156
Hipkins, Lewis, 129, 132, 134, 141, 143
Hipkins, Widow, 163
Hipsico, Jno., 149
Hix, Francis, 167
Hix, William, 167
Hodgkins, 173
Hodgkins, Danl., 163
Hodgkins, Perry Green, 169
Hodgkins, Ralph, 164, 172
Hogskins, Raphael, 169
Holbert, Danl., 164
Holden, Charles, 138
Holden, Chs. G., 147, 167
Holless, John, 47
Hollice, 160(2), 167(2)
Hollies, 157
Holliman, William, 146
Hollis, 58, 84, 85, 93, 144(2), 147(2), 151, 152
Hollis, John/Jno., 28, 49, 51, 56, 72, 106
Holmes, 52, 112, 120, 125, 127, 135, 136, 140, 161(2)
Holoman, Wm., 139
Homes, 30
Homes, Samuel, 78
Honesty, John, 163
Hooe, 117, 145, 153(2), 164
Hooe, John (and Col. and Gent.), 81, 138, 152(2), 153
Hooe, Robert Townshend/Robert T., 125, 141
Horsman, John, 163
Horseman, John, 134
Hoskins, Ralph, 134
Hough, John, 14, 22, 24, 36
House, John, 89, 94

House, Margaret (wife of John), 94
Howard, Samuel/Saml., 149, 162
Hubbal, John, 170
Hubbard, Jacob, 86
Hughes, Nathan, 53, 58
Hughs, John, 142
Hulls, John, 136
Hulls, Saml., 161
Hunt, Bazil/Bazzel, 147, 167(2)
Henry/Hy., 139, 146, 166
Hunt, James/Js., 138, 147, 167
Hunt, Pricilla/Priscilla, 146, 166
Hunter, 81, 118
Hunter, George (and Mr.), 114, 119, 128, 130, 134, 161, 163
Hunter, James, 156
Hunter, John/Jno., 45, 118(2), 134, 149, 161
Hunter, John Chapman/John C., 129, 134, 141, 143, 164
Hunter, William/Wm., 136, 146, 165
Hurdle, Richard, 107
Hurst, James/Js., 124, 134, 141, 148
Hurst, John, 32, 33, 52, 55, 65, 71, 72, 76, 77, 79, 81, 104, 113, 134(2)
Hurst, Peter, 167
Hurst, Sybel/Sybil, 147, 167
Hurt, John, 50
Husler, 165
Hussey, George, 162
Hussey, Henry, 133, 162
Hutcheson, Jere:, 146
Hutcheson, Walter, 146
Hutchinson, Benjamin, 41
Hutchison, Benjamin, 22
Hutchison, Daniel, 23
Hutchison, Jeremiah, 165
Hutchison, Joseph, 13, 26(2), 27, 36
Hutchison, Mr., 7
Hutchison, Walter, 165
Hynaman, Jacob, 161
Irmell, Paul, 60
Irmill, Paul, 38
Irvin, James, 165, 174
Isaac, Fielden, 133
Jackson, Alexander, 107

Jackson, Eleanor/Elinor, 44, 45, 48, 55
Jackson, John/Jno. (and Mr.), 3, 9, 13, 85, 114(2), 118, 124, 128, 130, 131(2), 134, 140, 141, 142, 144, 150, 159, 164, 174
Jackson, John Sr., 175
Jackson, John Jr., 173, 175
Jackson, Jonathan, 135, 144
Jackson, Joseph, 138
Jackson, Solomon, 138, 147
Jacobs, Benjamin, 108
Jacobs, Edward, 158, 171
Jacobs, Joseph, 28, 139
Jacobs, Jos. Jr., 145
Jacobs, Mordecai, 139, 146, 166
Javins, John, 137, 148, 166
Javins, Joseph/Jos., 137, 146, 165
Jefferson, 135
Jefferson, Jeremiah/Jere., 135, 149, 161
Jefferson, Jeremiah Jr., 161
Jefferson, John, 164
Jenkins, Charles/Ch., 134, 149, 161
Jenkins, Daniel (and Mr.), 91, 94, 95, 130, 131(2), 134, 164
Jenkins, Hanson, 161
Jenkins, James/Jas., 2, 20, 74, 76, 81(2)
Jenkins, John, 2, 44
Jenkins, Margaret/Margt., 148, 161
Jenkins, Samuel, 14, 20, 27, 35, 88, 94, 95, 134
Jenkins, Simon, 134, 148
Jenkins, Widow, 161
Jennings, Alexander, 3, 4, 7, 10
Jennings, Daniel, 85, 103
Johnson, Baldwin, 165
Johnson, Bennett, 138, 160
Johnson, Hanah, 91
Johnson, James, 147
Johnson, John (and shoemaker), 135, 139, 166
Johnson, John (son of James), 147
Johnson, John Jr., 137
Johnson, Jos., 166
Johnson, Lancelot/Launchley, 135, 160
Johnson, Samuel/Saml., 136, 161
Johnson, Seth, 52, 56(2)
Johnson, Strutton, 134
Johnson, Thomas/Thos., 135(2), 136, 163
Johnson, Walter, 135, 160
Johnson, William/Wm., 134, 135, 163, 165
Johnson, William/Wm. Jr., 90, 137
Johnston, Aaron, 147
Johnston, Archibald (and Capt.), 136, 146
Johnston, Bennett, 144
Johnston, G., 76
Johnston, George/Geo. (and Mr. and Gent.), 3(2), 6, 7, 20, 21(2), 22, 40(2), 45, 48, 51, 55, 57, 58, 59, 63, 64, 67, 68, 72, 73, 75
Johnston, John (son of Wm.), 144
Johnston, John (shoemaker), 146
Johnston, Lancelott, 144
Johnston, Samuel/Saml., 32, 39, 43, 67, 149
Johnston, Seth, 59(2)
Johnston, Walter, 144
Johnston, Wilfred, 136
Johnston, William/Wm., 75, 144, 146
Johnston, William Jr., 146
Jones, _____ (tailor), 136
Jones, _____, 163
Jones, Charles, 143, 169
Jones, Henry, 163
Jones, John, 123, 135, 144, 155, 163
Jones, Joseph, 42, 103
Jones, William/Will., 149, 165
Jordan, William, 165
Kean, Henry, 149
Keating, Wm., 149
Keaton, Wm., 136
Keen, 87
Keen, Francis, 147
Keen, James, 79, 108
Keen, William/Wm., 73, 143, 147
Keene, Francis/Frans., 138, 167
Keene, John, 169
Keene, Newton, 169
Keene, Wm., 138, 167
Keith, James, 83

Keller, Michael/Michl., 134, 152, 164
Kelley, Thomas, 41
Kelly, George/Geo., 149, 161
Kelly, Thomas, 26
Kennett, Zachariah, 107
Kenny, James, 133
Kent, Danl., 147, 167
Kent, Elijah, 145
Kent, Elisha, 164, 166
Kent, Jane, 147
Kent, John, 136, 147
Kidwell, Alexander, 164
Kidwell, Elijah, 147
Kidwell, Hezekiah, 147, 158, 167
Kidwell, Jesse, 134, 147
Kidwell, Joshua, 134, 164
Kidwell, Marshall P., 160
Kilgore, George, 95
Kincheloe, 173
Kincheloe, Cornelius, 138, 145(2), 164
Kincheloe, Hector, 164, 174
Kincheloe, Jesse, 174
Kincheloe, Wm., 72
Kinchloe, William, 58
King, Benjamin/Benja./Ben:, 108, 139, 145, 164, 164
King, Hargess/Hargiss/Hargus, 108, 114, 139, 145, 164
King, John, 139, 146, 166
King, Lawrence Suddith/Lawrence S., 145, 164
King, Samuel, 11, 53, 166
King, William/Wm., 138, 147, 160, 166
Kirby, John, 132
Kirby, John Jr., 162
Kirby, John Jr.'s father, 162
Kirby, Thomas/Thos., 135, 136, 144, 160
Kirk, Daniel, 138
Kirk, Grafton, 45, 48, 138, 147
Kirkbride, Joseph, 141
Kitchen, William, 1, 19, 23, 60, 61, 87
Knowland, Charles, 135
Korn, John, 161, 171
Lambert, Thomas/Thos./Tho., 135, 149, 161

Lampkin, George, 92, 108
Lander, Benjamin, 160
Lane, James, 8, 13, 29, 43(2),
Lane, James Jr., 37
Lane, Richard, 144, 160, 173(2)
Lane, William, 169
Lane, William Jr., 172, 174
Lanham, Edward, 128, 162
Lanham, Elisha, 164
Lannum, Edward, 130
Lasswell, Jacob, 8, 14
Lawson, Robert/Robt., 139, 146, 153, 165, 166
Lay, Abraham, 55
Layne, Henry, 134
Leake, Richard, 80
Lear, Tobias, 166
Leatherwood, Zacha., 164
Lee, Charles/Chas./Chs. (and Esq.), 127, 136, 149, 161
Lee, Col., 44
Lee, Hancock, 147, 153, 167, 169
Lee, Henry, 132, 164
Lee, Ludwell/Ludwells, 132, 137, 146, 149, 162, 165
Lee, Hon. Philip Ludwell, 49
Lee, Richard Bland, 169
Lee, Hon. Thomas, 1
Lee, Col. Thomas Ludwell, 95
Leonard, Richard/Richd., 150, 162
Lester, John/Jno., 134, 148
Lester, Widow, 134
Lewis, 5, 20, 118
Lewis, Daniel, 171
Lewis, Edward/Edwd., 136, 149
Lewis, Eliz. (infant), 76
Lewis, Fielding, 166
Lewis, John, 163
Lewis, Peter, 20
Lewis, Stephen (and Gent.), 1, 6, 11, 21, 22, 24, 28, 51
Lewis, Thomas (and Mr. and Gent.), 1, 2, 5, 25(2), 87, 94, 95(2), 104, 135, 144, 160
Lewis, Vincent, 3(2), 6, 7, 10, 11, 18, 29
Lewis, William/Wm., 134, 135, 145

Lidyard, Newbury, 22
Lightfoot, Richard, 74, 109
Lightfoot, Samuel/Saml., 149, 165
Lightfoot, Wm., 136
Linch, Daniel, 146
Lindsay, Mrs., 134
Lindsay, Opie, 135, 144, 160, 161
Lindsay, Susanna, 149
Lindsay, Thomas/Thos./Tho., 116, 119, 120, 134, 149, 154(2), 161, 168, 174
Lindsay, William/Wm., 118, 137, 140, 146, 150
Lindsey, Robert, 1, 68, 84, 92
Linton, John, 68
Linton, Moses (and Gent.), 5, 8, 13
Linton, William, 59, 62, 67, 79, 82, 85(2), 86(2)
Lister, John, 59, 65, 132, 150, 161, 162
Lister, Thomas, 95
Lister, William, 76, 79
Little, Charles/Chs. (and Gent.), 113, 117, 119, 120(2), 123, 124(2), 128, 133, 134, 141, 149, 156(3), 159, 161, 162, 174
Littlejohn, 85
Littlejohn, Samuel, 58, 84
Lloyd, Richd., 158
Logan, Barnaby, 138
Lomax, John, 82, 84, 88, 93, 102, 120
Lomax, Rachel (wife of John Lomax), 82
Lomax, Stephen, 161
Longdon, John, 161
Loofborough, 114, 116
Loofborough, David, 145
Loofborrow, 93
Loofborrow, David, 88
Loofbourrow, David, 138, 139
Loofburrow, 101, 103(2)
Loofburrow, David, 93, 108
Loudoun, Jesse, 149
Love, James, 160
Love, Samuel/Saml., 156, 169
Low, Charles, 167
Loyd, Henry, 137, 148, 166
Loyd, Richard, 163

Lucas, John, 1, 7, 9, 17, 31
Lucas, Thomas/Thos., 104, 105, 107(2), 108, 116, 139, 145, 164
Luckett, Ignatius, 135
Luffburrow, David, 166
Luke, John, 162
Lukehouse, Abram, 161
Lyles, William/Wm., 127, 133
Lyon, Daniel, 11
McAlister, Daniel, 133
McAtee, John, 160
McAtee, Samuel, 163
McCartee, Miss Molly, 165
McCarty, Col., 137
McCarty, Cornelius, 134, 148
McCarty, Daniel/Danl. (and Gent. and Col.), 2, 3, 5, 14, 15(2), 19, 24, 28, 35, 36, 47, 48, 57, 61, 62, 69, 73(2), 74, 75(2), 77(2), 81, 92, 93, 101, 102, 107, 109, 148(2), 159(2), 161, 165, 166
McCarty, Daniel Jr., 117, 137, 148
McCarty, James, 134
McCarty, Patrick, 148
McCarty, Wm., 135
McCason, Roger, 164
McClain, Alexander, 161
McConnell, Alexander, 161
McCormick, George, 105
McCormuck, George, 100
McCrea, Robert/Robt., 120, 149
McDaniel, Daniel, 161
McDaniel, James, 109
McDaniel, William, 104
McDonald, Alexander/Alexr., 123, 124, 136, 147, 154
McDonald, James, 93
McDonald, Miss, 163
McFarlin, John, 165
McFarling, Ignatius, 162
McGinnis, Lawrence, 137, 149, 166
McIntosh, John, 144, 160, 168
McIntosh, Laughlin, 160, 168
McKnight, William, 120, 139
McLain, 135
McLean, Samuel, 137

McNab, Chs., 166
McNabb, Charles/Chs., 139, 146
McNail, Hugh, 150
McNair, William, 163
McNamara, Dennis, 167
McNamarr, Dennis, 147
McNear, William, 163
McPherson, Burgess, 137, 147
McPherson, Edwd., 166
McPherson, Isaac, 141, 169
McPherson, Wm., 167
Macatie, Samuel, 133
Mackeltamare, Dennis, 138
Mackey, Tho., 144
Mackintosh, John, 135
Macknear, Wm., 158
Madden, 163, 164(2)
Madden, Michael, 161
Magie, Daniel, 161
Magruder, 107
Magruder, Thomas, 112(2)
Mandevill, 161
Manley, John H., 139, 146, 153
Manley, Mrs., 73, 74
Mansell, John, 135
Mansfield, John/Jno., 144, 160
Marl, 152
Marle, 132, 150, 162
Marmaduke, John, 51
Marshall, James, 173, 175
Marshall, John/Jno., 138, 145, 166
Marshall, Samuel, 173, 175
Marshall, Mr. Thomas, 17
Martin, Barton, 136
Martin, Henry, 160
Martin, James/Jas., 138, 144, 160
Martin, Noah, 144, 160(2)
Martin, Sampson, 135, 148
Martin, Saml., 160
Marton, Mrs., 137
Mason, 18
Mason, Ann, 44, 45, 55
Mason, Charles, 40
Mason, Col., 124, 136, 137, 149
Mason, George/Geo. (and Col. and Esq. and Gent.), 2, 3, 7, 9, 14, 19(2), 21, 24, 31, 38, 42, 44(2), 45, 52, 56, 61, 65, 71(2), 88, 101, 102, 107, 146, 148, 155, 159(2)
Mason, George/Geo. Jr., 74, 109(2), 126, 137, 148
Mason, George the Younger, 96
Mason, George/Geo. (Pohick), 137, 146, 153, 154, 165
Mason, John, 74, 162
Mason, Mrs., 165
Mason, Thomas, 165
Mason, Thomson/Thompson, 137, 149, 159, 165, 166
Massey, John, 132, 162, 170
Massey, Lee (and Revd. Mr. and Cl[erk]), 48, 83, 87, 103, 137, 146, 165
Masterson, Edward, 12, 17, 33
Masterson, Edward Sr., 18
Matney, James, 163
Matney, John, 163
Mattinly, James, 132(2)
Mattinly, John, 132
Maubley, Levin, 165
Mauzey, Peter, 145, 151
Mauzy, Peter, 76, 79, 122, 136
Mead, Samuel, 10, 28
Mead, William, 16, 41
Meara, John, 137, 146
Meeks, Anne, 172
Merchant, John, 161
Mercer, John (and Gent.), 20
Merryday, Wm., 133
Middleton, William/Wm., 134, 155, 173, 174, 175
Mill, George, 163
Millan, Thomas, 171
Millan, William, 173
Miller, Wm., 166
Mills, ____, 164
Mills, Daniel/Danl., 113(2), 136, 163
Mills, George/Geo., 136, 147
Mills, John, 113, 133, 162
Mills, Robert, 30, 33
Mills, Sarah (widow), 89, 109

Mills, William/Wm. (and bricklayer), 138, 139, 146, 147, 157, 165, 167
Minor, 16, 127
Minor, Ann, 132, 164
Minor, G., 153
Minor, George/Geo./Geo:, 121, 123, 125, 126, 128(2), 129, 132, 133, 141, 142, 156, 158(2), 159, 162, 164
Minor, George Jr., 162
Minor, Jemima, 162
Minor, John, 61, 65, 71, 74
Minor, Mrs., 133
Minor, Nicholas, 19, 20, 27, 32, 41, 42, 43
Mitchell, Adam, 13, 28
Mitchell, Wm., 170
Mittchel, William, 145
Moalds, Benjamin, 162
Molds, Benjn., 78
Mollerhorne, Robt., 146
Mollihon, Robert, 137
Moreland, Jacob Jr., 137
Morely, James, 135
Monroe, Andrew, 134, 147
Monroe, John, 62, 63, 67, 104
Monroe, Lawrence, 135, 149
Monroe, Sarah, 134, 147
Monroe, Spence/Spencer, 134, 147
Monroe, Thomas/Thos., 54, 90
Monroe, William, 147
Moody, Benjamin (and Mr.), 83(2), 100
Moody, Thomas, 132
Moon, Mrs. (Widow Moon), 136, 137
Moon, William/Wm., 104, 108, 114, 117, 139
Moone, John, 145
Moone, William, 145
Mooney, Nicholas, 134
Moor, James, 53
Moore, Cleon, 104, 118
Moore, Henry, 79, 84, 103
Moore, James, 57, 114
Moore, Jeremiah, 119, 129, 134, 143, 164, 168
Moore, Jesse, 134, 151, 152, 157, 167

Moore, John/Jno., 149, 164, 165
Moore, Manley, 165
Moore, William, 79, 164
Morgan, William, 161
Morris, Jacob, 33
Morris, Nehemiah/Nehe:, 138, 146, 149
Morris, Uriah, 145, 166
Morris, Zachh., 160
Morriss, Zachariah, 135
Morton, Jane, 148
Moseley, Mrs., 139
Moss, John, 23, 25, 116, 123, 128, 129, 130, 135, 141, 143, 149, 153, 155, 156, 160
Moss, John Jr., 42
Moss, Robert, 94, 100
Moulds, Benjamin/Benja., 132, 150
Mourland, John, 134
Moxley, Jacob, 133
Moxley, James, 144, 160
Moxley, John, 133, 164
Moxley, Joseph (and Gent.), 33, 35, 48, 49, 56, 57, 65, 71, 72, 79
Moxley, Richard/Richd., 32, 43, 51
Moxley, Samuel, 28, 36
Moxley, Thomas, 133
Moxley, William, 164
Moxly, Job, 164
Moxly, Patty, 133
Moxly, Spencer, 164
Moxly, Thomas, 164
Moxly, Thomas Jr., 164
Muir, Jno., 73
Mull, George, 40, 44
Mullichan, Jno., 149
Munday, William, 16
Munroe, Andrew, 158(2), 161, 168(2)
Munroe, Lawr., 161
Munroe, Sarah, 167
Munroe, Spencer, 167
Munroe, Wm., 167
Murray, Charles, 162
Murray, James, 127
Murray, John, 144, 160
Murr, John, 136
Muse, Thos., 132(2)

Nalley, Aaron, 147
Nally, Aaron, 134
Neale, Ann, 45
Neale, Christopher, 18, 33, 35, 37
Neale, Spencer, 63
Nelson, Richard/Richd., 10, 132
Nelson, Wm., 132
Netherton, 26
Newman, George, 174
Newton, Capt., 13
Nicholas, Geo., 166
Nicholas, Solomon, 79, 80
Nichols, George, 146
Nichols, James B., 166
Nicholson, Joseph, 132, 150, 162
Noddy, William, 172
Nolan, Charles, 160
Noland, 27
Noland, Archibald, 167
Noland, Charles, 144, 168
Noland, Jacob, 146
Noland, Jacob Jr., 146
Noland, Philip, 7
Norris, _____, 161
North, John, 24
O'Daniel, John/Jno., 29, 134, 148
O'Daniel, Stephen, 134, 148, 154
O'Neale, Charles, 27
Offield, Reisin, 134
Offord, Rezin, 148
Offutt, Reazin/Rezin, 129, 131, 154, 161
Ogden, Hugh, 164
Ogden, Thos., 136, 163
Ogden, Widow, 163
Ogden, Wm., 136
Ogdon, Tho., 147
Ogdon, William, 147
Ogelvay, George, 91
Omeara, Michael, 161
Orr, John, 144
Osborn, John, 16
Owens, Thos., 133
Owsley, Thomas, 30
Oxens, Thomas, 164
Padget, Benjamin, 160
Padget, Francis, 36, 43

Pagett, _____, 149
Pain, Abigail, 166
Pain, Geo., 166
Palmer, Thomas/Thos./Tho., 78, 87, 112(2), 135, 148, 156, 167
Pameroy, John, 138
Parker, Lawson, 137, 148, 166
Parle, William, 40
Parrot, Thomas (John Ward's hireling man), 100
Parrott, George, 65
Pasley, John, 134
Patterson, 117
Patterson, William, 158
Pattinson, Susannah, 109
Patton, James, 154
Payne, Alesworth, 146
Payne, Annanias, 132, 150, 162
Payne, Ebsworth, 137
Payne, Edward/Edwd./Edd./Ed. (and Mr.), 25, 25, 27, 36, 42, 43, 58, 71, 72, 73, 76(2), 84, 85, 87, 88, 100, 103
Payne, Francis, 150, 162
Payne, Jacob, 132, 150, 162
Payne, John/Jno., 135, 149, 160
Payne, Sanford, 89, 138, 145
Payne, Syrus, 137
Payne, Thomas/Tho., 132, 150
Payne, W., 128
Payne, William/Wm. (and Col. and Gent.), 10, 13, 47, 55, 65, 72, 73, 82, 84, 85, 101(2), 103, 107, 110, 113(3), 115, 118, 119, 120, 121, 123, 124, 125, 126, 128, 136, 141, 153(2), 156, 157, 159, 163(2), 168, 169, 170, 172, 174
Payne, William/Wm. Jr. (and Esq. and Gent.), 10, 13, 25, 46, 54, 57(2), 61(2), 72, 73, 100,
Peacock, John, 135
Peak, Humphrey/Humphry, 67, 80
Peak, John, 43, 51
Peak, Mary, 134
Peake, Humphrey, 56
Peake, John, 12, 20, 30, 32, 33, 43

Peake, Mary, 149, 166
Peake, Mrs., 137
Peake, William/Wm., 3, 28, 36, 46, 48(2), 49, 126, 137, 149, 156, 165
Peake, William Sr., 9
Peake, William's widow, 165
Pearson, Edwd., 146
Pearson, John, 133, 161
Pearson, Milkey, 133
Pearson, Simon, 57(2), 104, 133, 142, 161
Pearson, Thomas/Tho., 133, 150
Pell, Henry, 145, 164
Pelling, William, 157
Pendleton, James, 37
Pendall, Tho., 150
Pendoll, Thos., 132
Pennington, Isaac, 8
Peptisco, John, 134
Perry, Franklin, 27, 42, 56, 57, 74, 76, 81(2), 91, 134
Perry, Hanah, 100
Petitt, James, 166
Pettit, James, 149
Pettuth, John, 107
Peyton, 35
Peyton, Francis, 154, 166
Peyton, Henry (and Gent.), 20
Philips, John, 139, 146, 166
Philips, William, 163
Phillips, William, 172
Pilcher, Sarah, 135, 144
Piles, John, 16
Pindal, Thos., 162
Pinkstone, John, 88
Piper, David, 45, 46, 50, 71
Piper, Wm., 132, 133
Plumer, John, 63
Plummer, John, 55, 66
Pollard, Joseph, 73
Pollard, Thomas/Thos. (and Gent.), 71, 73, 83, 89, 95, 123(2), 126, 128, 135, 141, 143, 144, 160
Pollard, Thomas Jr., 160
Pollerd, Thomas Jr., 159
Pomery, John, 166

Pool, Peter, 147
Pool, Richard/Richd., 147, 165
Pool, Thos./Tho., 137, 148
Pool, Widow, 166
Pool, Wm., 137, 148
Popejoy, Ann, 135
Popejoy, Nathaniel, 91
Posey, 30, 32, 33, 44, 59, 73, 74, 117
Posey, Benjamin, 171
Posey, John, 33, 35, 40, 45, 73
Posten, Francis, 167
Poston, Francis, 134, 147
Potter, Edward/Edwd., 136, 144, 160
Potter, Joseph, 137, 146, 165
Potter, Sandford/Sanford, 138, 147, 167
Potter, William/Wm., 137, 148, 166
Potts, Dr., 161
Potts, Jno., 149
Poultney, 7, 69
Poultney, John, 37
Powell, _____ (Pohick), 167
Powell, Elisha, 88, 113
Powell, John, 108, 138, 139, 163
Powell, John (Pohick), 147
Powell, John/Jno. (son of William), 146, 166
Powell, Joseph, 133, 162, 163
Powell, Joseph (son of William), 164
Powell, Joseph Jr., 163
Powell, Micajah, 74
Powell, Robert (and Gent.), 113, 116, 125, 126, 133
Powell, William/Wm., 108, 145, 146, 155, 164
Pozey, John, 39
Pratt, Leonard, 65, 76
Presgrave, Wm., 134
Preston, _____, 165
Price, 99, 116(3), 126, 127, 128, 129, 135(2), 144, 147, 149, 151, 154, 160(2)
Price, Benoni, 149, 151, 155, 160
Price, Mrs., 135
Prichard, _____ (schoolmaster), 138
Prichard, Philip, 160
Pritchard, Lewis, 168

Pritchard, Philip, 135
Pritchart, Philip, 144
Proctor, John, 147, 167
Pshaw, Jacob, 26, 32, 38(2)
Pumroy, Jno., 147
Radcliff, John, 16
Ragan, Widow, 137
Ramey, Jacob, 8
Ramsay, 120, 133, 162
Ramsay, Dennis, 120
Ramsay, William/Wm. (and Gent.), 4, 14, 19(2), 21(2), 22, 25, 37, 69, 71, 73, 74, 81(2), 107(2)
Ramy, Sanford, 57
Rassau, Henry, 166
Ratcliff, Barbara/Barbary, 127, 135, 149
Ratcliff, John/Jno., 25, 30, 55, 56, 93, 102, 126, 127(2), 135, 149, 160
Ratcliff, Richard/Richd., 135, 143, 144, 156
Ratcliff, Robert, 157
Ratcliff, William, 144
Ratcliffe, John, 172
Ratcliffe, Richard/Richd., 157, 159, 160, 168(2), 169, 170, 174
Ratliff, John, 45
Rattle, James, 138
Rawland, Gilbert, 146
Read, John, 107
Reagan, Ann, 146, 165
Reagan, Michael, 35
Reardin, William, 18
Reardon, Henry/Hy., 123, 137, 146
Reardon, John, 85, 86(2)
Reardon, William/Will./Wm., 3, 19, 28, 36, 83, 87, 123, 137, 146
Reardon, Yelverton, 108, 139, 145, 164
Reed, Chs., 144
Reed, James, 132, 158, 170
Reed, John, 108, 139, 146
Reed, John Jr., 164
Reed, Nelson, 132
Reed, Sarah, 108, 139, 145, 164
Regan, Michael, 10, 16
Reid, James, 163

Reid, John, 163, 166
Reid, Joseph, 43
Reid, Nelson, 163
Remey, Jacob, 28, 31
Remy, Jacob, 13
Remy, Sandford/Sanford, 13, 45
Rerdon, Mrs., 165
Rhoby, Miles, 135
Rhoby, Young, 134
Rhodes, James, 81
Rhodes, John, 35, 47, 58, 63, 78, 91
Rice, Hopkins, 135
Richards, Caleb, 132, 150, 156, 162
Richards, James, 136, 153, 163, 174
Richards, William/Wm., 76, 136, 163
Richardson, David, 7, 37
Richster, John, 155, 165
Ricketts, 95, 115(4), 134, 136(2)
Ricketts, John T., 161
Riddle, Andrew, 47, 50(2), 51
Riddle, John, 154
Riddle, Joshua, 171
Rigg, Ben:, 137
Rigg, John/Jno., 138, 147, 167
Rigg, Widow, 138
Riley, Benjamin/Benj., 138, 147, 167
Riley, Chs., 147
Riley, Peter, 138
Roberts, John, 35
Roberts, Saml., 149
Robertson, James, 133, 142
Robertson, John, 132, 134
Robertson, John/Jno. C., 148, 167
Robey, Elijah, 147
Robey, Elisha, 167
Robey, Michael/Michl. H., 148, 160
Robinson, Jesse, 162
Robinson, John's widow, 162
Robinson, Mrs., 162
Robinson, Wm., 149
Roby, Elias, 160
Roby, Levy, 160
Rodgers, William, 160, 167
Rogers, Benjamin, 147
Rogers, Mary, 147, 167

Rogers, Widow, 138
Rogers, William/Wm. (and blacksmith), 124, 147, 163
Rogers, William Sr., 137
Rogers, William/Wm. Jr., 138, 147
Rollins, John, 162
Rollins, Luke, 165
Rooksberry, Ruben, 164
Rose, Francis, 134, 149, 161
Rose, Henry, 160
Rose, Isaac, 77, 91
Ross, H. (and Gent.), 74
Ross, Hector (and Gent.), 54, 59, 72, 75, 77(2)
Ross, William, 11, 16, 20
Roussaw, Henry, 138
Row, John, 139, 146
Rowland, Gilbert, 139, 166
Russau, Henry, 145
Russell, Anthony (and Gent.), 7, 10, 11, 18, 20, 29(2), 32
Russell, Charles, 173
Russell, Wm., 78, 162
Ruxby, Jacob, 132
Ryley, Peter, 108
Sanders, Daniel, 1
Sanders, Lewis, 86, 87, 88, 94
Sanders, Lewis Jr., 1, 19
Sanders, Thomas, 68
Sandford, Richd. Sr, 149
Sandford, Richd. Jr, 149
Sandford, Robert/Robt., 2, 13, 133, 149
Sands, Edmond/Edmund, 14, 16(2), 18, 24
Sanford, Elizabeth, 83, 93, 106
Sanford, Richard, 26, 41, 47, 94, 96, 124, 136
Sanford, Richard Jr., 136, 161
Sanford, Robert, 8, 21, 23, 41, 47
Sanford, William, 82, 83, 93, 112
Sangster, 84, 114(2), 116, 145, 146, 147, 152, 157
Sangster, Geo., 166
Sangster, John, 157
Sangster, Thomas, 84, 85(3), 86, 88, 95, 98(4), 100, 104, 105, 107, 145(3), 147, 150, 151
Sarter, John, 61
Sarter, John Peter, 30, 61
Sattridge, John, 132
Saunders, 117
Saunders, Benjamin, 144
Saunders, Lewis, 23, 134, 148
Saunders, Lewis Jr., 15
Saunders, Thomas, 70
Saven, James, 165
Savenus, Nicholas, 23
Sawkins, James, 172
Scisson, John, 160
Scisson, Robert, 144, 160
Scisson, William/Wm., 134, 163
Scot, Gustavus, 161
Scott, 114
Scott, Gustavous/Gustavus (and Esq.), 128, 134, 149
Scott, John/Jno. (and blacksmith), 132, 139, 146, 147, 150 , 162
Scott, Mr., 49, 130
Scott, Revd. Mr., 74
Scott, Sabret/Sabrett, 150, 162
Scott, Sebrek, 132
Scott, William, 112, 120
Scott, Zachariah/Zach., 150, 162
Scrivener, Richd., 137
Scutt, William, 19, 23, 29
Seale, John, 67(2), 68, 70, 77, 107, 110
Seal, John, 62, 64, 78
Seals, John, 68
Sebastian, Benjamin/Benjn. (and Gent.), 20, 26, 27, 31, 32(2), 36, 38, 41, 44, 45, 48, 51, 56, 61, 67(2), 68(2), 71, 72, 75, 125
Sebastian, Mr., 38, 71, 72, 73(2)
Sebastian, Nicholas, 134, 163
Seelig, Michael, 58
Selecman, Henry, 171
Self, Thomas, 30
Self, William/Williams, 47, 50
Sewell, 163
Sewell, James, 132
Sewell, Joseph, 163

Sewell, Robert, 132
Sewell, William, 25
Sexton, Daniel, 65
Sexton, John, 137, 146
Shaw, Thomas/Thos., 39, 41, 47(2), 53, 61, 71(2), 83, 90, 94, 96
Shearwood, Thomas, 132
Sherwood, Thomas/Tho., 150, 162
Shelton, Edward, 94, 100
Shelton, John, 38, 43
Shelvin, Edward, 80, 86
Shepherd, John, 115, 161
Shepard, Jno., 148
Sheppard, John, 119
Shepperd, John, 94, 134
Shilling, Jacob, 54
Shoemaker, Charles, 141
Shoemaker, Gedion, 149
Short, John, 146
Shortridge, John, 76, 114, 131, 140, 164, 173
Shortridge, William/Wm., 9, 56, 62(2), 72, 74, 86, 128
Shreeve, Samuel, 150
Shreve, John, 162
Shreve, Samuel, 142, 162
Shreiv, Samuel, 152
Shreive, Benjamin, 16
Simmonds, Wm., 139
Simmons, Elizabeth/Eliza., 161, 168
Simmons, Jeremiah, 158
Simmons, William, 146
Simons, Thomas, 51
Sims, Wm., 134
Simms, Alexander, 136, 161
Simms, Charles/Chs. (and Col.), 134, 146, 149, 161, 165
Simms, Craven/Cravin, 134, 149, 162
Simms, William/Wm., 161, 165, 166
Simms, Wm. Jr., 136
Simpson, 77
Simpson, Aaron, 138, 147, 167, 169
Simpson, Baxter, 108, 139, 145
Simpson, George/Geo., 59, 60, 68, 74, 75, 138, 146, 147, 167
Simpson, Gilbert, 117, 122, 137, 146, 165, 168(2)
Simpson, Gilbert Sr., 53
Simpson, James, 138, 147
Simpson, Jas. (son of George), 167
Simpson, John, 139, 161
Simpson, John (son of Thomas), 167
Simpson, John (son of William), 145
Simpson, Joseph, 128(2), 130, 131, 133
Simpson, Moses, 72, 108, 110, 145, 164
Simpson, Presley, 167
Simpson, Richard, 9, 116, 138, 141, 143, 145, 147, 164
Simpson, Richard Jr., 25
Simpson, Spence, 145
Simpson, Susanna, 138, 147, 167
Simpson, Thomas/Thos./Tho. (and shoemaker), 135, 144, 147, 164, 167
Simpson, William/Wm., 1, 28, 36, 50(2), 67, 74, 108, 116, 122, 136(2), 138, 139(2), 145, 152, 153, 155(2), 161, 166, 169, 172
Simpson, William (son of George), 146, 166
Simpson, William Jr., 145, 166
Simpson, William (brick), 146
Simpson, Wm. (buck), 137
Simpson, Will (son of Wm.), 138
Sinclair, Thomas, 168
Sisson, Robert, 135
Skidmore, Edward, 133, 162
Skidmore, William, 162
Skinner, John, 134, 144, 159, 166, 168, 173(2)
Skinner, Price, 148, 160(2)
Skinner, William/Wm., 135, 144, 160
Skinner, William/Wm. Jr., 135, 144
Slaughter, Mrs., 136, 149, 166
Smar, John 40
Smarr, John 36
Smith, 137
Smith, Alexander, 171
Smith, Augustine J., 160, 168
Smith, Charles, 117, 136, 143, 147, 154, 155, 163

Smith, George/Geo., 112, 114, 115, 118, 119, 128, 131, 132, 134, 140, 148, 173
Smith, John, 161
Smith, Joseph, 135, 163, 168
Smith, Leonard, 149
Smith, Mary, 144
Smith, Nathaniel/Nathan, 2, 9, 16, 21, 24(2), 163
Smith, Richd., 136
Smith, Samuel/Saml. (and shoemaker), 135, 144, 149, 151, 155, 160
Smith, Thomas, 15, 17, 19, 21,
Smith, Weathers Jr., 156
Smitherman, Samuel, 138, 147, 167
Smitherman, Samuel's sons, 147, 167
Smock, Robert, 159
Solomon, Geo., 133
Somer, John, 34
Songster, 124, 164, 166, 173, 174
Songster, Thomas/Tho., 122, 135, 138(3), 139(2), 164, 166(3), 167
Sorrell, Thomas, 12, 14, 20, 25
Southerd, Benjamin, 87
Sparrow, Matthew, 68
Speak, John, 165
Speak, Robert/Robt., 146, 165
Speak, William, 165
Speake, _____, 147
Speake, Robert/Robt., 71, 137
Spencer, Francis/Fran:, 137, 148, 166
Spencer, Wm., 148
Spilman, Js., 146
Spinks, Chandler, 135, 151, 163
Spinks, Gerrard/Gerrd., 115, 136, 144, 155(2), 160
Spirlin, Hanson, 164
Spirlin, Jeremiah, 164
Sprag, John/Jno., 146, 166
Spragg, John, 139
Spur, James, 42
Spurlin, Jere:, 132
Spurr, James, 35
Stanhope, William, 174
Starke, William, 7, 22,

Stark, William, 8(2), 10, 24, 29, 31
Starks, William, 20
Stearns, Balty, 63
Steel, John, 92
Steel, William, 144, 160
Stephens, 5
Stephens, Joseph, 12, 25, 30, 40, 55
Stephens, Richard, 19, 43
Stephens, Robert, 11, 19
Stephens, William, 147
Steptoe, 30, 73
Steptoe, Mr. James, 15(2)
Stevens, Joseph, 47, 48
Steward, James, 77
Stewart, John, 139
Stiel, Wm., 135
Stith, Buckner, 137, 149
Stone, Caleb, 139, 146, 166
Stone, Casandra, 137, 146
Stone, Daniel/Danl., 137, 148, 166
Stone, Elie, 144
Stone, Francis, 139, 143, 146, 166
Stone, John, 140, 150
Stone, John S., 166
Stone, Samuel/Saml., 137, 143, 147, 167
Stone, William, 59, 61, 112(2)
Stonemier, Agness, 99, 100
Storer, Daniel, 134
Storey, Danl., 148
Strother, Christopher, 4, 7
Stuart, David (and Gent. and Dr.), 123, 126, 128, 135, 144, 150, 160, 162, 168
Stuart, Hugh, 169
Stuart, James, 81
Stuart, John, 172
Stuart, Mrs., 145, 166
Suddeth, Benj., 147
Suddeth, Benjamin Jr., 145
Suddith, Benja./Ben:, 138, 139
Suddith, James (son of Silent), 167
Suddith, Lawrence, 139
Suddith, Silent, 167
Sumers, 25
Sumers, John, 26, 95

Summers, 54(2), 115, 116(2), 134, 149(2), 160, 161
Summers, Francis/Frans., 1, 28, 29, 43, 88, 90, 91, 99, 104, 108, 116, 135, 149, 161
Summers, Francis Jr., 82, 106(2), 161
Summers, George/Geo., 135, 144, 159, 160, 166, 170
Summers, John, 17, 19, 22(2), 46, 54(2), 64, 65, 82(2), 97, 99, 113(2), 151, 153, 155
Summers, John (son of Francis), 163
Summers, John Jr., 53, 126
Summers, John 3d, 136, 163
Summers, Mary, 163
Summers, Simon, 115, 122, 134, 151, 153, 154, 162, 174
Summers, Widow, 134, 143
Summers, William, 63, 91, 104, 113(2), 142, 154
Swain, Archibald, 132, 150, 162
Swain, Benjamin/Benja., 132, 150, 162
Swallow, John, 135
Swallow, Zach., 147
Swallow, Zepheniah, 137
Sweet, John, 3
Swink, 164
Swink, William/Wm., 128, 130, 133
Talbert, Rhodam, 164
Talbot, Benjamin, 60(2)
Talbot, Daniel, 44, 77
Talbot, Samuel, 60
Talbot, William, 44, 45, 50
Talbott, Benjamin/Benja./Benjn., 58(2), 67, 72, 86, 87, 88, 90, 94, 95, 110, 144
Talbott, Daniel, 90(2)
Talbott, John, 78
Talbott, Mary M., 134
Talbott, Samuel, 64, 91, 95, 104, 107
Talbott, William, 80
Talbut, 113
Talbut, Benjamin, 122, 160
Talbut, Daniel, 34
Talbut, Levy, 162

Talbut, McKinzie, 161
Talbut, Samuel, 57
Talbut, William, 57, 161
Talbutt, 150
Talbutt, Benjamin/Ben:, 112(2), 134
Talbutt, Mary M., 147
Talbutt, McKinzie/McKenzy/McKinsey/McKinzey, 121, 135, 149, 155
Talbutt, Wm., 135, 149
Tanner, Paul, 164
Tascox, James, 138
Tayloe, Col., 3, 13
Taylor, Edward, 135, 145, 160
Taylor, Edward Jr., 135, 145
Taylor, Francis, 135, 148
Taylor, Harry, 53
Taylor, James, 145
Taylor, Jesse, 120, 162
Taylor, Sarah, 160
Taylor, Thomas/Thos., 129, 132, 133
Taylor, Vincent, 133
Templeman, 26
Templeman, John, 2, 7, 12, 25, 27, 29
Terret, William Henry, 6
Terrett, 149, 161
Terrett, William Henry/William H./Wm. Henry/Wm. Hy./Wm. H./WHy./ W. H. (and Capt. and Mr. and Gent.), 21, 23(2), 33, 35, 112, 119, 120, 126(2), 127, 133, 151, 159(2), 162
Thomas, David, 135
Thomas, Daniel, 10, 16, 24
Thomas, James, 132
Thomas, James Jr., 161
Thomas, John, 133, 162
Thomas, Robert, 10, 26
Thomas, Wm., 133
Thompson, 107
Thompson, Charles, 162
Thompson, Cornelius, 138, 145
Thompson, Daniel/Danl., 134, 147, 167
Thompson, Edward, 10, 16
Thompson, James, 167
Thompson, John, 148, 160, 167
Thompson, Joseph, 104, 133

Thompson, Richard, 96, 162
Thompson, Samuel, 134, 147, 167
Thompson, Saml. Jr., 147, 167
Thompson, William/Wm., 18, 107, 117, 137, 138, 139, 146, 150, 152, 159, 165, 171
Thorn, Michael, 165
Thrift, Charles/Chas. (and Mr.), 37, 41, 44(2), 47, 49, 56, 62, 118, 119, 128, 130, 131(2), 134, 139, 140, 145, 164
Thrift, George/Geo., 32, 36, 37, 40, 44, 125, 133, 156, 161, 163
Thrift, Hamilton, 148, 161, 164
Thrift, Jeremiah/Jere:, 132, 150, 162
Thrift, William, 152, 162
Thrifton, Benjamin, 167
Throgmorton, Thomas, 95
Tillet, George, 166
Tillet, Jas., 166
Tillet, Samuel, 13, 27
Tillett, George, 138
Tillett, George Jr., 145
Tillett, James, 138, 145
Tillett, John/Jno., 93, 104, 139, 146
Tillett, John Jr., 138
Tillett, Samuel, 25, 60, 95, 100
Toulson, 148
Towers, John, 158, 172
Tramel, Garrard, 25
Tramel, John, 24
Tramel, William, 42
Tramell, Garrard/Gerard/Gerrard/Gerrd., 26, 73, 115, 161
Tramell, John, 24, 28
Tramell, William, 35, 42
Trammel, Garrard, 33
Trammel, John, 37
Trammel, John Jr., 22
Trammell, 54, 116
Trammell, Garrard/Gerrard /Gerard/ Gerrd., 29, 54, 55, 57(2), 61, 66, 81(3), 105, 134, 149,
Trammell, John, 28, 36, 40, 71, 76, 161
Trammell, Samson/Sampson, 42, 163

Trammell, Sampson Jr., 134
Trammell, Thomas/Thos./Tho., 134, 149, 161
Trammell, William, 2, 45
Trammill, Gerrard Jr., 65
Trammill, Sampson (and Mr.), 130
Triplet, Thomas 44
Triplett, Geo., 167
Triplett, Mrs., 165
Triplett, Sarah (and Mrs.), 137, 146
Triplett, Thomas, 32, 52, 73, 77, 79, 80, 91, 165
Triplett, William/Wm., 73, 79, 91, 93, 109, 146, 156, 165, 166
Triplett, Wm. Sr., 137
Triplett, Wm. Jr. 137, 146
Tristler, Lewis, 149
Tucker, Charles/Chs., 149, 165
Tucker, Geo., 162
Tucker, Widow, 133(2), 162(2)
Tuel, William, 172
Tull, Jacob, 166
Turbe.vill, Mr., 57
Turbeville, Mr., 56
Turberville, 106
Turberville, Corbin, 164
Turberville, George R. L., 169
Turberville, John (and Mr.), 128, 130, 133, 140, 164
Turley, 47
Turley, James, 27
Turley, John/Jno. (and Gent.), 1, 18, 71
Turley, Paul, 5, 27, 95
Turley, Peter, 27
Turley, Sampson (and Gent.), 8, 13, 27(2), 36, 37, 45, 89, 103, 108, 114, 117, 139, 145
Turner, 158
Turner, Charles, 82, 84, 91, 93
Turner, Ellzey, 173
Turner, Fielding (and Gent.), 1, 7, 10, 23, 26, 28(2), 29, 36
Turner, Thomas, 160
Turner, Thomas E., 160
Turner, Vincent, 144, 160

Turner, William/Wm., 61, 76(2), 92, 122, 134, 135, 144, 147, 152, 158, 160, 167
Tustar, Lewis, 135
Tuttle, John Baptist, 107
Tuttle, Thomas, 15
Tyler, Charles, 4, 8, 15, 42, 45, 54, 59,
Tyler, Henry/Hy., 137, 146
Tyler, Wm., 137, 146
Tylor, Henry, 165
Tylor, Wm., 165
Urton, John, 49, 62
Valandingham, Dorson, 70
Valentine, Jacob, 145, 160
Vallandigham, Dorson, 76
Vallandingham, Dorson, 68
Veale, Bond, 9
Vermillion, Hanson/Henson, 148, 161, 170
Vernon, 123
Vernon, John, 146
Vestal, 102(2), 111
Vestel, 30
Vinayard, Stephen, 44
Vinayard, Stephen's son, 44
Vincent, Robert, 150, 162
Vineyard, James, 78
Violet, Edward, 5, 8, 12, 15, 43
Violett, 18
Violett, Edward, 53, 57, 90
Violett, Hugh, 165
Violett, John, 133
Violett, Wm., 137
Voilett, _____ 144
Voilett, William/Will., 146, 165
Vowell, John, 158
Wade, Valinda/Verlina/Verlinda, 17, 33, 35
Wadlington, Thomas, 15(2)
Wadlington, William, 40
Wagener, Col., 137
Wagener, Peter (and Gent.), 18, 19, 21, 23, 28, 36, 37, 38, 41, 42, 45(2), 47, 48(2), 49, 50, 54, 61, 64, 65(2), 67, 69, 72, 77, 83, 84, 89, 92, 93, 101, 104, 117, 137, 146, 165

Wagener, Peter Sr., 80
Wagener, Peter Jr. (and Gent.), 77, 81, 101, 102, 104, 109(2), 110
Wailes, Andrew, 84, 108
Wales, Andrew, 91
Walker, James, 163
Walker, John, 164
Walker, Wm., 134
Waller, John (and Mr.), 94, 95(2), 105, 107
Ward, 153, 154
Ward, John, 68, 90, 100, 107, 123, 124, 138, 143, 147, 153, 167
Ward, John's hireling man Thomas Parrot, 100
Ward, Jonathan, 133, 162
Ward, Josiah, 161
Ward, Mary, 147, 167
Ward, Widow, 138
Ward, William/Wm., 127, 136, 154
Ward, Zachariah/Zach., 137, 146, 165, 174
Warden, Nathaniel, 164
Warden, Mrs. Sarah, 145
Warner, Joseph/Jos., 137, 147, 165
Warthin, Edward, 165
Warthin, Henry, 165
Washington, Bushrod Gent., 140
Washington, Corbin, 157
Washington, Edward/Edd./Edwd., 13, 14, 18, 49, 67, 87, 139, 141, 146, 152(2), 159, 166, 167, 169, 174
Washington, George, Esq. (and Gent. and Gen. Washington/the General), 51, 52, 59, 88, 103, 127, 137(2), 148(2), 156, 166(4), 167
Washington, George/Geo. A., 137, 148
Washington, Laurence/Lawrence, 137, 141, 146, 152, 165
Washington, Lund, 75, 113, 117, 137, 146
Washington, Lund Jr., 137, 146
Washington, Maj., 149
Washington, Mrs., 165
Washington, Will., 150

Washington, William H., 162, 171
Waters, 58
Waters, Elizabeth, 58, 60, 65
Waters, William/Wm., 134, 163
Wathen, Edward, 89
Watson, 134
Watson, Benja., 132
Watson, John, 41
Watson, Josiah, 161
Waugh, James, 123, 138, 145, 151, 166, 169
Waugh, Tyler/Tylor, 138, 145
Webb, James, 133
Weeden, John, 144
Weeden, Saml., 144
Weedon, Samuel, 135, 160
Weightman, Richard, 161, 169
Wells, Cornelius, 146, 150, 165
West, 134
West, Ann, 138
West, Benjamin/Benj., 145, 166
West, George (and Mr.), 17, 80
West, H., 71, 75
West, Hugh (and Mr. and Gent.), 2(2), 9, 18, 21, 22, 28, 31(4), 32(2), 33, 38(2), 41(2), 42, 47, 51(2), 52, 56(4), 59, 60, 61, 63, 67, 68, 71, 76, 77
West, Hugh Sr., 18
West, Hugh Jr. (and Gent.), 17, 20, 24, 28
West, John/Jno. (and Mr. and Gent.), 6, 21, 26, 28, 31, 34, 35, 37, 51, 53, 55, 69, 108, 120, 125, 130, 133, 149, 161(2)
West, John/Jno. Jr. (and Esq. and Gent.), 35, 39, 41, 47, 54, 68, 69(2), 71, 82, 86, 101, 103, 108, 134
West, Joseph, 17
West, Mrs., 133, 135, 149, 160
West, Roger/Rogers, 123, 126, 137, 149, 154, 155, 159, 166
West, Sarah, 138, 145, 166
West, Sybell/Sibell (widow), 33, 38
West, Capt. T., 120
West, Thomas/Thos./Tho., 112, 121, 133, 136, 143, 149(2), 158, 161
West, William, 3, 4(2), 7, 13, 22, 23, 29, 35, 41, 145, 160
Weston, William/Wm., 137, 146
Whaley, Gilson, 131
Whaley, James, 30
Whaley, William/Wm., 138, 145, 167
Wheaton, Samuel, 161
Wheeler, Drummond, 139, 146, 155
Wheeler, Ignatius, 172
Wheeler, Richd., 139, 146
Wheeler, Saml., 148
Wheler, Richd., 159
Wherry, Jesse, 161
Whery, Jesse, 159
White, Abednego, 144
White, Benja., 134
White, John, 169
Whiting, 133
Whiting, Carlyle, 162(2)
Whiting, Francis, 80, 83
Wickliff, Robert/Robt., 138, 145, 164
Wickliffe, Robert, 173
Wickliff, Susanna (widow), 81
Wigginton, Roger, 7, 17
Wiggs, John, 138
Wiggs, Ann, 137, 146
Wilcoxen, Margaret/Margt., 148, 167
Wilcoxin, Rezin, 168
Wiley, George/Geo., 146, 165
Wiley, James, 142, 152, 161, 168
Wilkinson, John, 160
Willet, Edward, 45
Willett, Edward, 48
William/Williams, 6, 8, 14, 26, 102(2), 111
Williams, Aler/Alier/Ayler, 133, 136, 144, 164
Williams, Bazil, 144, 160
Williams, Bozwell, 138
Williams, George/Geo., 94, 95, 134, 150, 158, 162, 164
Williams, Edward/Edwd., 53, 147
Williams, Elijah, 75
Williams, Jeremiah/Jere:, 95, 113(2), 134

Williams, James, 108
Williams, John/Jno., 78, 85, 132, 164
Williams, John Lock, 107
Williams, Joseph, 108, 145
Williams, Oliver, 160
Williams, Owen (and Gent.), 20, 35, 42, 48, 49, 60, 65
Williams, Thomas/Thos., 26, 139
Williams, Walter, 6
Williams, William/Wm., 16, 62, 80, 134, 137, 146, 165
Williamson, Jane, 147, 167
Williamson, Jesse, 137, 146, 165
Williamson, Widow, 138
Willis, Mrs., 6
Willit, Edward, 43
Wilson, John, 132, 149, 161, 173
Wilson, Geo., 149
Wilson, Mrs., 137
Windsor, Mrs., 167
Windsor, Richard, 167
Windsor, Thomas 73, 87, 138, 160
Winsor, Thos., 147
Winsor, Thomas Jr., 144
Wingate, Henry, 138
Winn, George, 124
Wise, Peter, 88, 102, 120, 123
Wise, John, 149
Wood, Elijah, 132
Wood, John/Jno., 134, 149, 161, 168
Wood, Joseph/Jos:, 135, 144, 160
Woodard, Jeremiah, 166
Woodard, Thomas, 151, 160
Woodward, Joseph, 110
Woodward, Thomas/Thos., 138, 144
Woodyard, Henry, 166
Woodyard, Jere:, 139, 146
Woolbright, Barnaby, 135, 144
Woolbright, Jacob, 135
Woolbright, Samuel/Saml., 138, 144, 160
Workman, John, 171
Worthin, Edward, 137
Worthin, Henry, 137
Worthing, Edwd., 146
Worthing, Hy., 146
Wren, Col., 130, 131, 155, 159(2), 163
Wren, James (and Gent.), 46, 55(2), 62, 67, 71, 100, 112, 118, 119, 120, 123, 127, 130, 133, 134, 149, 151, 153, 154, 161, 162, 167
Wren, James Jr., 135
Wren, John/Jno., 86, 87, 90, 104, 113, 122, 124, 133, 134, 143, 144, 160(2)
Wren, Mr., 57
Wren, Thomas/Thos., 1, 33, 38, 49, 55(2), 56, 62(2), 67, 72, 105
Wren, William/Wm., 115, 125, 134, 149, 161
Wrenn, James, 128
Wrenn, John, 156
Wrenn, Thomas, 107, 174
Wright, Chs., 146
Wright, Charles's mother, 146
Wright, Thomas, 162
Wyley, George, 137
Yates, 173, 174
Yeamon, John, 139
Yearman, John Jr., 164
Yeoman, John, 145
Yeoman, John Jr., 145
Yost, Wm., 149
Young, David, 55, 59, 67, 79, 84
Young, Dorotha/Dorothy, 86, 93, 102
Young, William, 140, 160
Youst, Wm., 135

Water Features: Creeks, Falls, Rivers, Runs, Springs, and Other Features

Accotink, 2, 8, 14, 22, 25(2), 27, 29, 30(3), 32(2), 34, 40, 42, 43, 46, 47, 52, 60, 61, 62(2), 74, 80, 82, 85(2), 86, 87, 88, 92, 93, 94, 97, 99, 100, 102, 105, 106, 107, 113(2), 115(2), 116(2), 117(3), 137(2), 143, 144(2), 146(2), 165(2)
Back Lick/Black Lick, 5, 22, 43, 46, 53, 57, 80, 86, 88, 93, 94, 97, 99(2),

100(2), 117, 126, 127, 136(2), 163
Bayly's Mill Run, 139
Beaver Dam, 14
Brickens Branch, 103
Bridge Branch (near the old Court House), 95, 112
Broad Run, 14, 19, 20, 24(4), 28(2), 29, 33, 36, 37, 39
Major Broadwater's Spring Branch, 66(2)
Buck Horn, 112
Bull Run, 48, 49, 58(2), 67, 81, 95, 104, 107, 108, 117(2), 138, 145(3), 157(2), 164(2), 173, 174
    Wolf Shoals/Wolf Run Shoals on, 145, 157(2), 164 (see also Occoquan)
Cameron Run, 93, 118, 126, 127, 137, 149(2), 154, 155, 166, 170(2) (NOTE: Some entries under "Cameron" in the Miscellaneous section may also refer to Cameron Run)
The Creek (with Dawson's ford), 41
Carpenters Spring, 72(2)
Courthouse Run, 134, 148, 149, 161
Cub Run, 1, 13, 29, 30, 31, 41
Davis's Branch, 20
Difficult Run/Branch/Creek, 1, 6, 9, 11(2), 12, 19, 30(3), 37(2), 39, 44, 50, 53, 54, 55, 57, 62, 66, 71, 74, 76, 77(2), 81(2), 85, 87, 88, 93(2), 94(3), 95(3), 101(2), 103(2), 108, 112(2), 114, 115(2), 119, 120, 121(2), 124, 126, 128, 130, 131, 133, 134, 135, 141, 142(2), 145, 148, 153, 154, 156, 161, 164(2), 166
    Mouth of 101
Dogue/Doeg Run/Creek, 5, 8(2), 32(2), 33, 39, 44, 46, 52(2), 54, 59(2), 61, 62, 69, 75, 85, 86, 90, 96, 97, 102, 105, 107(2), 117(4), 136, 137, 146, 149, 165(2)
Lewis Ellzey's Meadow Branch, 119, 120

Falls:
    Falls of Occoquan, 30(2)
    The Falls, 2, 7, 12, 47, 68, 72, 77, 120, 121, 15667, 105, 159(2), 202
    Falls of Potomac, 96
    Great Falls of Potomac, 112, 121, 124, 128, 129, 130, 131(5), 139, 140, 141, 163
    Little Falls (of Potomac), 81, 103, 118, 119, 140, 142, 162
NOTE: See also Falls road, Little Falls road, Old Falls road, Falls mill)
Flat-Lick Run, 174
Four Mile Run/Creek, 4, 18, 26, 30, 46, 52, 62, 63, 68(2), 69(3), 77, 87, 91, 92, 99, 101, 113, 114, 115(2), 129, 132(2), 141, 142, 150(2), 162(3), 170
Frying Pan Run, 98, 169
Giants Castle Branch, 123
Giles Run, 59(2), 102
Goose Creek, 2, 3(4), 6(3), 7, 8, 10(2), 11, 14, 16, 19, 20, 24(2), 29(2), 30(2), 32(2), 38, 42, 44
    Bank of Goose Creek (digging bank down at the ferry), 3, 6, 8
    Beaver Dam branch/fork of, 3, 5, 26, 41
    North branch of, 17
    Upper fork of, 6(2)
Gum Spring, 15, 32, 33, 59, 97, 117(2), 126, 137, 149, 165, 166(2)
Branch near the Hop-yard, 120
Holmes/Homes Run, 30, 52, 112, 120, 125, 135, 136, 140, 161(2)
    North run of, 127
Branch near George Hunter's line, 128
Hunting Creek, 4(2), 8, 15, 21, 32, 38, 46, 54, 70, 81, 86(2), 88, 90, 93, 96, 97(2), 102, 107, 136
    North Run of, 116(2)
Great Hunting Creek, 61
    Run of, 19
Little Hunting Creek, 29, 39, 40, 43, 44, 69
Indian Run, 126

Great Limestone/Great Limestone Spring, 7, 27
Limestone Run, 3, 6
Johnimore, 89
Little River, 7, 10, 20, 26(2), 27, 29, 30, 32, 35
Lubber Branch, 141, 142
Maddams Branch, 95
Mine Branch, 171
Occoquan/Occoquon, 2(2), 3, 6, 7, 9, 14, 19(3), 21, 24, 28, 30(2), 31, 36, 38, 42, 44(2), 47, 50(2), 51, 52, 56, 61, 65, 71(2), 84(2), 85, 98, 105, 107, 108, 114, 117, 139, 141, 152(3), 153(2), 167, 169
    Wolf Run Shoals on, 85, 107, 114, 116, 139(2) (see also Bull Run)
Old Courthouse Run, 134, 161
Pimmets/Pimmits Run, 26, 65, 82, 84, 92, 96, 97, 100, 118, 119(2), 130, 141
Piney Branch/Run, 26, 27, 30, 76, 77, 100, 116, 121, 122, 136
P.k. [Pohick?] Creek, 146
Pohick Run/Creek, 2(2), 14, 19, 23, 32, 62(2), 74, 77, 84, 85, 92, 93, 96, 101, 102, 104(2), 105, 109, 117(2), 123, 144, 153, 154, 155, 159(2), 165
The pond on the road from William Gardner's to Ravensworth road, 73
The ponds, 87, 114, 115
The Ponds on the road from Cameron to Falls Church, 112
Ponds on Pohick road, 138(2), 147(2), 167(2)
Popeshead Run, 24, 25(2), 30, 37, 58, 83, 87, 88, 92, 168, 169, 170
    Mouth of, 27, 168
Potomac River, 30, 37, 39
Poultney's Run, 69
      Quarter Branch, 54
The River bank (Potomac), 37
The River, 17, 132, 150, 158
Rock Creek, 18, 49, 63, 69, 89, 93, 100
Rocky Run, 169
Great Rocky Run, 13
Little Rocky Run, 13(2), 37, 42
Sandy Run, 110
Sasse Branch, 49
Scott's Run, 114, 128, 130
Segalon Run, 41
Sugarland Run, 3, 6, 14, 19, 20, 24, 37, 39, 52
Tuscarora Run, 42
White Marsh, 165
White Oak Spring, 135
Wolf Trap Run, 2, 104, 118, 122
Wolf Run, 37, 92, 145
Wolf Run Shoals/ Wolf Shoals, 85, 107, 114, 116, 139(2), 145, 157(2), 164 (also see entries under Occoquan and Bull Run)

Roads

Roads are cross-indexed to all locations and persons mentioned. Descriptions have been standardized to aid in identifying roads and to simplify the preparation of this index. This was a necessity since many road descriptions changed slightly in the orders as different landmarks were cited. Various roads went under similar general descriptions and the reader should bear this in mind when determining the identify of each road.

Roads from Accotink to Pohick:
- Road from Pohick to Accotink and crossroad, thence the road to the warehouse from the crossroad (and parts thereof) 2, 14
- Road from Accotink to Pohick Run (including the fork of the road at Boggess's and the fork to Mr. Barnes's) 32
- Roads from Pohick/Pohick Run to Accotink 62, 74, 93, 117
- Road from Pohick to Accotink in the Back road 85
- River side road from Pohick to Accotink 105
- Sections of the Colchester road: from the Turnpike gate to Dogue Run; from Dogue Run to Accotink; from Accotink to Pohick Run; from Pohick Run to Colchester 165

Roads from Accotink to Dogue Run:
- Road from Dogue Run to Accotink/Accotink Run 8, 32, 52, 62, 102
- Road from Accotink to Dogue Run on the Back road 85
- Road from Dogue Run to Accotink (the lower road) 107
- Road from the lower side of Accotink Run on the River side road to Dogue Run 117
- Road from Dogue Run at Mrs. Moon's to Accotink 137
- Road from Accotink, on the Alexandria road, to the Parish line at Dogue Run 146
- Sections of the Colchester road: from the Turnpike gate to Dogue Run; from Dogue Run to Accotink; from Accotink to Pohick Run; from Pohick Run to Colchester 165

Road from Accotink to Cameron 22

Road from Accotink to Rocky Run Chapel 25

Road from Accotink to Popeshead 25, 87, 88

Road from Accotink 27

Road from Little Hunting Creek to Accotink/Accotink Run 29, 40, 43

Road from Accotink to the old school house 34, 42, 60

Road from Accotink/Accotink Run to the Back Lick 46, 80, 86, 93, 94, 100, 117

Road from Accotink to the lower end of Mr. Barns's plantation 47

Road from the ford over Accotink to the upper end of Barns's quarter 61

Road from the school house near John Summers's to Accotink Run 82

Road on the upper side of Accotink 92

Fork road leading to the Back Lick and Alexandria from the road from Accotink to the Back Lick 97

Road from Accotink to the school house near John Summers's (with intersection of the road at Price's leading from Accotink to Alexandria) 99

Road from the fork of the road near Francis Summers Jr.'s to Accotink Run 106

Road from the great road by Daniel Mills's to William Payne's old mill on Accotink Run 113(2)

Road from Ricketts's to Accotink 115

Road from Accotink to the Ox road 115

Road from Accotink above Price's to the Ox road 116

Road from Price's to Accotink 116

Road from Accotink to Colchester 137

Road from the neighborhood of Occoquan Forrest crossing Accotink at John Ward's mill and down to Alexandria 143

Road from Accotink on the Turnpike road to Price's tavern 144

Road from Hollis's old field to Accotink, on the Turnpike road 144

Road from Colchester ferry, along the road to Alexandria to Accotink Run 146

Road from Accotink, on the Alexandria road, to the Parish line at Dogue Run 146

Road from Georgetown ferry to Adams's mill 113

Road from the rock hole on Four Mile Creek to John Carlyle and Robert Adam Gent.'s mills 68(2)

Road from the Georgetown road at R. Adams's to Avery's road 121(2)

Road from Robert Adam's mill to Avory's road 132

Road from the Red house to Samuel Adams's mill on Pimmets Run, and from thence to the road leading from the Little Falls to Alexandria (road beginning near the Red house, thence to Pimmets Run and the mill, thence through Samuel Adams's plantation to the long branch of Pimmets Run, along John Butcher's fence to Mathew Earp's, and across the Church road to the Alexandria road) 118, 119

Road from Samuel Adams's mill, to the place where Benjamin Sebastian formerly lived 125

Road from William Adams's house to Hunting Creek 4

Road from Cameron to Mr. William Adams's/William Adams's Gent.'s house/path/plantation 19, 63, 66, 69, 80(2), 82, 83

Road from William Adams's Gent. to Gerrard Trammell's 66

Road from Mr. William Adam's to the old Court House 80

Road from Mr. William Adams's to Pimmets Run 82, 84, 92

Road from Mr. William Adams's/William Adams's Gent. to Pimmets Run (including fork[?] road from the Alexandria road to the ferry at Georgetown) 96, 97, 100

Road from Gerrard Trammell's bridge towards William Adams's 105

Road from Little River to the widow Adams's at Goose Creek 32

Road from the ferry road at the town of Alexandria to Hunting Creek ford and from thence to Four Mile Run ford 4

Road from Alexandria to Rockey Run Chapel 17, 18

Streets and public landings in Alexandria 21

Roads between Alexandria and Cameron:
- Road from the lower Cross Street in Alexandria to the road that leads to Cameron (not cleared) 27, 29
- Road from Alexandria to Cameron/Cameron Run 46, 66, 69, 80, 93, 96, 98, 105
- Road from Alexandria to Cameron (fork at Cameron with the roads leading up and down the Country) 98
- Road from Alexandria to Cameron Run and to the forks of the road at Cameron hill 118

- Road from Cameron into <u>Alexandria</u> (entrance road into <u>Alexandria</u> at King Street) 120

Streets in <u>Alexandria</u>:
Cameron Street 37
The Lower Cross Street, 27, 29
Duke Street, 120
King's Street 37, 120
Princess Street, 120
Queen's Street 94, 96

Road from <u>Alexandria</u> to the plantation of John Carlyle (running between the plantations of Baldwin Dade and Townshend Dade) 41

Road from <u>Alexandria</u> to Four Mile Run 46, 52, 63, 91, 96, 114

Road from Robert Boggess's to <u>Alexandria</u> 46

Road from <u>Alexandria</u> to the Falls warehouse 48, 76

Old road from the Falls warehouse to <u>Alexandria</u> 49

Road from James Donaldson's house into the main road leading to <u>Alexandria</u> 51, 53

Road from the old Court House road near John Monroe's, to intersect with the road to <u>Alexandria</u> near Balty Stearns's 62, 63

Road (from Colchester to <u>Alexandria</u> ordinary) at the hill below Boggess's 69, 72

Road from the Falls Church to <u>Alexandria</u> 71, 113

Road from the Falls Church to the road leading from <u>Alexandria</u> to the Falls warehouse 79, 114, 132

Road from Court's ferry to the road from <u>Alexandria</u> to Colchester 84

Road from Hunting Creek to <u>Alexandria</u> 86

Road from the termination of Queen's Street to the road leading from <u>Alexandria</u> to Georgetown (road along Mr. Carlyle's ditch on a parallel line with Queen Street to the county road from Georgetown to <u>Alexandria</u>) 94, 96

Road from Mr. William Adams's/William Adams's Gent. to Pimmets Run (including fork[?] road from the <u>Alexandria</u> road to the ferry at Georgetown) 96, 97, 100

Fork road leading to the Back Lick and Alexandria from the road from Accotink to the Back Lick 97

Road dividing at Boggess's (into the River side and Back roads to Alexandria) 97

Road from Accotink to the school house near John Summers's (with intersection of the road at Price's leading from Accotink to Alexandria) 99

Road from the Alexandria road to the Four Mile Run (including fork leading to Georgetown and the Falls warehouse) 99

Road from the forks of the road at the Falls Church to the old road from Rock Creek to Alexandria 100

Road from the Alexandria road to the Ox road near the Piney Branch 100

Main road from Alexandria to Difficult Bridge 101, 103

Road from Colchester to Boggess's mill on Pohick (road to leave the main road from Colchester to Alexandria above Giles Run and go through Mr. Bayly's land and George Mason's land above the plantation where Joseph Garberry lives then to cross the Church road at the Widow Atcheson's plantation and through her plantation to the mill) 102

Main roads from Vestal's and Williams's Gaps leading to Alexandria and Colchester 102(2), 111

Road from Hereford's ferry to the old Church and to the road leading from Colchester to Alexandria 109, 116

Road from the road leading out of the Alexandria road about three miles above the old Court House to the Great Falls of Potomac 112

Road from the Ponds (on the road from Cameron to the Falls Church) to Alexandria 112

Road from the bridge at Alexandria to Four Mile Run 115

Road from Ricketts's to the Alexandria road 115

Road from the Red house to Samuel Adams's mill on Pimmets Run, and from thence to the road leading from the Little Falls to Alexandria (road beginning near the Red house, thence to Pimmets Run and the mill, thence through Samuel Adams's plantation to the long branch of Pimmets Run, along John Butcher's fence to Mathew Earp's, and across the Church road to the Alexandria road) 118, 119

Road from Alexandria to Difficult bridge near Capt. W. H. Terrett's (road from Princess Street passing near Baldwin Dade's old tobacco house, and Ramsay's quarter, and to a white oak in the road above Capt. W. H. Terrett's old field) 119, 120

Road from Alexandria to Lewis Ellzey's meadow branch and the ford over Holmes Run (road from Duke Street crossing the gut below John West's meadow, passing Dutch Caty's house and Captain T. West's field to the road at the mouth of Col. Gilpin's lane, thence with the road to a branch near the hop-yard, and along the side of the hill to Holmes Run opposite the mouth of Messers Bird and Hawkins's lane) 119, 120

Road from Alexandria to Georgetown ferry 119

Road from the forks of Avery's road to the road leading from Alexandria to Georgetown 121

Road from the Great Falls of Potomac to Alexandria (road from the lower ford over Difficult Creek/Difficult Run via the Great road, near Edward Lanham's/Lannum's field, thence through the lands of William Swink, Joseph Simpson, Gustavus Scott Esqr. [rented by Daniel Jenkins], John Jackson, and Robert Carter, Esq. [rented by Charles Thrift], to the place on Scott's Run where William Shortridge's old mill stood, thence through the lands of Turberville, Scott, George Hunter, John Jackson, across Pimmets Run, through the lands of Richard Conway, Philip Darrell, William Gunnell, Mr. John West, Mr. Sampson Trammell, and Col. James Wren, to the Great road leading from Leesburg to Alexandria, where Col. Wren's store stands, thence by the Great road [Turnpike road] from the store to Alexandria) 121, 124(2), 128, 129, 130, 131(3), 139, 140, 141

See also:
Divisions of the road from the Great Falls of Potomac to Alexandria:
- from Col. Wren's store to the Falls road 131, 155
- from the Falls road to Difficult 131

See also:
Road from Charles Thrift's near the Little Falls road to the new road leading from the Great Falls to Alexandria (road from the Little Falls road to the Great Falls road, running through Charles Thrift's land/Robert Carter's land rented by Thrift to the Great Falls road at John Jackson's land; alteration of the road running through John Turberville's land) 131, 140

Road from John Ward's mill to the road that leads to Alexandria 123

Road from the Colchester road to the road to Alexandria (road from the Colchester road near John Ward's mill, thence through William Rogers's plantation to Richard Sanford's fence, and to the road leading to Alexandria opposite to Capt. Harper's fence) 124

Turnpike road from Alexandria to Georgetown ferry 126

Turnpike road from Alexandria to Difficult 126

Turnpike road from Alexandria to Newgate 126

Road from the main road at or near Price's to Hepburn and Dundas's mill on Back Lick Run and thence into the main road leading to Alexandria (road beginning at the dividing line between the lands of William Fitzhugh and the heirs of Henry Fitzhugh on the road to Newgate, thence to the Colchester road and through/adjoining the dividing line of John and Barbara Ratcliff, the lands of Barbara Ratcliff and the land of the heirs of James Murray to Hepburn and Dundas's mill, and thence through Hepburn and Dundas's land, and the lands of David and Samuel Arell, Minor, Charles Lee Esq., to William Bird's fence on the Newgate road, and along the said road to Alexandria) 126, 127

Road from the County line to the White Oak Spring (part of the Turnpike road leading to Alexandria) 135

Road from Back Lick Run to where it intersects the road from Colchester to Alexandria at the end of John Hereford Jr.'s lane 136

Road from the forks of the road near Martin Cockburn's to where it intersects the road from Alexandria to Colchester (the Alexandria road), near Bayly's Mill Run/Bayley's old mill 139, 148

Road from the landing opposite Charles Shoemaker and Joseph Kirkbride's mill upon Occoquan River to the town of Alexandria 141

Road from Georgetown ferry beginning at Lubber Branch and extending to Dowdall's corner on the hill by Four Mile Run (running on the line between Thomas Darne and Samuel Shreve and then straight until it intersects the Georgetown road below the Alexandria road) 141, 142

Road from the neighborhood of Occoquan Forrest crossing Accotink at John Ward's mill and down to Alexandria 143

Road from the County line on the Turnpike road leading to Alexandria and to the Ox road 144

Road from Colchester ferry, along the road to Alexandria to Accotink Run 146

Road from Accotink, on the Alexandria road, to the Parish line at Dogue Run 146

Road from the Alexandria road at John Hereford's lane, to Price's tavern on the old Court House road 147

Road from the Turnpike (near Alexandria) to Four Mile Run 150, 162

Road from Alexandria to John Hooe's ferry on Occoquan, near the old warehouse landing 152(2), 153(2)

Road from Roger West's house into the main road leading to Alexandria, at the ford of Cameron Run 154, 155

(Alteration of) the road from Jesse Moore's blacksmith shop to Alexandria 157

(Alteration of) Ox road from the corner between Simpson Halley and Richard Ratcliffe to the Alexandria road 157(2)

Road from the head of Pohick Creek to the road leading from Alexandria to the Colchester road 159(2)

Sections of the Turnpike road:
- from the County line on the Turnpike road leading from Newgate to Alexandria as far as the Ox road 160
- from the Ox road, along said Turnpike road to the fork of the road at Hollice's old field near Price Skinner's 160
- from Hollice's old field in the fork of the road Price Skinner's to Price's tavern 160
- from Price's tavern to the forks of the road above Summers's 160

Road from the Widow Tucker's to the Turnpike [to?] Alexandria 162

Road from Ellzey's Church road opposite Richard Ratcliffe's land, passing by William Payne's mill, and from thence to the Alexandria road 168, 169

Road from the Ellicot and Co. bridge on Occoquan into the road leading from Washington's shop to Robert Boggess's mill and from thence to the intersection with the Alexandria Post road 169

Road from Fox's mill to Alexandria 169

Road from the Beaver Dam of Goose Creek to the school house above Anderson's and from thence to the top of the Blue Ridge 41

Road from the main road at or near Price's to Hepburn and Dundas's mill on Back Lick Run and thence into the main road leading to Alexandria (road beginning at the dividing line between the lands of William Fitzhugh and the heirs of Henry Fitzhugh on the road to Newgate, thence to the Colchester road and through/adjoining the dividing line of John and Barbara Ratcliff, the lands of Barbara Ratcliff and the land of the heirs of James Murray to Hepburn and Dundas's mill, and thence through Hepburn and Dundas's land, and the lands of David and Samuel Arell, Minor, Charles Lee Esq., to William Bird's fence on the Newgate road, and along the said road to Alexandria) 126, 127

Road from the old Court House to the Falls warehouse (road coming from Bull Run through the Court House corn field, Mr. Scott's land and John Urton's land, into the old Falls road at the head of Sasse Branch, thence along the old roads by Thomas Ashbury's, through the Honourable Philip Ludwell Lee's plantation, and to the Falls) 48, 49

Road from Little River via Goose Creek to Ashby's Gap on the Blue Ridge 10

Road from the main road to William Ashford's house 35, 36, 37

Road from William Ashford's to John Hollis's 49

Road from William Ashford's [to the?] old Court House 54

Road from Colchester to Boggess's mill on Pohick (road to leave the main road from Colchester to Alexandria above Giles Run and go through Mr. Bayly's land and George Mason's land above the plantation where Joseph Garberry lives then to cross the Church road at the Widow Atcheson's plantation and through her plantation to the mill) 102

NOTE: Entries for Avery and Avory are combined

Road from the forks of Avery's road to the road leading from Alexandria to Georgetown 121

Road from Avory's road to the Falls 121

Road from the Georgetown road at R. Adams's to Avery's road 121(2)

Road from the entrance of Avory's road near Marle's house to the Turnpike road leading to the Falls Church 132

Road from Robert Adam's mill to Avory's road 132

Road from Avory's road to the Mountain road that leads to the Falls landing 132

Road from Avery's road to where it intersects the road leading from Fendall and Hipkins's mill 156

Road from Four Mile Run below the old mill to Avery's road 162

Road from Avery's road to the Little Falls road 162

Awbrey's road to Cameron 4

Road from the head of Dogue Creek into the Back Lick road (the Middle road) 5

Road from the Black [Back?] Lick to the Church road 22

Road from the Back Lick road to the county road 43

Road from Accotink/Accotink Run to the Back Lick 46, 80, 86, 93, 94, 100, 117

Road from the Back Lick Run to Falls Church (road from the Back Lick to the Church, going through Simon Pearson's plantation and round Edmund Butler's fence); (including intersection at Price's, leading from thence to the Church) 53, 57(2), 88, 99(2), 100

Fork road leading to the Back Lick and Alexandria from the road from Accotink to the Back Lick 97

Road from the main road at or near Price's to Hepburn and Dundas's mill on Back Lick Run and thence into the main road leading to Alexandria (road beginning at the dividing line between the lands of William Fitzhugh and the heirs of Henry Fitzhugh on the road to Newgate, thence to the Colchester road and through/adjoining the dividing line of John and Barbara Ratcliff, the lands of Barbara Ratcliff and the land of the heirs of James Murray to Hepburn and Dundas's mill, and thence through Hepburn and Dundas's land, and the lands of David and Samuel Arell, Minor, Charles Lee Esq., to William Bird's fence on the Newgate road, and along the said road to Alexandria) 126, 127

Road from Nicholas Fitzhugh's gate to Back Lick Run 136

Road from Back Lick Run to where it intersects the road from Colchester to Alexandria at the end of John Hereford Jr.'s lane 136

Back Lick road from Price's ordinary to the fork of the road above Nicholas Fitzhugh's gate 151

Road from the forks of the road below Fitzhugh's along the Back Lick road to Back Lick Run 163

Road from the Back Lick Run to the intersection of the Colchester road at Hereford's 163

Back road 30(2)

Back road from Dogue Run to Great Hunting Creek 61

Road from Pohick to Accotink in the Back road 85

Road from Accotink to Dogue Run on the Back road 85

Road dividing at Boggess's (into the River side and Back roads to Alexandria) 97

Forks of the Back and River side roads on the road from Hunting Creek to the Gum Spring 97

NOTE: Entries for Balendine, Ballandine, and Ballendine are combined

Road from the old road at the upper side of Ballendine's farm to Difficult 85

Road from the Widow Brewster's landing to Ballandine's mill 173

Road from the Ox road to Mr. John Balendine's mill 49

Road from the Ox road to John Ballendine's ford 50

Road from John Ballendine's mill at the Falls landing to or near the mouth of Difficult Run (road from the Falls mill to the mouth of Difficult Run) 74, 77

Road from Accotink to Pohick Run (including the fork of the road at Boggess's and the fork to Mr. Barnes's) 32

Road from Accotink to the lower end of Mr. Barns's plantation 47

Road from the ford over Accotink to the upper end of Barns's quarter 61

Road from the forks of the road near Martin Cockburn's to where it intersects the road from Alexandria to Colchester (the Alexandria road), near Bayly's Mill Run/Bayley's old mill 139, 148

Road from Colchester to Boggess's mill on Pohick (road to leave the main road from Colchester to Alexandria above Giles Run and go through Mr. Bayly's land and George Mason's land above the plantation where Joseph Garberry lives then to cross the Church road at the Widow Atcheson's plantation and through her plantation to the mill) 102

Road from John Hooe's ferry on Occoquan to the lands of Samuel Bayly 153

Road from Samuel Bayly's land to Ward's mill 153

Road from the Ox road at Charles Beach's/Beache's/Beeche's (fence) to the Ponds on the Pohick road 138, 147, 167

Road from Sugarland Run to the Chapel on the Beaver Dam 14

Road from Little River to the Beaver Dam fork of Goose Creek 26

Road from the Beaver Dam of Goose Creek to the school house above Anderson's and from thence to the top of the Blue Ridge 41

Road from the cross roads at Hodgkins's to the mouth of Beard's lane, on Lee's land 173

Road from Alexander Henderson, Gent.'s plantation on Bull Run into the main road (road from Henderson's plantation through/adjoining the lands of Yelverton Reardon, William Moon, Marmaduke Beckwith, James Keen, Sampson Turley, William Simpson, and Peter Ryley to the main road) 104, 107, 108

Road through Thomas Pollard's plantation (Road between Joseph Pollard's and Mr. Beckwith's plantations) 71, 73

Road from Hunting Creek to Belhaven 32

Road from Ellzey's Church road to the old Court House road (at Bennett's) 148, 167

Road from the old Court House road opposite Joseph Bennett's plantation, to where it intersects the Ox road 135

Road from Berryman's quarter to the Church 11

Road from the forks of the road above Summers's to Bride's (Bird's?) bridge 149

The road from Bird's (Bride's?) bridge to the Turnpike 149

Road from Alexandria to Lewis Ellzey's meadow branch and the ford over Holmes Run (road from Duke Street crossing the gut below John West's meadow, passing Dutch Caty's house and Captain T. West's field to the road at the mouth of Col. Gilpin's lane, thence with the road to a branch near the hop-yard, and along the side of the hill to Holmes Run opposite the mouth of Messers Bird and Hawkins's lane) 119, 120

Road from the main road at or near Price's to Hepburn and Dundas's mill on Back Lick Run and thence into the main road leading to Alexandria (road beginning at the dividing line between the lands of William Fitzhugh and the heirs of Henry Fitzhugh on the road to Newgate, thence to the Colchester road and through/adjoining the dividing line of John and Barbara Ratcliff, the lands of Barbara Ratcliff and the land of the heirs of James Murray to Hepburn and Dundas's mill, and thence through Hepburn and Dundas's land, and the lands of David and Samuel Arell, Minor, Charles Lee Esq., to William Bird's fence on the Newgate road, and along the said road to Alexandria) 126, 127

Road from the fork of the road at Blackburn's fence to Difficult Run (Col. Broadwater's mill road) 164

Road from Edward Blackburn's to Charles Broadwater's mill on Difficult Run to intersect with the old Lawyer's road (road from Blackburn's fence through Walter Garrett's plantation and through the land of Col. Thomas Ludwell Lee to the mill) 94, 95

Road from Edward Blackburn's to the Church road near Mr. Benjamin Moody's 100

Road from the corner of Richard Scott Blackburn's fence to where it intersects with the old Court House road 134

Road from Little River via Goose Creek to Ashby's Gap on the Blue Ridge 10

Road from Edward Thompson's to the top of the Blue Ridge 16

Road from (Peter?) Lewis's mill to the County line at the Blue Ridge 20

Lower road from Goose Creek to top of the [Blue] ridge 32

Road from the Beaver Dam of Goose Creek to the school house above Anderson's and from thence to the top of the Blue Ridge 41

Ox road where it intersects at Thomas Sangster's (intersection leading from Colchester to the [Blue Ridge?] Mountains) 98

Road from Accotink to Pohick Run (including the fork of the road at Boggess's and the fork to Mr. Barnes's) 32

Road (from Colchester to Alexandria ordinary) at the hill below Boggess's 69, 72

Road dividing at Boggess's (into the River side and Back roads to Alexandria) 97

Road from Colchester to Boggess's mill on Pohick (road to leave the main road from Colchester to Alexandria above Giles Run and go through Mr. Bayly's land and George Mason's land above the plantation where Joseph Garberry lives then to cross the Church road at the Widow Atcheson's plantation and through her plantation to the mill) 102

Road from the Ponds on the Pohick road, crossing the Stage road to the intersection of the road leading from Boggess's to Gen. Washington's mill 167

Road at Boggess's 172

Road from Fitzhugh's quarter on Ravensworth to Robert Boggess's 12

Road to Robert Boggess's mill 15

Road from Robert Boggess's to Alexandria 46

Road from the fork of the road by Robert Boggess's to Dogue Run 59

Road from Robert Boggess's to Gen. Washington's mill 137, 166

Road from Pohick new Church to intersect the road from Robert Boggess's to Gen. (George) Washington's mill (road from Robert Boggess's to the General's [George Washington's] mill, and from that road to Pohick Church) 137, 148

Road from the Ellicot and Co. bridge on Occoquan into the road leading from Washington's shop to Robert Boggess's mill and from thence to the intersection with the Alexandria Post road 169

Road from Colchester to Robert Boggess Jr.'s mill 81

Road from Mathew Boswell's shop on Elzey's Church road to the mouth of the Widow Summers's lane 143

Road from the Widow Brewster's landing to Ballandine's mill 173

Road from the Widow Brewster's, to the intersection of Fairfax's mill road, on the line of Fairfax 175

Road from the Turnpike road to the Ox road, near the Brick Church 145

Road from Bricken's Branch to the Little Falls 103

Road from the forks of the road above Summers's to Bride's (Bird's?) bridge 149

The road from Bird's (Bride's?) bridge to the Turnpike 149

Road from the bridge/Dirt bridge/Bridge Branch above/near the old Court House to Difficult bridge/Difficult Run 88, 94, 95

Road from the run above the old Court House to where it intersects the road below the Court House (road from Bridge Branch to James Collins's old field and then into the old road) 112(2)

Road from Broad Run to the new Church 20

Road between Goose Creek and Broad Run 29

Road from the old Court House to Difficult Run, from Difficult Run to Sugarland Run and from Sugarland Run to Broad Run 37, 39

Road crossing the bridge made by John Trammel Jr./Broad Run bridge 22, 24

Road to Broad Run Chapel and from thence to Vestal's Gap 30

Road from Mr. Sebastian's quarter to Maj. Charles Broadwater's quarter 71, 72, 73

Road from Edward Blackburn's to Charles Broadwater's mill on Difficult Run to intersect with the old Lawyer's road (road from Blackburn's fence through Walter Garrett's plantation and through the land of Col. Thomas Ludwell Lee to the mill) 94, 95

Road from Col. Broadwater's mill to the old Court House 150

Road from the fork of the road at Blackburn's fence to Difficult Run (Col. Broadwater's mill road) 164

Extension of the Parish line from its present termination at Col. Broadwater's mill dam, along the Lawyer's road until it intersects the Ox road, and thence along the Ox road to Frying Pan Run and to the present County line 169

Roads from the Falls Church to Maj. Broadwater's Spring Branch and from Broadwater's Spring Branch to the Ox road 66

Road by Garrard Tramel's/Tramell's plantation (by Pimmets Run, and the orchard of Templeman's plantation, and to Mr. Broadwater's plantation) 25, 26

Road to Hereford's ferry (via Mr. Ferguson's cleared ground opposite to Mrs. Bronaugh's corn field) 45

Road from Brown's ordinary to John Hollis's 28

Road between Little Rockey Run and Brown's ordinary 42

Road from Brown's ordinary to Colchester 75

Mountain road (portion between Richard Brown's and Little Rockey Run) 13, 37

Road from Buck horn bridge to the top of the hill on the west side of Holmes Run 112

Road from Bull Run to the old Court House 48

Road from the old Court House to the Falls warehouse (road coming from Bull Run through the Court House corn field, Mr. Scott's land and John Urton's land, into the old Falls road at the head of Sasse Branch, thence along the old roads by Thomas Ashbury's, through the Honourable Philip Ludwell Lee's plantation, and to the Falls) 48, 49

Road from the Ox road to Bull Run 58

(Second) Road from the Ox road to Bull Run 58

Road that goes through Benjamin Talbott's plantation from the old Court House to Bull Run 67

Road from Thomas Sangster's/Songster's to the Wolf Shoals/Wolf Run Shoals (on Bull Run/Occoquan) 85, 116, 139, 145, 164

Road from Thomas Sangster's to Thomas Throgmorton's mill on Bull Run 95

Road from Alexander Henderson, Gent.'s plantation on Bull Run into the main road (road from Henderson's plantation through/adjoining the lands of Yelverton Reardon, William Moon, Marmaduke Beckwith, James Keen, Sampson Turley, William Simpson, and Peter Ryley to the main road) 104, 107, 108

Road from Alexander Henderson's plantation on Bull Run by Sampson Turley's to the Ox road 117

Road from Bull Run at Col. John Hooe's old mill to the Mountain road near Thomas Sangster's/Songster's (road from the Mountain road to Bull Run near Hooe's mill) 117, 138, 145, 164

Road from the Ox road above William Simpson's to where it intersects the road from Sangster's to Bull Run 145

Road from the Wolf Run Shoals to the shallow ford on Bull Run 157(2)

Road from Yates's ford on Bull Run, to Songster's (through the lands of Hector Kincheloe and Jesse Kincheloe the heirs of Daniel Kincheloe, and Robert Wickliffe) 173, 174

Road from the Red house to Samuel Adams's mill on Pimmets Run, and from thence to the road leading from the Little Falls to Alexandria (road beginning near the Red house, thence to Pimmets Run and the mill, thence through Samuel Adams's plantation to the long branch of Pimmets Run, along John Butcher's fence to Mathew Earp's, and across the Church road to the Alexandria road) 118, 119

Road from the Back Lick Run to Falls Church (road from the Back Lick to the Church, going through Simon Pearson's plantation and round Edmund Butler's fence); (including intersection at Price's, leading from thence to the Church) 53, 57(2), 88, 99(2), 100

Road from Cameron to the top of Cameron hill 47(2)

Road from the forks of the road near Francis Summers Jr.'s to the foot of Cameron hill 106

Road from the forks of the road at Cameron hill to the mile tree 116

Road from Alexandria to Cameron Run and to the forks of the road at Cameron hill 118

NOTE: The Fairfax County records frequently use the term "Cameron" for both Cameron Run and the vicinity (including probable references to Cameron ordinary), and often do not specify which feature is being cited; entries for both Cameron and Cameron Run are combined in this index

Awbrey's road to Cameron 4

Road (from Cameron?) to the run of Great Hunting Creek 19

Road from Cameron to Mr. William Adams's/William Adams's Gent.'s house/path/plantation 19, 63, 66, 69, 80(2), 82, 83

Road from Accotink to Cameron 22

Roads between Alexandria and Cameron:
- Road from the lower Cross Street in Alexandria to the road that leads to Cameron (not cleared) 27, 29
- Road from Alexandria to Cameron/Cameron Run 46, 66, 69, 80, 93, 96, 98, 105
- Road from Alexandria to Cameron (fork at Cameron with the roads leading up and down the Country) 98
- Road from Alexandria to Cameron Run and to the forks of the road at Cameron hill 118
- Road from Cameron into Alexandria (entrance road into Alexandria at King Street) 120

Road from the old school house to the road at Cameron 33

Road from the bridge at Cameron to John Summers's 46

Road from Cameron to the top of Cameron hill 47(2)

Road from Cameron to Mr. Daniel French's plantation 63, 66

Road from Cameron to the old Court House road 80

(Fork?) road from Cameron to the Falls Church 97

(Fork?) road from Cameron to John Summers's 97

Road from Cameron to the Mountain road 101, 103

Road from Dogue Run to Cameron 105, 117, 149

Road from the Ponds (on the road from Cameron to the Falls Church) to Alexandria 112

Road from the Gum Spring to Cameron/Cameron Run 117, 126, 149, 166

Road from Cameron Run into the Turnpike road 127

Road from Cameron Run or Creek by Clish to where General Washington's new road intersects the River side old road 137

Road from Roger West's house into the main road leading to Alexandria, at the ford of Cameron Run 154, 155

Road from Hepburn and Dundas's mill to the Colchester road near Cameron Run 170(2)

Road from Mark Canton's near Little River to Peyton's mill path 35

Road from Alexandria to the plantation of John Carlyle (running between the plantations of Baldwin Dade and Townshend Dade) 41

Road from the rock hole on Four Mile Creek to John Carlyle and Robert Adam Gent.'s mills 68(2)

Road from the termination of Queen's Street to the road leading from Alexandria to Georgetown (road along Mr. Carlyle's ditch on a parallel line with Queen Street to the county road from Georgetown to Alexandria) 94, 96

Road by Carpenter's Spring and Mr. George Johnston's plantation 72(2)

Road from the Great Falls of Potomac to Alexandria (road from the lower ford over Difficult Creek/Difficult Run via the Great road, near Edward Lanham's/Lannum's field, thence through the lands of William Swink, Joseph Simpson, Gustavus Scott Esqr. [rented by Daniel Jenkins], John Jackson, and Robert Carter, Esq. [rented by Charles Thrift], to the place on Scott's Run where William Shortridge's old mill stood, thence through the lands of Turberville, Scott, George Hunter, John Jackson, across Pimmets Run, through the lands of Richard Conway, Philip Darrell, William Gunnell, Mr. John West, Mr. Sampson Trammell, and Col. James Wren, to the Great road leading from Leesburg to Alexandria, where Col. Wren's store stands, thence by the Great road [Turnpike road] from the store to Alexandria) 121, 124(2), 128, 129, 130, 131(3), 139, 140, 141

Road from Charles Thrift's near the Little Falls road to the new road leading from the Great Falls to Alexandria (road from the Little Falls road to the Great Falls road, running through Charles Thrift's land/Robert Carter's land rented by Thrift to the Great Falls road at John Jackson's land; alteration of the road running through John Turberville's land) 131, 140

Road from Sugarland Run to the Chapel on the Beaver Dam 14

Road from the Sugar Lands to the new Chapel 18

Road to Broad Run Chapel and from thence to Vestal's Gap 30

Road from the Honourable Thomas Lee's quarter to the Church 1

Road from Berryman's quarter to the Church 11

Road from the Mountain road by Martha Hall's to the Church 11

Road from James Hamilton's to the Church 13

Road from the ferry to the Church road 18

Road from Broad Run to the new Church 20

Road from Sugarland Run to the new Church 20

The Church road 22, 27, 30

Road from the Black (Back?) Lick to the Church road 22

Road from the fork of the old road near Benjamin Talbott's to the Church road near Lewis Sanders's (road from the forks of the old Court House road near Benjamin Talbott's to the Church road near Lewis Sanders's, for people to go to the new Church built by Mr. Edward Payne and to the mill, also a nearer and most convenient way to Colchester and Dumfries; road to go through the plantations of Benjamin Talbott, William Kitchen, Edward Washington and John Cotton); (also proposed was an alternate, and more difficult route through/near the plantations of Benoni Hally, William Kitchen, Mrs. Mary Ferguson, Edward Washington and John Cotton) 86, 87, 88

Road from the upper Church to the new Church 93

Road from Hollis's to the new Church 93

Road from the fork of the old Court House road to the old Church road near Lewis Sanders's/Saunders's 94, 117

Road from Edward Blackburn's to the Church road near Mr. Benjamin Moody's 100

Road from Colchester to Boggess's mill on Pohick (road to leave the main road from Colchester to Alexandria above Giles Run and go through Mr. Bayly's land and George

Mason's land above the plantation where Joseph Garberry lives then to cross the Church road at the Widow Atcheson's plantation and through her plantation to the mill) 102

Road from Hereford's ferry to the old Church and to the road leading from Colchester to Alexandria 109, 116

Road from the mile tree to the Church 116

Road from the new Church to the River side road 117

Road from the Red house to Samuel Adams's mill on Pimmets Run, and from thence to the road leading from the Little Falls to Alexandria (road beginning near the Red house, thence to Pimmets Run and the mill, thence through Samuel Adams's plantation to the long branch of Pimmets Run, along John Butcher's fence to Mathew Earp's, and across the Church road to the Alexandria road) 118, 119

Road from Piney Branch into the Ox road near the Church 121, 122

NOTE: For roads associated with a specific chapel or church (i.e., Falls Church, Pohick Church, Rocky Run Chapel/Church), please see the entries for the particular chapel or church

Road from the old Court House to the road leading from the Chestnut tree to the Falls warehouse 67

Road to Clifton's ferry 52, 53

Road from the Post in the Main road to the ferry at Mr. William Clifton's 17

Road from Cameron Run or Creek by Clish to where General Washington's new road intersects the River side old road 137

Road from the forks of the road near Martin Cockburn's to where it intersects the road from Alexandria to Colchester (the Alexandria road), near Bayly's Mill Run/Bayley's old mill 139, 148

Road from William Hall Jr.'s house to Major Catesbys Cocke's road where it intersects 1

Road from Major Cocke's road to Cub Run 1

NOTE: Entries for Cockerell, Cockrell, Cockerill and Cockrill are combined

Road from the crossroads at Cockerill's old place along Ellzey's Church road to the Parish line 163

Road from the crossroads near Cockrill's to the forks of the road below Fitzhugh's 163

244

Road from the Falls Church to the crossroad near Joseph Cockerill's/Cockrell's old place 105, 107, 163

Bridle way from Colchester to Occoquan warehouse 48, 49

Road from the Ox road to Colchester 48, 67, 68

Road from the Ox road to Colchester (alteration made by going through John Ford's plantation, along an old mill path, and into the Ox road at the Wolf Pit hill) 50

Road from Pohick/Pohick Church to Colchester 60, 62, 101, 104

Road (from Colchester to Alexandria ordinary) at the hill below Boggess's 69, 72

Roads from William Gardner's to the place where the road to Colchester leaves the Ox road and from that place to Colchester warehouse 74

Road from Brown's ordinary to Colchester 75

Road from Colchester to Robert Boggess Jr.'s mill 81

Road from Court's ferry to the road from Alexandria to Colchester 84

Road from the fork of the old road near Benjamin Talbott's to the Church road near Lewis Sanders's (road from the forks of the old Court House road near Benjamin Talbott's to the Church road near Lewis Sanders's, for people to go to the new Church built by Mr. Edward Payne and to the mill, also a nearer and most convenient way to Colchester and Dumfries; road to go through the plantations of Benjamin Talbott, William Kitchen, Edward Washington and John Cotton); (also proposed was an alternate, and more difficult route through/near the plantations of Benoni Hally, William Kitchen, Mrs. Mary Ferguson, Edward Washington and John Cotton) 86, 87, 88

Road from Loofburrow's to Colchester 93, 103

Road from the Loudoun County line to Thomas Sangster's (and fork road at Thomas Sangster's leading to Colchester) 98

Road from Colchester up the Ox road (angle of the road up the Ox road) 98

Road from the Ox road to the Forge (angle of the road leading to the Forge and Colchester) 98

Ox road where it intersects at Thomas Sangster's (intersection leading from Colchester to the [Blue Ridge?] Mountains) 98

Main roads from Colchester to Loudoun County and the Ox Road from Loofburrow's to Difficult 101

Main roads from Vestal's and Williams's Gaps leading to Alexandria and Colchester 102(2), 111

Road from Colchester to Boggess's mill on Pohick (road to leave the main road from Colchester to Alexandria above Giles Run and go through Mr. Bayly's land and George Mason's land above the plantation where Joseph Garberry lives then to cross the Church road at the Widow Atcheson's plantation and through her plantation to the mill) 102

Road from Hereford's ferry to the old Church and to the road leading from Colchester to Alexandria 109, 116

Road from Colchester/Colchester warehouse to Wolf Pit hill 117, 139

Road from the Colchester road to the road to Alexandria (road from the Colchester road near John Ward's mill, thence through William Rogers's plantation to Richard Sanford's fence, and to the road leading to Alexandria opposite to Capt. Harper's fence) 124

Road from the main road at or near Price's to Hepburn and Dundas's mill on Back Lick Run and thence into the main road leading to Alexandria (road beginning at the dividing line between the lands of William Fitzhugh and the heirs of Henry Fitzhugh on the road to Newgate, thence to the Colchester road and through/adjoining the dividing line of John and Barbara Ratcliff, the lands of Barbara Ratcliff and the land of the heirs of James Murray to Hepburn and Dundas's mill, and thence through Hepburn and Dundas's land, and the lands of David and Samuel Arell, Minor, Charles Lee Esq., to William Bird's fence on the Newgate road, and along the said road to Alexandria) 126, 127

Road to Colchester 133

Road from Back Lick Run to where it intersects the road from Colchester to Alexandria at the end of John Hereford Jr.'s lane 136

Road from the Turnpike gate down the Colchester road to Dogue Run at Mrs. Moon's 136

Road from Accotink to Colchester 137

Road from the forks of the road near Martin Cockburn's to where it intersects the road from Alexandria to Colchester (the Alexandria road), near Bayly's Mill Run/Bayley's old mill 139, 148

Road from the White Oak Harbor/old White Oak Harbor, on the Ox road to Colchester (at the old sign post) 146, 166

Road from Colchester ferry, along the road to Alexandria to Accotink Run 146

Colchester road (portion from Samuel Smith's lane, through Benoni Price's plantation) 155

Road from John Summers's house to the Colchester road 155

Road from the head of Pohick Creek to the road leading from Alexandria to the Colchester road 159(2)

Road from the Back Lick Run to the intersection of the Colchester road at Hereford's 163

Sections of the Colchester road: from the Turnpike gate to Dogue Run; from Dogue Run to Accotink; from Accotink to Pohick Run; from Pohick Run to Colchester 165

Road from Hepburn and Dundas's mill to the Colchester road near Cameron Run 170(2)

Road from (John?) Hunter's fence to Collins's fence (road to begin at William Deneale's corner on Wolf Trap (Run) and thence with Lewis's line and thence along the line of Deneale and Fairfax to their corner near Collins's fence) 118(2)

Road from the run above the old Court House to where it intersects the road below the Court House (road from Bridge Branch to James Collins's old field and then into the old road) 112(2)

Road from the Mountain road by Col. Colvill's Tuskarora quarter to the ford of Goose Creek by Samuel King's 11

Road turned at Thomas Colvill Gent.'s plantation 55

Road from the Great Falls of Potomac to Alexandria (road from the lower ford over Difficult Creek/Difficult Run via the Great road, near Edward Lanham's/Lannum's field, thence through the lands of William Swink, Joseph Simpson, Gustavus Scott Esqr. [rented by Daniel Jenkins], John Jackson, and Robert Carter, Esq. [rented by Charles Thrift], to the place on Scott's Run where William Shortridge's old mill stood, thence through the lands of Turberville, Scott, George Hunter, John Jackson, across Pimmets Run, through the lands of Richard Conway, Philip Darrell, William Gunnell, Mr. John West, Mr. Sampson Trammell, and Col. James Wren, to the Great road leading from Leesburg to Alexandria, where Col. Wren's store stands, thence by the Great road [Turnpike road] from the store to Alexandria) 121, 124(2), 128, 129, 130, 131(3), 139, 140, 141

Road from the fork of the old road near Benjamin Talbott's to the Church road near Lewis Sanders's (road from the forks of the old Court House road near Benjamin Talbott's to the Church road near Lewis Sanders's, for people to go to the new Church built by Mr. Edward Payne and to the mill, also a nearer and most convenient way to

Colchester and Dumfries; road to go through the plantations of Benjamin Talbott, William Kitchen, Edward Washington and John Cotton); (also proposed was an alternate, and more difficult route through/near the plantations of Benoni Hally, William Kitchen, Mrs. Mary Ferguson, Edward Washington and John Cotton) 86, 87, 88

Road from (Peter?) Lewis's mill to the County line at the Blue Ridge 20

Road from Popeshead to the County line 58

Road from the County line to the fork of the road by Waters's ordinary 58

Road from the County line to the White Oak Spring (part of the Turnpike road leading to Alexandria) 135

Road from the County line on the Newgate road to David Loofbourrow's 138

Road from the County line on the Turnpike road leading to Alexandria and to the Ox road 144

Road from the County line on the Mountain road, to Thomas Sangster's 145

Sections of the Turnpike road:
- from the County line on the Turnpike road leading from Newgate to Alexandria as far as the Ox road 160
- from the Ox road, along said Turnpike road to the fork of the road at Hollice's old field near Price Skinner's 160
- from Hollice's old field in the fork of the road Price Skinner's to Price's tavern 160
- from Price's tavern to the forks of the road above Summers's 160

Road from the County line at the Turnpike road to Thomas Songster's 166

Road from the Ox road at the Wood Cutting to Occoquan and the County line leading to the mill 167

Extension of the Parish line from its present termination at Col. Broadwater's mill dam, along the Lawyer's road until it intersects the Ox road, and thence along the Ox road to Frying Pan Run and to the present County line 169

Road from Court's ferry to Pohick Church 84

Road from Court's ferry to the road from Alexandria to Colchester 84

Prince William road between the Ox road and where it comes into the road to the Court House 1

Road that leads from the Court House to Mrs. Willis's 6

Road from the Court House to Williams's Gap 8, 14

Road from the Court House/old Court House to the Ox road 9, 57, 105, 107, 110, 114

Road from the Court House to the Ox road near Capt. Lewis Ellzey's 11

Old Court House road 22

Road from the old Court House to the Mountain road 32, 33

Road from the old Court House to Difficult Run, from Difficult Run to Sugarland Run and from Sugarland Run to Broad Run 37, 39

Road from the old Court House to Turley's mill 47

Road from Bull Run to the old Court House 48

Road from the old Court House to the Falls warehouse (road coming from Bull Run through the Court House corn field, Mr. Scott's land and John Urton's land, into the old Falls road at the head of Sasse Branch, thence along the old roads by Thomas Ashbury's, through the Honourable Philip Ludwell Lee's plantation, and to the Falls) 48, 49

Road from William Ashford's [to the?] old Court House 54

Road from the old Court House to the fork of the road by John Ratcliff's plantation 55, 56

Road from the old Court House road near John Monroe's, to intersect with the road to Alexandria near Balty Stearns's 62, 63

Road from below the old Court House to Difficult 62

Road from Difficult to the old Court House 66

Road that goes through Benjamin Talbott's plantation from the old Court House to Bull Run 67

Road from the old Court House to the road leading from the Chestnut tree to the Falls warehouse 67

Road from the old Court House to the Falls road 68, 133

Road out of the old Court House road near Benoni Halley's to the Ravensworth road near John Hollis's 72

Road from the old Court House (via fence and line of John Mason's fence and Mr. Scott's line to the Falls road) to the Falls warehouse 72, 74

Road from Cameron to the old Court House road 80

Road from Mr. William Adam's to the old Court House 80

Road from the fork of the road leading to John Summers's to the old Court House road 82

Road from the old Court House to William Shortridge's 86

Road from the fork of the old road near Benjamin Talbott's to the Church road near Lewis Sanders's (road from the forks of the old Court House road near Benjamin Talbott's to the Church road near Lewis Sanders's, for people to go to the new Church built by Mr. Edward Payne and to the mill, also a nearer and most convenient way to Colchester and Dumfries; road to go through the plantations of Benjamin Talbott, William Kitchen, Edward Washington and John Cotton); (also proposed was an alternate, and more difficult route through/near the plantations of Benoni Hally, William Kitchen, Mrs. Mary Ferguson, Edward Washington and John Cotton) 86, 87, 88

Road from the bridge/Dirt bridge/Bridge Branch above/near the old Court House to Difficult bridge/Difficult Run 88, 94, 95

Road from the fork of the old Court House road to the old Church road near Lewis Sanders's/Saunders's 94, 117

Road from the old school house near Francis Summers's to the old Court House (with intersection of the road leading to the old Court house road) 99

Road from the run above the old Court House to where it intersects the road below the Court House (road from Bridge Branch to James Collins's old field and then into the old road) 112(2)

Road from the road leading out of the Alexandria road about three miles above the old Court House to the Great Falls of Potomac 112

Road from the forks of the road above Summers's to the road that leads to the old Court House 115

Road from the old Court House to the road leading to the Falls 120

Road from the old Court House to Doctor Dick's mill 124

Road from the corner of Richard Scott Blackburn's fence to where it intersects with the old Court House road 134

Road from the old Court House to William Turner's tavern 134

Road from the old Court House road opposite Joseph Bennett's plantation, to where it intersects the Ox road 135

Road from the Alexandria road at John Hereford's lane, to Price's tavern on the old Court House road 147

Road from William Turner's tavern to the Parish line on the old Court House road 147, 158, 167

Road from Ellzey's Church road to the old Court House road (at Bennetts) 148, 167

Road from Col. Broadwater's mill to the old Court House 150

Road from the old Court House to the Red house 156

Road from the old Court House to the Parish line 164

Old Court House road from the mouth of Popeshead Run to the corner of John Gibson's fence at the road leading to the Newgate road 168, 169

Road from the Falls Church to Deakins's near the Court House Run 134

Road from Deakins's near the old Court House Run, to Difficult 134

Road from Difficult bridge (on the Turnpike road) to the old Court House Run/Court House Run 148, 161

Road from the Court House Run to the Falls Church 149, 161

Road from Major Cocke's road to Cub Run 1

Road from the lower side of Little Rockey Run to Cub Run 13

Road from Alexandria to the plantation of John Carlyle (running between the plantations of Baldwin Dade and Townshend Dade) 41

Road from Alexandria to Difficult bridge near Capt. W. H. Terrett's (road from Princess Street passing near Baldwin Dade's old tobacco house, and Ramsay's quarter, and to a white oak in the road above Capt. W. H. Terrett's old field) 119, 120

Road from Georgetown ferry beginning at Lubber Branch and extending to Dowdall's corner on the hill by Four Mile Run (running on the line between Thomas Darne and Samuel Shreve and then straight until it intersects the Georgetown road below the Alexandria road) 141, 142

Road from the Great Falls of Potomac to Alexandria (road from the lower ford over Difficult Creek/Difficult Run via the Great road, near Edward Lanham's/Lannum's field, thence through the lands of William Swink, Joseph Simpson, Gustavus Scott Esqr. [rented by Daniel Jenkins], John Jackson, and Robert Carter, Esq. [rented by Charles Thrift], to the place on Scott's Run where William Shortridge's old mill stood, thence through the lands of Turberville, Scott, George Hunter, John Jackson, across Pimmets Run, through the lands of Richard Conway, Philip Darrell, William Gunnell, Mr. John West, Mr. Sampson Trammell, and Col. James Wren, to the Great road leading from Leesburg to Alexandria, where Col. Wren's store stands, thence by the Great road [Turnpike road] from the store to Alexandria) 121, 124(2), 128, 129, 130, 131(3), 139, 140, 141

Road from Davis's Branch to Peter Lewis's mill 20

Road from Scott's Run to Allen Davis's lane 114

Road from Edward Thompson's to Thomas Davis's 10

Road from Thomas Davis's/Thomas Davis's house to Goose Creek 10, 19, 20

Road from Thomas Kelley's road below William Ellzey's plantation to Segalon Run, across the creek at Dawson's ford and continuing to Nicholas Minor's 41

Road from the Falls Church to Deakins's near the Court House Run 134

Road from Deakins's near the old Court House Run, to Difficult 134

Road from (John?) Hunter's fence to Collins's fence (road to begin at William Deneale's corner on Wolf Trap (Run) and thence with Lewis's line and thence along the line of Deneale and Fairfax to their corner near Collins's fence) 118(2)

Road from William Deneale's mill to the road leading from Hollis's old field to Thomas Sangster's 151, 152

Road from the old Court House to Doctor Dick's mill 124

Road from the back line to the forks of the road above Dickens's 115

Road from the forks of the road above Dickens's to Ricketts's 115

NOTE: Entries for <u>Difficult bridge</u>, <u>Difficult Branch</u>, <u>Difficult Run</u>, and <u>Difficult</u> [Run] are combined

Road from Bond Veale's to the upper side of <u>Difficult Run</u> 9

Road from <u>Difficult Run</u> to Sugarland Run 19

Road from the old Court House to <u>Difficult Run</u>, from <u>Difficult Run</u> to Sugarland Run and from Sugarland Run to Broad Run 37, 39

Road from the Falls warehouse up to the Falls road leading to <u>Difficult</u> 44

Road from <u>Difficult</u> to James Dozier's 44

Road from <u>Difficult Branch</u> to Capt. Lewis Ellzey's 57

Road from below the old Court House to <u>Difficult</u> 62

Road from <u>Difficult</u> to the old Court House 66

Road from John Ballendine's mill at the Falls landing to or near the mouth of <u>Difficult Run</u> (road from the Falls mill to the mouth of <u>Difficult Run</u>) 74, 77

Road from young Gerrard Trammell's to Tolston mill upon <u>Difficult</u> 81

Road from John Hurst's plantation above the Falls Church to Tolston mill upon <u>Difficult</u> 81

Road from the old road at the upper side of Ballendine's farm to <u>Difficult</u> 85

Road from the bridge/Dirt bridge/Bridge Branch above/near the old Court House to <u>Difficult bridge</u>/<u>Difficult</u> 88, 94, 95

Road from Edward Blackburn's to Charles Broadwater's mill on <u>Difficult Run</u> to intersect with the old Lawyer's road (road from Blackburn's fence through Walter Garrett's plantation and through the land of Col. Thomas Ludwell Lee to the mill) 94, 95

Road from Benjamin Talbott's to Mr. Thomas Lewis's mill on <u>Difficult Run</u> (road from Benjamin Talbott's plantation thence along the Main road to the end of Samuel Talbott's fence thence through Mr. John Waller's plantation, and along the old mill path to Mr. Thomas Lewis's mill on <u>Difficult Run</u>) 94, 95

Main road from Alexandria to <u>Difficult bridge</u> 101, 103

Main roads from Colchester to Loudoun County and the Ox Road from Loofburrow's to <u>Difficult</u> 101

Road from Benjamin Talbott's plantation to the Loudoun County line at Amos Fox's mill on <u>Difficult</u> 112(2)

Road from <u>Difficult</u> to Sangster's shop/Songster's 114, 124

Road from the forks at <u>Difficult</u> to the Red house 115

Road from Gullatt's fence to <u>Difficult</u> 115

Road from Alexandria to <u>Difficult bridge</u> near Capt. W. H. Terrett's (road from Princess Street passing near Baldwin Dade's old tobacco house, and Ramsay's quarter, and to a white oak in the road above Capt. W. H. Terrett's old field) 119, 120

Road from the Great Falls of Potomac to Alexandria (road from the lower ford over <u>Difficult Creek/Difficult Run</u> via the Great road, near Edward Lanham's/Lannum's field, thence through the lands of William Swink, Joseph Simpson, Gustavus Scott Esqr. [rented by Daniel Jenkins], John Jackson, and Robert Carter, Esq. [rented by Charles Thrift], to the place on Scott's Run where William Shortridge's old mill stood, thence through the lands of Turberville, Scott, George Hunter, John Jackson, across Pimmets Run, through the lands of Richard Conway, Philip Darrell, William Gunnell, Mr. John West, Mr. Sampson Trammell, and Col. James Wren, to the Great road leading from Leesburg to Alexandria, where Col. Wren's store stands, thence by the Great road [Turnpike road] from the store to Alexandria) 121, 124(2), 128, 129, 130, 131(3), 139, 140, 141

Turnpike road from Alexandria to <u>Difficult</u> 126

Divisions of the road from the Great Falls of Potomac to Alexandria:
- from Col. Wren's store to the Falls road 131, 155
- from the Falls road to <u>Difficult</u> 131

Roads from the forks of the road near the old Farm to <u>Difficult</u> 133

Road from Deakins's near the old Court House Run, to <u>Difficult</u> 134

Road from <u>Difficult bridge/Difficult Run</u> on the Ox road, to Songster's/Thomas Sangster's/Songster's shop 135, 145, 166

Road from <u>Difficult bridge</u> (on the Turnpike road) to the old Court House Run/Court House Run 148, 161

Road from the Traff hill to Amos Fox's mill on <u>Difficult Run</u> 156

Road from Madden's shop to <u>Difficult Run</u> 164

Road from the fork of the road at Blackburn's fence to <u>Difficult Run</u> (Col. Broadwater's mill road) 164

Road from <u>Difficult</u> mill to the forks of the road at the farm 114

Road from the bridge/<u>Dirt bridge</u>/Bridge Branch above/near the old Court House to Difficult bridge/Difficult 88, 94, 95

Road from the head of <u>Dogue Creek</u> into the Back Lick road (the Middle road) 5

Roads from Accotink to <u>Dogue Run</u>:
- Road from <u>Dogue Run</u> to Accotink/Accotink Run 8, 32, 52, 62, 102
- Road from Accotink to <u>Dogue Run</u> on the Back road 85
- Road from <u>Dogue Run</u> to Accotink (the lower road) 107
- Road from the lower side of Accotink Run on the River side road to <u>Dogue Run</u> 117
- Road from <u>Dogue Run</u> at Mrs. Moon's to Accotink 137
- Road from Accotink, on the Alexandria road, to the Parish line at Dogue Run 146
- Sections of the Colchester road: from the Turnpike gate to <u>Dogue Run</u>; from <u>Dogue Run</u> to Accotink; from Accotink to Pohick Run; from Pohick Run to Colchester 165

Road from <u>Dogue Run</u> to the Gum Spring, and from the spring to Posey's ferry 32, 33, 59, 117

Road from <u>Dogue Run</u> to (John) Posey's/Pozey's ferry landing and from thence to Little Hunting Creek 39, 44

Road from Hunting Creek to <u>Dogue Run</u> 46, 54, 90, 107

Road between Holmes Run and <u>Dogue Run</u> 52

Road from the fork of the road by Robert Boggess's to <u>Dogue Run</u> 59

Back road from <u>Dogue Run</u> to Great Hunting Creek 61

Road from <u>Dogue Run</u> to Little Hunting Creek 69

Road from the fork of the road below William Johnston's to <u>Dogue Run</u> 75

Road from <u>Dogue Run</u> to Cameron 105, 117, 149

Road from the forks of the road below Pattersons to <u>Dogue Run</u> 117

Road from the Turnpike gate down the Colchester road to <u>Dogue Run</u> at Mrs. Moon's 136

Road from James Donaldson's house into the main road leading to Alexandria 51, 53

Way by Mr. Wren's plantation and through Gerrard Trammell's old field (to be established a bridle way for James Donaldson, Gent.) 57

Road from William Donaldson's to Summers's old school house 54(2)

Road from Georgetown ferry beginning at Lubber Branch and extending to Dowdall's corner on the hill by Four Mile Run (running on the line between Thomas Darne and Samuel Shreve and then straight until it intersects the Georgetown road below the Alexandria road) 141, 142

Road from Difficult to James Dozier's 44

Road from Dozer's to the head of Giles Run 59

Road from Capt. Lewis Ellzey's to James Ingo Dozer's 57

Road from the road near William West's to the road that leads to Dumfries 23

Road from the fork of the old road near Benjamin Talbott's to the Church road near Lewis Sanders's (road from the forks of the old Court House road near Benjamin Talbott's to the Church road near Lewis Sanders's, for people to go to the new Church built by Mr. Edward Payne and to the mill, also a nearer and most convenient way to Colchester and Dumfries; road to go through the plantations of Benjamin Talbott, William Kitchen, Edward Washington and John Cotton); (also proposed was an alternate, and more difficult route through/near the plantations of Benoni Hally, William Kitchen, Mrs. Mary Ferguson, Edward Washington and John Cotton) 86, 87, 88

Road from the main road at or near Price's to Hepburn and Dundas's mill on Back Lick Run and thence into the main road leading to Alexandria (road beginning at the dividing line between the lands of William Fitzhugh and the heirs of Henry Fitzhugh on the road to Newgate, thence to the Colchester road and through/adjoining the dividing line of John and Barbara Ratcliff, the lands of Barbara Ratcliff and the land of the heirs of James Murray to Hepburn and Dundas's mill, and thence through Hepburn and Dundas's land, and the lands of David and Samuel Arell, Minor, Charles Lee Esq., to William Bird's fence on the Newgate road, and along the said road to Alexandria) 126, 127

Road from Hepburn and Dundas's mill to the Colchester road near Cameron Run 170(2)

Road from Alexandria to Lewis Ellzey's meadow branch and the ford over Holmes Run (road from Duke Street crossing the gut below John West's meadow, passing Dutch Caty's house and Captain T. West's field to the road at the mouth of Col. Gilpin's lane, thence with the road to a branch near the hop-yard, and along the side of the hill to Holmes Run opposite the mouth of Messers Bird and Hawkins's lane) 119, 120

Road from the Red house to Samuel Adams's mill on Pimmets Run, and from thence to the road leading from the Little Falls to Alexandria (road beginning near the Red house, thence to Pimmets Run and the mill, thence through Samuel Adams's plantation to the long branch of Pimmets Run, along John Butcher's fence to Mathew Earp's, and across the Church road to the Alexandria road) 118, 119

Road from the Ellicot and Co. bridge on Occoquan into the road leading from Washington's shop to Robert Boggess's mill and from thence to the intersection with the Alexandria Post road 169

Road from Mathew Boswell's shop on Elzey's Church road to the mouth of the Widow Summers's lane 143

Road from Ellzey's Church road to the old Court House road (at Bennetts) 148, 167

Road from the Ox road to the Parish line on Ellzey's Church road 148, 154, 167

Road from the crossroads at Cockerill's old place along Ellzey's Church road to the Parish line 163

Road from Ellzey's Church road opposite Richard Ratcliffe's land, passing by William Payne's mill, and from thence to the Alexandria road 168, 169

Road from the Court House to the Ox road near Capt. Lewis Ellzey's 11

Road from Difficult Branch to Capt. Lewis Ellzey's 57

Road from Capt. Lewis Ellzey's to James Ingo Dozer's 57

Road from Alexandria to Lewis Ellzey's meadow branch and the ford over Holmes Run (road from Duke Street crossing the gut below John West's meadow, passing Dutch Caty's house and Captain T. West's field to the road at the mouth of Colonel Gilpin's lane, thence with the road to a branch near the hop-yard, and along the side of the hill to Holmes Run opposite the mouth of Messers Bird and Hawkins's lane) 119, 120

Road from Thomas Kelley's road below William Ellzey's plantation to Segalon Run, across the creek at Dawson's ford and continuing to Nicholas Minor's 41

Road from (John?) Hunter's fence to Collins's fence (road to begin at William Deneale's corner on Wolf Trap (Run) and thence with Lewis's line and thence along the line of Deneale and Fairfax to their corner near Collins's fence) 118(2)

Road from the Widow Brewster's, to the intersection of <u>Fairfax's</u> mill road, on the line of <u>Fairfax</u> 175

Road from the main road near the <u>Falls Church</u> 49

Road from the Back Lick Run to <u>Falls Church</u> (road from the Back Lick to the Church, going through Simon Pearson's plantation and round Edmund Butler's fence); (including intersection at Price's, leading from thence to the Church) 53, 57(2), 88, 99(2), 100

Road from the <u>Falls Church</u> to the Ox road 60

Road from the <u>Falls Church</u> to Rock Creek 63

Roads from the <u>Falls Church</u> to Maj. Broadwater's Spring Branch and from Broadwater's Spring Branch to the Ox road 66

Road from the <u>Falls Church</u> to Alexandria 71, 113

Road from the <u>Falls Church</u> to the road leading from Alexandria to the Falls warehouse 79, 114, 132

Road from John Hurst's plantation above the <u>Falls Church</u> to Tolston mill upon Difficult 81

(Fork?) road from Cameron to the <u>Falls Church</u> 97

Road from the forks of the road at the <u>Falls Church</u> to the old road from Rock Creek to Alexandria 100

Road from the <u>Falls Church</u> to the crossroad near Joseph Cockerill's/Cockrell's old place 105, 107, 163

Road from the Ponds (on the road from Cameron to the <u>Falls Church</u>) to Alexandria 112

Road from the <u>Falls Church</u> to Fitzhugh's gate 113

Road from the <u>Falls Church</u> to Trammell's 116

Road from the North run of Holmes to the <u>Falls Church</u> 127

Road from Georgetown to the <u>Falls Church</u> 129

Road from the entrance of Avory's road near Marle's house to the Turnpike road leading to the <u>Falls Church</u> 132

Road from the <u>Falls Church</u> to (the white oak at) Widow Tucker's 133, 162

Road from the Falls Church to Deakins's near the Court House Run 134

Road from the Falls Church by Ricketts's to the Ox road 134

Road from the Falls Church to the Little Falls of Potomac 142, 163

Road from the Court House Run to the Falls Church 149, 161

Road from the Falls Church to the forks of the road near the Falls 156

NOTE: Entries for Falls, Great Falls, and Little Falls are grouped below

the Falls road 17

Road from the Falls warehouse up to the Falls road leading to Difficult 44

Road from the Falls warehouse to the Falls landing 44, 107, 110(2)

Road from Alexandria to the Falls warehouse 48, 76

Road from the old Court House to the Falls warehouse (road coming from Bull Run through the Court House corn field, Mr. Scott's land and John Urton's land, into the old Falls road at the head of Sasse Branch, thence along the old roads by Thomas Ashbury's, through the Honourable Philip Ludwell Lee's plantation, and to the Falls) 48, 49

Old road from the Falls warehouse to Alexandria 49

Road (fork?) from the Falls road to the Sugar Land road 52

Road from Trammell's old field to the Falls road 54

Road from the old Court House to the road leading from the Chestnut tree to the Falls warehouse 67

Road from the old Court House to the Falls road 68, 133

Road from the old Court House (via fence and line of John Mason's fence and Mr. Scott's line to the Falls road) to the Falls warehouse 72, 74

Road from John Ballendine's mill at the Falls landing to or near the mouth of Difficult Run (road from the Falls mill to the mouth of Difficult Run) 74, 77

Road from the forks of the road to Four Mile Run leading to the Falls 77

Road from the Falls Church to the road leading from Alexandria to the <u>Falls</u> warehouse 79, 114, 132

Road from the Alexandria road to the Four Mile Run (including fork leading to Georgetown and the <u>Falls</u> warehouse) 99

Road from the <u>Falls</u> warehouse to the Red house 103

Road from the <u>Falls</u> warehouse to the fork of the road by Turberville's quarter 106

Road from the Red house to the <u>Falls</u> landing 114, 132

Road from the old Court House to the road leading to the <u>Falls</u> 120

Road from Avory's road to the <u>Falls</u> 121

Road from Avory's road to the Mountain road that leads to the <u>Falls</u> landing 132

Road from the Falls Church to the forks of the road near the <u>Falls</u> 156

Road from the road leading out of the Alexandria road about three miles above the old Court House to the <u>Great Falls</u> of Potomac 112

Road from the <u>Great Falls</u> of Potomac to Alexandria (road from the lower ford over Difficult Creek/Difficult Run via the Great road, near Edward Lanham's/Lannum's field, thence through the lands of William Swink, Joseph Simpson, Gustavus Scott Esqr. [rented by Daniel Jenkins], John Jackson, and Robert Carter, Esq. [rented by Charles Thrift], to the place on Scott's Run where William Shortridge's old mill stood, thence through the lands of Turberville, Scott, George Hunter, John Jackson, across Pimmets Run, through the lands of Richard Conway, Philip Darrell, William Gunnell, Mr. John West, Mr. Sampson Trammell, and Col. James Wren, to the Great road leading from Leesburg to Alexandria, where Col. Wren's store stands, thence by the Great road [Turnpike road] from the store to Alexandria) 121, 124(2), 128, 129, 130, 131(3), 139, 140, 141

Road from Charles Thrift's near the Little Falls road to the new road leading from the <u>Great Falls</u> to Alexandria (road from the Little Falls road to the <u>Great Falls</u> road, running through Charles Thrift's land/Robert Carter's land rented by Thrift to the <u>Great Falls road</u> at John Jackson's land; alteration of the road running through John Turberville's land) 131, 140

Divisions of the road from the <u>Great Falls</u> of Potomac to Alexandria
    from Col. Wren's store to the <u>Falls</u> road 131, 155
    from the <u>Falls</u> road to Difficult 131

Road from Col. Wren's tavern up the <u>Great Falls</u> road to Madden's shop 163

Road from Hunters old store house to the Little Falls 81

Road from Bricken's Branch to the Little Falls 103

Road from the Little Falls to Four Mile Run 113

Road from the Red house to Samuel Adams's mill on Pimmets Run, and from thence to the road leading from the Little Falls to Alexandria (road beginning near the Red house, thence to Pimmets Run and the mill, thence through Samuel Adams's plantation to the long branch of Pimmets Run, along John Butcher's fence to Mathew Earp's, and across the Church road to the Alexandria road) 118, 119

Road from Charles Thrift's near the Little Falls road to the new road leading from the Great Falls to Alexandria (road from the Little Falls road to the Great Falls road, running through Charles Thrift's land/Robert Carter's land rented by Thrift to the Great Falls road at John Jackson's land; alteration of the road running through John Turberville's land) 131, 140

Road from the Falls Church to the Little Falls of Potomac 142, 163

Road from the Sycamore near the Little Falls warehouse to the river 158

Road from Avery's road to the Little Falls road 162

Road from Madden's shop to the Little Falls warehouse 164

Road from Difficult mill to the forks of the road at the farm 114

Roads from the forks of the road near the old Farm to Difficult 133

Road from Avery's road to where it intersects the road leading from Fendall and Hipkins's mill 156

Road to Hereford's ferry (via Mr. Ferguson's cleared ground opposite to Mrs. Bronaugh's corn field) 45

Road from the fork of the old road near Benjamin Talbott's to the Church road near Lewis Sanders's (road from the forks of the old Court House road near Benjamin Talbott's to the Church road near Lewis Sanders's, for people to go to the new Church built by Mr. Edward Payne and to the mill, also a nearer and most convenient way to Colchester and Dumfries; road to go through the plantations of Benjamin Talbott, William Kitchen, Edward Washington and John Cotton); (also proposed was an alternate,

and more difficult route through/near the plantations of Benoni Hally, William Kitchen, Mrs. Mary Ferguson, Edward Washington and John Cotton) 86, 87, 88

Road from the ferry road at the town of Alexandria to Hunting Creek ford and from thence to Four Mile Run ford 4

Road from the ferry to the Church road 18

The ferry road 30(2)

The ferry road (George Mason's?) 159

NOTE: For roads associated with specific ferries (i.e., Clifton's, Court's, Georgetown, Hereford's/Herryford's, Holland Point, Hooe's, Mason's, Noland's, Posey's/Pozey's, Poultney's, Rock Creek), please see the entries for the particular name

Road from Fitzhugh's quarter on Ravensworth to Robert Boggess's 12

Road from the Falls Church to Fitzhugh's gate 113

Road turned in Fitzhugh's lane 141, 143

Road from Ward's mill to the road leading from Fitzhugh's quarters near Mrs. French's 154

Road from the forks of the road below Fitzhugh's, to the intersection of the Turnpike road at Frazier's old place 163

Road from the forks of the road below Fitzhugh's along the Back Lick road to Back Lick Run 163

Road from the crossroads near Cockrill's to the forks of the road below Fitzhugh's 163

Road from the main road at or near Price's to Hepburn and Dundas's mill on Back Lick Run and thence into the main road leading to Alexandria (road beginning at the dividing line between the lands of William Fitzhugh and the heirs of Henry Fitzhugh on the road to Newgate, thence to the Colchester road and through/adjoining the dividing line of John and Barbara Ratcliff, the lands of Barbara Ratcliff and the land of the heirs of James Murray to Hepburn and Dundas's mill, and thence through Hepburn and Dundas's land, and the lands of David and Samuel Arell, Minor, Charles Lee Esq., to William Bird's fence on the Newgate road, and along the said road to Alexandria) 126, 127

Road from Nicholas Fitzhugh's gate to Back Lick Run 136

Back Lick road from Price's ordinary to the fork of the road above Nicholas Fitzhugh's gate 151

Road from Price's ordinary to the forks of the road above Nicholas Fitzhugh's plantation 154

Road from the main road at or near Price's to Hepburn and Dundas's mill on Back Lick Run and thence into the main road leading to Alexandria (road beginning at the dividing line between the lands of William Fitzhugh and the heirs of Henry Fitzhugh on the road to Newgate, thence to the Colchester road and through/adjoining the dividing line of John and Barbara Ratcliff, the lands of Barbara Ratcliff and the land of the heirs of James Murray to Hepburn and Dundas's mill, and thence through Hepburn and Dundas's land, and the lands of David and Samuel Arell, Minor, Charles Lee Esq., to William Bird's fence on the Newgate road, and along the said road to Alexandria) 126, 127

Road from the Ox road to Colchester (alteration made by going through John Ford's plantation, along an old mill path, and into the Ox road at the Wolf Pit hill) 50

Road from the Ox road to the Forge (angle of the road leading to the Forge and Colchester) 98

Road from the ferry road at the town of Alexandria to Hunting Creek ford and from thence to Four Mile Run ford 4

Road from Four Mile Run to Mason's ferry opposite to Rock Creek 18

Road from Alexandria to Four Mile Run 46, 52, 63, 91, 96, 114

Road from the rock hole on Four Mile Creek to John Carlyle and Robert Adam Gent.'s mills 68(2)

Road from Four Mile Run to Rock Creek/Rock Creek ferry 69, 115

Road from the forks of the road to Four Mile Run leading to the Falls 77

Road from Four Mile Run to Georgetown ferry 87, 150, 162

Road from the Alexandria road to the Four Mile Run (including fork leading to Georgetown and the Falls warehouse) 99

Road from the Little Falls to Four Mile Run 113

Road from the bridge at Alexandria to Four Mile Run 115

Road from the ferry opposite Georgetown to Four Mile Run 132

The road from Four Mile Run to the Turnpike gate 132

Road from Georgetown ferry beginning at Lubber Branch and extending to Dowdall's corner on the hill by Four Mile Run (running on the line between Thomas Darne and Samuel Shreve and then straight until it intersects the Georgetown road below the Alexandria road) 141, 142

Road from the Turnpike (near Alexandria) to Four Mile Run 150, 162

Road from Four Mile Run below the old mill to Avery's road 162

Road from the Turnpike road at Hallace's/Hollice's/Hollis's old field to where it intersects the Pohick road at John Fowler's plantation/fence 138, 147, 167

Road from Fox's mill to Alexandria 169

Road from Benjamin Talbott's plantation to the Loudoun County line at Amos Fox's mill on Difficult Run 112(2)

Road from the Traff hill to Amos Fox's mill on Difficult Run 156

Road from Ricketts's to Frazer's 136

Road from the forks of the road below Fitzhugh's, to the intersection of the Turnpike road at Frazier's old place 163

Road from Cameron to Mr. Daniel French's plantation 63, 66

Road proposed to be altered by Daniel French 90

Road from Ward's mill to the road leading from Fitzhugh's quarters near Mrs. French's 154

Road from Thomas Sangster's to Frying Pan (leading out from the Mountain road) 98

Extension of the Parish line from its present termination at Col. Broadwater's mill dam, along the Lawyer's road until it intersects the Ox road, and thence along the Ox road to Frying Pan Run and to the present County line 169

Road from Colchester to Boggess's mill on Pohick (road to leave the main road from Colchester to Alexandria above Giles Run and go through Mr. Bayly's land and George Mason's land above the plantation where Joseph Garberry lives then to cross the Church road at the Widow Atcheson's plantation and through her plantation to the mill) 102

Road from William Garner's/Gardner's to Ravensworth road (from the Ox road via the pond and Steptoe's quarter) 72, 73

Roads from William Gardner's to the place where the road to Colchester leaves the Ox road and from that place to Colchester warehouse 74

Road from Pimmet's Run to Nicholas Garrett's plantation 65

Road from Edward Blackburn's to Charles Broadwater's mill on Difficult Run to intersect with the old Lawyer's road (road from Blackburn's fence through Walter Garrett's plantation and through the land of Col. Thomas Ludwell Lee to the mill) 94, 95

Road from Four Mile Run to Georgetown ferry 87, 150, 162

Road from the termination of Queen's Street to the road leading from Alexandria to Georgetown (road along Mr. Carlyle's ditch on a parallel line with Queen Street to the county road from Georgetown to Alexandria) 94, 96

Road from Mr. William Adams's/William Adams's Gent. to Pimmets Run (including fork[?] road from the Alexandria road to the ferry at Georgetown) 96, 97, 100

Road from the Alexandria road to the Four Mile Run (including fork leading to Georgetown and the Falls warehouse) 99

Road from Georgetown ferry to Adams's mill 113

Road from Alexandria to Georgetown ferry 119

Road from the forks of Avery's road to the road leading from Alexandria to Georgetown 121

Road from the Georgetown road at R. Adams's to Avery's road 121(2)

Turnpike road from Alexandria to Georgetown ferry 126

Road from Georgetown to the Falls Church 129

Road from the ferry opposite Georgetown to Four Mile Run 132

Road from the Turnpike gate at the intersection of the Georgetown road to the white oak at the Widow Tucker's 133

Road from Georgetown ferry beginning at Lubber Branch and extending to Dowdall's corner on the hill by Four Mile Run (running on the line between Thomas Darne and Samuel Shreve and then straight until it intersects the Georgetown road below the Alexandria road) 141, 142

the Georgetown road 162

Road from John Gest's house to the road by William Tramell's 35

Old Court House road from the mouth of Popeshead Run to the corner of John Gibson's fence at the road leading to the Newgate road 168, 169

Road from Giles Run to Occoquan warehouse 59

Road from Dozer's to the head of Giles Run 59

Road from Colchester to Boggess's mill on Pohick (road to leave the main road from Colchester to Alexandria above Giles Run and go through Mr. Bayly's land and George Mason's land above the plantation where Joseph Garberry lives then to cross the Church road at the Widow Atcheson's plantation and through her plantation to the mill) 102

Road from Alexandria to Lewis Ellzey's meadow branch and the ford over Holmes Run (road from Duke Street crossing the gut below John West's meadow, passing Dutch Caty's house and Captain T. West's field to the road at the mouth of Col. Gilpin's lane, thence with the road to a branch near the hop-yard, and along the side of the hill to Holmes Run opposite the mouth of Messers Bird and Hawkins's lane) 119, 120

Road from the glebe of Cameron Parish to Goose Creek Chapel 33

Road from Herryford's (Hereford's) ferry to the Church of Pohick and into the main road by William Godfrey's 9

Road from the glebe of Cameron Parish to Goose Creek Chapel 33

Road from Sugarland Run to Goose Creek/Goose Creek ferry and from Goose Creek to Limestone Run (including digging down the banks at Goose Creek ferry) and parts thereof 3, 6, 8, 24

Road from Williams's Gap to the upper fork of Goose Creek 6

Road from the upper fork of Goose Creek to Elisha Hall's 6

Road from the inhabitants of Little River and Goose Creek down to the road by Mr. Hutchison's 7

Road from Little River via Goose Creek to Ashby's Gap on the Blue Ridge 10

Road from Thomas Davis's/Thomas Davis's house to Goose Creek 10, 19, 20

Road from the Mountain road by Col. Colvill's Tuskarora quarter to the ford of Goose Creek by Samuel King's 11

Road from Little River to the Beaver Dam fork of Goose Creek 26

Road between Little River and Goose Creek 29

Road between Goose Creek and Broad Run 29

Road from Little River to the widow Adams's at Goose Creek 32

Lower road from Goose Creek to top of the [Blue] ridge 32

Road from the Beaver Dam of Goose Creek to the school house above Anderson's and from thence to the top of the Blue Ridge 41

Road (from Goose Creek?) near Tuscarora Run 42

Road from the bridge at Grayson's mill to Occoquan ferry 6

Road by Benjamin Grayson's mill 4

Road near Pohick (near Col. Grayson's mill dam) 23

Road from the great road by Daniel Mills's to William Payne's old mill on Accotink Run 113(2)

Road from the Great Falls of Potomac to Alexandria (road from the lower ford over Difficult Creek/Difficult Run via the Great road, near Edward Lanham's/Lannum's field, thence through the lands of William Swink, Joseph Simpson, Gustavus Scott Esqr. [rented by Daniel Jenkins], John Jackson, and Robert Carter, Esq. [rented by Charles Thrift], to the place on Scott's Run where William Shortridge's old mill stood, thence through the lands of Turberville, Scott, George Hunter, John Jackson, across Pimmets Run, through the lands of Richard Conway, Philip Darrell, William Gunnell, Mr. John West, Mr. Sampson Trammell, and Col. James Wren, to the Great road leading from Leesburg to Alexandria, where Col. Wren's store stands, thence by the Great road [Turnpike road] from the store to Alexandria) 121, 124(2), 128, 129, 130, 131(3), 139, 140, 141

Road by James Green's plantation 85

Road through James Richards's plantation to Grymes's mill (Grymes's mill road) 153, 174

Road from Gerrard Tramell's to Gullatt's fence 115

Road from Gullatt's fence to Difficult 115

Road from the Gum Spring to Hunting Creek 15

Forks of the Back and River side roads on the road from Hunting Creek to the Gum Spring 97

Road from Dogue Run to the Gum Spring, and from the spring to Posey's ferry 32, 33, 59, 117

Road from the Gum Spring to Cameron/Cameron Run 117, 126, 149, 166

Road from General (George) Washington's ferry to his mill, from thence to his tumbling dam, thence along his new road to intersect the riverside old road above the Gum Spring (and parts thereof: includes road from the Gum Spring to General Washington's mill) 137, 166

Road from the Great Falls of Potomac to Alexandria (road from the lower ford over Difficult Creek/Difficult Run via the Great road, near Edward Lanham's/Lannum's field, thence through the lands of William Swink, Joseph Simpson, Gustavus Scott Esqr. [rented by Daniel Jenkins], John Jackson, and Robert Carter, Esq. [rented by Charles Thrift], to the place on Scott's Run where William Shortridge's old mill stood, thence through the lands of Turberville, Scott, George Hunter, John Jackson, across Pimmets Run, through the lands of Richard Conway, Philip Darrell, William Gunnell, Mr. John West, Mr. Sampson Trammell, and Col. James Wren, to the Great road leading from Leesburg to Alexandria, where Col. Wren's store stands, thence by the Great road [Turnpike road] from the store to Alexandria) 121, 124(2), 128, 129, 130, 131(3), 139, 140, 141

Road from the upper fork of Goose Creek to Elisha Hall's 6

Road from the Mountain road by Martha Hall's to the Church 11

Road from William Hall Jr.'s house to Major Catesbys Cocke's road where it intersects 1

NOTE: Hallace: see Hollis, etc.

NOTE: Entries for Halley and Hally are combined

Road from the forks of the road below Sangster's to the Mountain road by Hally's 114

Road out of the old Court House road near Benoni Halley's to the Ravensworth road near John Hollis's 72

Road from the fork of the old road near Benjamin Talbott's to the Church road near Lewis Sanders's (road from the forks of the old Court House road near Benjamin Talbott's to the Church road near Lewis Sanders's, for people to go to the new Church built by Mr. Edward Payne and to the mill, also a nearer and most convenient way to Colchester and Dumfries; road to go through the plantations of Benjamin Talbott, William Kitchen, Edward Washington and John Cotton); (also proposed was an alternate,

and more difficult route through/near the plantations of Benoni Hally, William Kitchen, Mrs. Mary Ferguson, Edward Washington and John Cotton) 86, 87, 88

Road on the line between James Hally and John Halley 124

(Alteration of) Ox road from the corner between Simpson Halley and Richard Ratcliffe to the Alexandria road 157(2)

Road from James Hamilton's to the Church 13

Road by James Hamilton's plantation 26

Road from the Colchester road to the road to Alexandria (road from the Colchester road near John Ward's mill, thence through William Rogers's plantation to Richard Sanford's fence, and to the road leading to Alexandria opposite to Capt. Harper's fence) 124

Road from William Hartshorne's mill to the Turnpike road 151

Road from Alexandria to Lewis Ellzey's meadow branch and the ford over Holmes Run (road from Duke Street crossing the gut below John West's meadow, passing Dutch Caty's house and Captain T. West's field to the road at the mouth of Col. Gilpin's lane, thence with the road to a branch near the hop-yard, and along the side of the hill to Holmes Run opposite the mouth of Messers Bird and Hawkins's lane) 119, 120

Road from Alexander Henderson, Gent.'s plantation on Bull Run into the main road (road from Henderson's plantation through/adjoining the lands of Yelverton Reardon, William Moon, Marmaduke Beckwith, James Keen, Sampson Turley, William Simpson, and Peter Ryley to the main road) 104, 107, 108

Road from Occoquan above the Wolf Run Shoals to Alexander Henderson's mill 114

Road from Alexander Henderson's plantation on Bull Run by Sampson Turley's to the Ox road 117

Road from the main road at or near Price's to Hepburn and Dundas's mill on Back Lick Run and thence into the main road leading to Alexandria (road beginning at the dividing line between the lands of William Fitzhugh and the heirs of Henry Fitzhugh on the road to Newgate, thence to the Colchester road and through/adjoining the dividing line of John and Barbara Ratcliff, the lands of Barbara Ratcliff and the land of the heirs of James Murray to Hepburn and Dundas's mill, and thence through Hepburn and Dundas's land, and the lands of David and Samuel Arell, Minor, Charles Lee Esq., to William Bird's fence on the Newgate road, and along the said road to Alexandria) 126, 127

Road from Hepburn and Dundas's mill to the Colchester road near Cameron Run 170(2)

Road from the forks of the road above Herbert's to the north run of Hunting Creek 116

Road from Herryford's (Hereford's) ferry to the Church of Pohick and into the main road by William Godfrey's 9

Road to Hereford's ferry (via Mr. Ferguson's cleared ground opposite to Mrs. Bronaugh's corn field) 45

Road from Pohick Church to Hereford's ferry 46, 109

Road from Hereford's ferry to the old Church and to the road leading from Colchester to Alexandria 109, 116

Road from the Back Lick Run to the intersection of the Colchester road at Hereford's 163

Road from Back Lick Run to where it intersects the road from Colchester to Alexandria at the end of John Hereford Jr.'s lane 136

Road from the Alexandria road at John Hereford's lane, to Price's tavern on the old Court House road 147

Road from Avery's road to where it intersects the road leading from Fendall and Hipkins's mill 156

Road from the cross roads at Hodgkins's to the mouth of Beard's lane, on Lee's land 173

The road from Pohick old Church (old Pohick Church) to Holland Point ferry 137, 148

NOTE: Entries for Hallace, Hollies, Hollis and Hollice are combined

Road from the Ox road to Hollis's old field (by Samuel Littlejohn's fence) 58

Road that leads from Hollis's old field to Sangster's 84, 85

Road from Hollis's to the new Church 93

Road from the Ox road near Sangster's/Thomas Songster's to Hallace's/Hollies'/Hollis's/Hollice's old field (at the Turnpike road) 138, 147, 157, 167

Road from the Turnpike road at Hallace's/Hollice's/Hollis's old field to where it intersects the Pohick road at John Fowler's plantation/fence 138, 147, 167

Road from Hollis's old field to Accotink, on the Turnpike road 144

Road from the Ox road on the Turnpike road to Hollis's old fields, opposite to where another road intersects the Turnpike road leading from the Ox road 144

Road from William Deneale's mill to the road leading from Hollis's old field to Thomas Sangster's 151, 152

Sections of the Turnpike road:
- from the County line on the Turnpike road leading from Newgate to Alexandria as far as the Ox road 160
- from the Ox road, along said Turnpike road to the fork of the road at Hollice's old field near Price Skinner's 160
- from Hollice's old field in the fork of the road Price Skinner's to Price's tavern 160
- from Price's tavern to the forks of the road above Summers's 160

Road from Brown's ordinary to John Hollis's 28

Road from William Ashford's to John Hollis's 49

Road out of the old Court House road near Benoni Halley's to the Ravensworth road near John Hollis's 72

Road from where John Hollis formerly lived to Pohick warehouse 106

Road between Holmes Run and Dogue Run 52

Road from Buck horn bridge to the top of the hill on the west side of Holmes Run 112

Road from Alexandria to Lewis Ellzey's meadow branch and the ford over Holmes Run (road from Duke Street crossing the gut below John West's meadow, passing Dutch Caty's house and Captain T. West's field to the road at the mouth of Col. Gilpin's lane, thence with the road to a branch near the hop-yard, and along the side of the hill to Holmes Run opposite the mouth of Messers Bird and Hawkins's lane) 119, 120

Road from the North run of Holmes to the Falls Church 127

Road from Price's tavern to Holmes Run 135

Road from Holmes Run, to the Turnpike gate, where William Simpson lives 136

Road from Holmes Run to the Stone bridge below the Turnpike 161

Road from the forks of the road above Summers's to Holmes Run 161

Road from Bull Run at Col. John Hooe's old mill to the Mountain road near Thomas Sangster's/Songster's (road from the Mountain road to Bull Run near Hooe's mill) 117, 138, 145, 164

Road from Alexandria to John Hooe's ferry on Occoquan, near the old warehouse landing 152(2), 153(2)

Road from John Hooe's ferry on Occoquan to the lands of Samuel Bayly 153

Road from Hunter's old store house to the Little Falls 81

Road from the Great Falls of Potomac to Alexandria (road from the lower ford over Difficult Creek/Difficult Run via the Great road, near Edward Lanham's/Lannum's field, thence through the lands of William Swink, Joseph Simpson, Gustavus Scott Esqr. [rented by Daniel Jenkins], John Jackson, and Robert Carter, Esq. [rented by Charles Thrift], to the place on Scott's Run where William Shortridge's old mill stood, thence through the lands of Turberville, Scott, George Hunter, John Jackson, across Pimmets Run, through the lands of Richard Conway, Philip Darrell, William Gunnell, Mr. John West, Mr. Sampson Trammell, and Col. James Wren, to the Great road leading from Leesburg to Alexandria, where Col. Wren's store stands, thence by the Great road [Turnpike road] from the store to Alexandria) 121, 124(2), 128, 129, 130, 131(3), 139, 140, 141

Road from (John?) Hunter's fence to Collins's fence (road to begin at William Deneale's corner on Wolf Trap (Run) and thence with Lewis's line and thence along the line of Deneale and Fairfax to their corner near Collins's fence) 118(2)

Road from the ferry road at the town of Alexandria to Hunting Creek ford and from thence to Four Mile Run ford 4

Road from William Adams's house to Hunting Creek 4

Road from Dogue Run on the Back road to Hunting Creek (Back road from Hunting Creek to Dogue Run; Road from Hunting Creek to Dogue Run on the back road [with fork[?] roads leading from Hunting Creek]) 8, 86, 96, 97

Road from the Gum Spring to Hunting Creek 15

Road from Hunting Creek to Belhaven 32

Road from Hunting Creek to Dogue Run 46, 54, 90, 107

Road from Hunting Creek to Alexandria 86

Forks of the Back and River side roads on the road from Hunting Creek to the Gum Spring 97

Road from the north run of Hunting Creek to the forks of the road above (Francis?) Summers's 116

Road from the forks of the road above Herbert's to the north run of Hunting Creek 116

Road (from Cameron?) to the run of Great Hunting Creek 19

Back road from Dogue Run to Great Hunting Creek 61

Road from Little Hunting Creek to Accotink/Accotink Run 29, 40, 43

Road from Dogue Run to (John) Posey's/Pozey's ferry landing and from thence to Little Hunting Creek 39, 44

Road from Dogue Run to Little Hunting Creek 69

Road from John Hurst's plantation above the Falls Church to Tolston mill upon Difficult 81

Road from the mouth of John Hurst's lane to the upper part of Talbot's fence 113

Road from the inhabitants of Little River and Goose Creek down to the road by Mr. Hutchison's 7

Road from the Great Falls of Potomac to Alexandria (road from the lower ford over Difficult Creek/Difficult Run via the Great road, near Edward Lanham's/Lannum's field, thence through the lands of William Swink, Joseph Simpson, Gustavus Scott Esqr. [rented by Daniel Jenkins], John Jackson, and Robert Carter, Esq. [rented by Charles Thrift], to the place on Scott's Run where William Shortridge's old mill stood, thence through the lands of Turberville, Scott, George Hunter, John Jackson, across Pimmets Run, through the lands of Richard Conway, Philip Darrell, William Gunnell, Mr. John West, Mr. Sampson Trammell, and Col. James Wren, to the Great road leading from Leesburg to Alexandria, where Col. Wren's store stands, thence by the Great road [Turnpike road] from the store to Alexandria) 121, 124(2), 128, 129, 130, 131(3), 139, 140, 141

Road from Charles Thrift's near the Little Falls road to the new road leading from the Great Falls to Alexandria (road from the Little Falls road to the Great Falls road, running through Charles Thrift's land/Robert Carter's land rented by Thrift to the Great Falls road at John Jackson's land; alteration of the road running through John Turberville's land) 131, 140

Road from the Great Falls of Potomac to Alexandria (road from the lower ford over Difficult Creek/Difficult Run via the Great road, near Edward Lanham's/Lannum's field, thence through the lands of William Swink, Joseph Simpson, Gustavus Scott Esqr. [rented by Daniel Jenkins], John Jackson, and Robert Carter, Esq. [rented by Charles Thrift], to the place on Scott's Run where William Shortridge's old mill stood, thence through the lands of Turberville, Scott, George Hunter, John Jackson, across Pimmets Run, through the lands of Richard Conway, Philip Darrell, William Gunnell, Mr. John West, Mr. Sampson Trammell, and Col. James Wren, to the Great road leading from Leesburg to Alexandria, where Col. Wren's store stands, thence by the Great road [Turnpike road] from the store to Alexandria) 121, 124(2), 128, 129, 130, 131(3), 139, 140, 141

Road from Sampson Turley's plantation to the Mountain road (road from Sampson Turley's plantation on Johnimore into the Mountain road) 27, 89

Road by Carpenter's Spring and Mr. George Johnston's plantation 72(2)

Road from the fork of the road below William Johnston's to Dogue Run 75

Road from Pohick Church to the Ox road at Keen's hill 83, 87

Road from Alexander Henderson, Gent.'s plantation on Bull Run into the main road (road from Henderson's plantation through/adjoining the lands of Yelverton Reardon, William Moon, Marmaduke Beckwith, James Keen, Sampson Turley, William Simpson, and Peter Ryley to the main road) 104, 107, 108

Road from Thomas Kelley's road below William Ellzey's plantation to Segalon Run, across the creek at Dawson's ford and continuing to Nicholas Minor's 41

Road from Yates's ford on Bull Run, to Songster's (through the lands of Hector Kincheloe and Jesse Kincheloe the heirs of Daniel Kincheloe, and Robert Wickliffe) 173, 174

Road from the Mountain road by Col. Colvill's Tuskarora quarter to the ford of Goose Creek by Samuel King's 11

Road from the landing opposite Charles Shoemaker and Joseph Kirkbride's mill upon Occoquan River to the town of Alexandria 141

Road from the fork of the old road near Benjamin Talbott's to the Church road near Lewis Sanders's (road from the forks of the old Court House road near Benjamin Talbott's to the Church road near Lewis Sanders's, for people to go to the new Church built by Mr. Edward Payne and to the mill, also a nearer and most convenient way to Colchester and Dumfries; road to go through the plantations of Benjamin Talbott, William Kitchen, Edward Washington and John Cotton); (also proposed was an alternate,

and more difficult route through/near the plantations of Benoni Hally, William Kitchen, Mrs. Mary Ferguson, Edward Washington and John Cotton) 86, 87, 88

Road from the new road between Col. Tayloe's quarter and James Lane's, and thence on the lower side of Lane's plantation to the Mountain road a little above Capt. Newton's quarter on Great Rockey Run 13

Road from the Great Falls of Potomac to Alexandria (road from the lower ford over Difficult Creek/Difficult Run via the Great road, near Edward Lanham's/Lannum's field, thence through the lands of William Swink, Joseph Simpson, Gustavus Scott Esqr. [rented by Daniel Jenkins], John Jackson, and Robert Carter, Esq. [rented by Charles Thrift], to the place on Scott's Run where William Shortridge's old mill stood, thence through the lands of Turberville, Scott, George Hunter, John Jackson, across Pimmets Run, through the lands of Richard Conway, Philip Darrell, William Gunnell, Mr. John West, Mr. Sampson Trammell, and Col. James Wren, to the Great road leading from Leesburg to Alexandria, where Col. Wren's store stands, thence by the Great road [Turnpike road] from the store to Alexandria) 121, 124(2), 128, 129, 130, 131(3), 139, 140, 141

Road from Edward Blackburn's to Charles Broadwater's mill on Difficult Run to intersect with the old Lawyer's road (road from Blackburn's fence through Walter Garrett's plantation and through the land of Col. Thomas Ludwell Lee to the mill) 94, 95

Extension of the Parish line from its present termination at Col. Broadwater's mill dam, along the Lawyer's road until it intersects the Ox road, and thence along the Ox road to Frying Pan Run and to the present County line 169

Road from the cross roads at Hodgkins's to the mouth of Beard's lane, on Lee's land 173

Road from the main road at or near Price's to Hepburn and Dundas's mill on Back Lick Run and thence into the main road leading to Alexandria (road beginning at the dividing line between the lands of William Fitzhugh and the heirs of Henry Fitzhugh on the road to Newgate, thence to the Colchester road and through/adjoining the dividing line of John and Barbara Ratcliff, the lands of Barbara Ratcliff and the land of the heirs of James Murray to Hepburn and Dundas's mill, and thence through Hepburn and Dundas's land, and the lands of David and Samuel Arell, Minor, Charles Lee Esq., to William Bird's fence on the Newgate road, and along the said road to Alexandria) 126, 127

Road from the old Court House to the Falls warehouse (road coming from Bull Run through the Court House corn field, Mr. Scott's land and John Urton's land, into the old Falls road at the head of Sasse Branch, thence along the old roads by Thomas Ashbury's, through the Honourable Philip Ludwell Lee's plantation, and to the Falls) 48, 49

Road from the Honourable Thomas Lee's quarter to the Church 1

Road from Edward Blackburn's to Charles Broadwater's mill on Difficult Run to intersect with the old Lawyer's road (road from Blackburn's fence through Walter Garrett's plantation and through the land of Col. Thomas Ludwell Lee to the mill) 94, 95

Road from Ricketts's to the Leesburg road 115, 136

Road from the Great Falls of Potomac to Alexandria (road from the lower ford over Difficult Creek/Difficult Run via the Great road, near Edward Lanham's/Lannum's field, thence through the lands of William Swink, Joseph Simpson, Gustavus Scott Esqr. [rented by Daniel Jenkins], John Jackson, and Robert Carter, Esq. [rented by Charles Thrift], to the place on Scott's Run where William Shortridge's old mill stood, thence through the lands of Turberville, Scott, George Hunter, John Jackson, across Pimmets Run, through the lands of Richard Conway, Philip Darrell, William Gunnell, Mr. John West, Mr. Sampson Trammell, and Col. James Wren, to the Great road leading from Leesburg to Alexandria, where Col. Wren's store stands, thence by the Great road [Turnpike road] from the store to Alexandria) 121, 124(2), 128, 129, 130, 131(3), 139, 140, 141

Road from Marle's to the Turnpike road leading to Leesburg (road from Marl's to the Leesburg road) 150, 152, 162

Road from (John?) Hunter's fence to Collins's fence (road to begin at William Deneale's corner on Wolf Trap (Run) and thence with Lewis's line and thence along the line of Deneale and Fairfax to their corner near Collins's fence) 118(2)

Road from Daniel Lewis's mill to the Turnpike road, at the Mine Branch 171

Road from (Peter?) Lewis's mill to the County line at the Blue Ridge 20

Road from Davis's Branch to Peter Lewis's mill 20

Road from Benjamin Talbott's to Mr. Thomas Lewis's mill on Difficult Run (road from Benjamin Talbott's plantation thence along the Main road to the end of Samuel Talbott's fence thence through Mr. John Waller's plantation, and along the old mill path to Mr. Thomas Lewis's mill on Difficult Run) 94, 95

Road from Vincent Lewis's road to William West's house 3

Road from Vincent Lewis's road to Col. Tayloe's quarter 3

Road from Sugarland Run to Goose Creek/Goose Creek ferry and from Goose Creek to Limestone Run (including digging down the banks at Goose Creek ferry) and parts thereof 3, 6, 8, 24

Road from the Great Limestone Spring to Poultney's ferry 7

Road from Shanadore road to the Great Limestone 27

Road from the Ox road to Hollis's old field (by Samuel Littlejohn's fence) 58

Road from Samuel Littlejohn's to Occoquan new road 84, 85

Road from the inhabitants of Little River and Goose Creek down to the road by Mr. Hutchison's 7

Road from Little River via Goose Creek to Ashby's Gap on the Blue Ridge 10

Road from Little River to the Beaver Dam fork of Goose Creek 26

Road from the fork of the roads above Netherton's to Little River 26

Road from the Piney Branch to Little River 27

Road between Little River and Goose Creek 29

Road from Little River to the widow Adams's at Goose Creek 32

Road from Mark Canton's near Little River to Peyton's mill path 35

NOTE: Loofburrow and variant spellings are combined

Main roads from Colchester to Loudoun County and the Ox Road from Loofburrow's to Difficult 101

(Main) road from the Loudoun County line to Loofburrow's/Loofborough's 103(2), 114

Road from Loofburrow's to Colchester 93, 103

Road from the County line on the Newgate road to David Loofbourrow's 138

Road from the (upper part of) Wolf Pit hill to Loofborough's/David Loofbourrow's 116, 139

Road from the Loudoun County line to Popeshead Run 92

Road from the Loudoun County line to Thomas Sangster's (and fork road at Thomas Sangster's leading to Colchester) 98

Road from the Loudoun County line to Thomas Sangster's 98, 100

Main roads from Colchester to Loudoun County and the Ox Road from Loofburrow's to Difficult 101

(Main) road from the Loudoun County line to Loofburrow's/Loofborough's 103(2), 114

Road from Benjamin Talbott's plantation to the Loudoun County line at Amos Fox's mill on Difficult Run 112(2)

Road from the Ox road to the Loudoun County line 116

Road from Samuel Love's mill on Rocky Run, into the Turnpike road 169

Road from Georgetown ferry beginning at Lubber Branch and extending to Dowdall's corner on the hill by Four Mile Run (running on the line between Thomas Darne and Samuel Shreve and then straight until it intersects the Georgetown road below the Alexandria road) 141, 142

Road from the crossroad where Ricketts lives to the Maddams branch 95

Road from Col. Wren's tavern up the Great Falls road to Madden's shop 163

Road from Madden's shop to Difficult Run 164

Road from Madden's shop to the Little Falls warehouse 164

Road from near Mrs. Manley's plantation to Posey's ferry 73, 74

Road from the entrance of Avory's road near Marle's house to the Turnpike road leading to the Falls Church 132

Road from Marle's to the Turnpike road leading to Leesburg (road from Marl's to the Leesburg road) 150, 152, 162

Road from Four Mile Run to Mason's ferry opposite to Rock Creek 18

Road from Colchester to Boggess's mill on Pohick (road to leave the main road from Colchester to Alexandria above Giles Run and go through Mr. Bayly's land and George Mason's land above the plantation where Joseph Garberry lives then to cross the Church road at the Widow Atcheson's plantation and through her plantation to the mill) 102

The ferry road (George Mason's?) 159

Road at George Mason, Jr.'s mill 109

Main road over the mill race at George Mason Jr.'s mill on Pohick 109

Road from the old Court House (via fence and line of John Mason's fence and Mr. Scott's line to the Falls road) to the Falls warehouse 72, 74

Road from Vernon's mill on Pohick to the stand or meeting house 123

Road from the mile tree to the Church 116

Road from the forks of the road at Cameron hill to the mile tree 116

Road from the great road by Daniel Mills's to William Payne's old mill on Accotink Run 113(2)

Road from Daniel Lewis's mill to the Turnpike road, at the Mine Branch 171

Road from the main road at or near Price's to Hepburn and Dundas's mill on Back Lick Run and thence into the main road leading to Alexandria (road beginning at the dividing line between the lands of William Fitzhugh and the heirs of Henry Fitzhugh on the road to Newgate, thence to the Colchester road and through/adjoining the dividing line of John and Barbara Ratcliff, the lands of Barbara Ratcliff and the land of the heirs of James Murray to Hepburn and Dundas's mill, and thence through Hepburn and Dundas's land, and the lands of David and Samuel Arell, Minor, Charles Lee Esq., to William Bird's fence on the Newgate road, and along the said road to Alexandria) 126, 127

Road from Nicholas Minor's to the road to Noland's ferry 27

Road from Thomas Kelley's road below William Ellzey's plantation to Segalon Run, across the creek at Dawson's ford and continuing to Nicholas Minor's 41

Road from the old Court House road near John Monroe's, to intersect with the road to Alexandria near Balty Stearns's 62, 63

Road from Edward Blackburn's to the Church road near Mr. Benjamin Moody's 100

Road from the Turnpike gate down the Colchester road to Dogue Run at Mrs. Moon's 136

Road from Dogue Run at Mrs. Moon's to Accotink 137

Road from Alexander Henderson, Gent.'s plantation on Bull Run into the main road (road from Henderson's plantation through/adjoining the lands of Yelverton Reardon, William Moon, Marmaduke Beckwith, James Keen, Sampson Turley, William Simpson, and Peter Ryley to the main road) 104, 107, 108

(Alteration of) the road from Jesse Moore's blacksmith shop to Alexandria 157

Road from the Mountain road by Col. Colvill's Tuskarora quarter to the ford of Goose Creek by Samuel King's 11

Road from the Mountain road by Martha Hall's to the Church 11

Mountain road (portion between Richard Brown's and Little Rockey Run) 13, 37

Road from the new road between Col. Tayloe's quarter and James Lane's, and thence on the lower side of Lane's plantation to the Mountain road a little above Capt. Newton's quarter on Great Rockey Run 13

Road from Sumers's/John Summers's to the Mountain road near Rocky Run Chapel 22, 25

Road from Popeshead on the new road to the Mountain road 25

Road from the post in the Mountain road to Piney Branch 26

Road from Sampson Turley's plantation to the Mountain road (road from Sampson Turley's plantation on Johnimore into the Mountain road) 27, 89

Road from the Mountain road to Sampson Turley's and a road out of that to the mouth of Popeshead 27

Road from the old Court House to the Mountain road 32, 33

The Mountain road 36

Fork of the Ox road and the Mountain road 52, 53

Road from the fork of the road that leads to Prince William into the Mountain road 61, 64

Road from Thomas Sangster's to Occoquan (leading out from the Mountain road) 86, 98, 105, 107

Road from John Pinkstone's to the Mountain road 88

Road from Thomas Sangster's to Frying Pan (leading out from the Mountain road) 98

Road from Cameron to the Mountain road 101, 103

Road from the forks of the road below Sangster's to the Mountain road by Hally's 114

Road from Bull Run at Col. John Hooe's old mill to the Mountain road near Thomas Sangster's/Songster's (road from the Mountain road to Bull Run near Hooe's mill) 117, 138, 145, 164

Road from Avory's road to the Mountain road that leads to the Falls landing 132

Road from the County line on the <u>Mountain</u> road, to Thomas Sangster's 145

Road from the main road at or near Price's to Hepburn and Dundas's mill on Back Lick Run and thence into the main road leading to Alexandria (road beginning at the dividing line between the lands of William Fitzhugh and the heirs of Henry Fitzhugh on the road to Newgate, thence to the Colchester road and through/adjoining the dividing line of John and Barbara Ratcliff, the lands of Barbara Ratcliff and the land of the heirs of James <u>Murray</u> to Hepburn and Dundas's mill, and thence through Hepburn and Dundas's land, and the lands of David and Samuel Arell, Minor, Charles Lee Esq., to William Bird's fence on the Newgate road, and along the said road to Alexandria) 126, 127

Road from the fork of the roads above <u>Netherton's</u> to Little River 26

Road from the main road at or near Price's to Hepburn and Dundas's mill on Back Lick Run and thence into the main road leading to Alexandria (road beginning at the dividing line between the lands of William Fitzhugh and the heirs of Henry Fitzhugh on the road to <u>Newgate</u>, thence to the Colchester road and through/adjoining the dividing line of John and Barbara Ratcliff, the lands of Barbara Ratcliff and the land of the heirs of James Murray to Hepburn and Dundas's mill, and thence through Hepburn and Dundas's land, and the lands of David and Samuel Arell, Minor, Charles Lee Esq., to William Bird's fence on the <u>Newgate</u> road, and along the said road to Alexandria) 126, 127

Turnpike road from Alexandria to <u>Newgate</u> 126

Road from the County line on the <u>Newgate</u> road to David Loofbourrow's 138

Sections of the Turnpike road:
- from the County line on the Turnpike road leading from <u>Newgate</u> to Alexandria as far as the Ox road 160
- from the Ox road, along said Turnpike road to the fork of the road at Hollice's old field near Price Skinner's 160
- from Hollice's old field in the fork of the road Price Skinner's to Price's tavern 160
- from Price's tavern to the forks of the road above Summers's 160

Old Court House road from the mouth of Popeshead Run to the corner of John Gibson's fence at the road leading to the <u>Newgate</u> road 168, 169

Road from the new road between Col. Tayloe's quarter and James Lane's, and thence on the lower side of Lane's plantation to the Mountain road a little above Capt. <u>Newton's</u> quarter on Great Rockey Run 13

Road from Nicholas Minor's to the road to <u>Noland's</u> ferry 27

Road from Pohick/Pohick Run to <u>Occoquan</u> ferry 2, 19, 84, 117

Road from the bridge at Grayson's mill to Occoquan ferry 6

Road from the Ox road to Occoquan ferry 28(2)

Road from Occoquan ferry road to the Ox road 36

Bridle way from Colchester to Occoquan warehouse 48, 49

Road from Giles Run to Occoquan warehouse 59

Road from Samuel Littlejohn's to Occoquan new road 84, 85

Road from Thomas Sangster's/Songster's to the Wolf Shoals/Wolf Run Shoals (on Bull Run/Occoquan) 85, 116, 139, 145, 164

Road from Thomas Sangster's to Occoquan (leading out from the Mountain road) 86, 98, 105, 107

Road from Occoquan above the Wolf Run Shoals to Alexander Henderson's mill 114

Road from the landing opposite Charles Shoemaker and Joseph Kirkbride's mill upon Occoquan River to the town of Alexandria 141

Road from Alexandria to John Hooe's ferry on Occoquan, near the old warehouse landing 152(2), 153(2)

Road from John Hooe's ferry on Occoquan to the lands of Samuel Bayly 153

Road from the Ox road at the Wood Cutting to Occoquan and the County line leading to the mill 167

Road from the Ellicot and Co. bridge on Occoquan into the road leading from Washington's shop to Robert Boggess's mill and from thence to the intersection with the Alexandria Post road 169

Road from the neighborhood of Occoquan Forrest crossing Accotink at John Ward's mill and down to Alexandria 143

Road between Ravensworth and the Ox Road 1

Road from the Prince William road to the Ox Road 1

Prince William road between the Ox Road and where it comes into the road to the Court House 1

Road from the Court House/old Court House to the Ox road 9, 57, 105, 107, 110, 114

Road from the Court House to the Ox road near Capt. Lewis Ellzey's 11

Road from the Ox road to Occoquan ferry 28(2)

Ox road 30(2), 41

Road from Occoquan ferry road to the Ox road 36

Road from the Ox road to Colchester 48, 67, 68

Road from the Ox road to Mr. John Balendine's mill 49

Road from the Ox road to John Ballendine's ford 50

Road from the Ox road to Colchester (alteration made by going through John Ford's plantation, along an old mill path, and into the Ox road at the Wolf Pit hill) 50

Fork of the Ox road and the Mountain road 52, 53

Road from the Ox road to Bull Run 58

(Second) Road from the Ox road to Bull Run 58

Road from the Ox road to Hollis's old field (by Samuel Littlejohn's fence) 58

Road from the Falls Church to the Ox road 60

Roads from the Falls Church to Maj. Broadwater's Spring Branch and from Broadwater's Spring Branch to the Ox road 66

Road from William Garner's/Gardner's to Ravensworth road (from the Ox road via the pond and Steptoe's quarter) 72, 73

Roads from William Gardner's to the place where the road to Colchester leaves the Ox road and from that place to Colchester warehouse 74

Road from the ford of Piney Branch into the Ox road above the Pohick road 76, 77

Road from Pohick Church to the Ox road at Keen's hill 83, 87

Road from the Ponds to the Ox road 87, 115

Road from Colchester up the Ox road (angle of the road up the Ox road) 98

Road from the Ox road to the Forge (angle of the road leading to the Forge and

Colchester) 98

Ox road where it intersects at Thomas Sangster's (intersection leading from Colchester to the [Blue Ridge?] Mountains) 98

Road from the Alexandria road to the Ox road near the Piney Branch 100

Main roads from Colchester to Loudoun County and the Ox road from Loofburrow's to Difficult 101

Road from Accotink to the Ox road 115

Road from Accotink above Price's to the Ox road 116

Road from the Ox road to the Loudoun County line 116

Road from Alexander Henderson's plantation on Bull Run by Sampson Turley's to the Ox road 117

Road from Piney Branch into the Ox road near the Church 121, 122

Road from the Falls Church by Ricketts's to the Ox road 134

Road from the old Court House road opposite Joseph Bennett's plantation, to where it intersects the Ox road 135

Road from Difficult bridge/Difficult Run on the Ox road, to Songster's/Thomas Sangster's/Songster's shop 135, 145, 166

Road from the Ox road near Sangster's/Thomas Songster's to Hallace's/Hollies'/Hollis's/Hollice's old field (at the Turnpike road) 138, 147, 157, 167

Road from the Ox road at Charles Beach's/Beache's/Beeche's (fence) to the Ponds on the Pohick road 138, 147, 167

Road from the Ox road above William Simpson's to where it intersects the road from the Wolf Run Shoals to Thomas Songster's by Sampson Turley's plantation 139

Road from the County line on the Turnpike road leading to Alexandria and to the Ox road 144

Road from the Ox road on the Turnpike road to Hollis's old fields, opposite to where another road intersects the Turnpike road leading from the Ox road 144

Road from the Turnpike road to the Ox road, near the Brick Church 145

Road from the Ox road above William Simpson's to where it intersects the road from Sangster's to Bull Run 145

Road from Sangster's/Thomas Songster's shop, on the Ox road, to the White Oak Harbor 146, 166

Road from the White Oak Harbor/old White Oak Harbor, on the Ox road to Colchester (at the old sign post) 146, 166

Road from the Ox road to the Parish line on Ellzey's Church road 148, 154, 167

(Alteration of) the Ox road near William Simpson's house and running through his land 155(2)

(Alteration of) Ox road from the corner between Simpson Halley and Richard Ratcliffe to the Alexandria road 157(2)

Sections of the Turnpike road:
- from the County line on the Turnpike road leading from Newgate to Alexandria as far as the Ox road 160
- from the Ox road, along said Turnpike road to the fork of the road at Hollice's old field near Price Skinner's 160
- from Hollice's old field in the fork of the road Price Skinner's to Price's tavern 160
- from Price's tavern to the forks of the road above Summers's 160

Road from the Ox road at the Wood Cutting to Occoquan and the County line leading to the mill 167

Extension of the Parish line from its present termination at Col. Broadwater's mill dam, along the Lawyer's road until it intersects the Ox road, and thence along the Ox road to Frying Pan Run and to the present County line 169

Road from Accotink, on the Alexandria road, to the Parish line at Dogue Run 146

Road from William Turner's tavern to the Parish line on the old Court House road 147, 158, 167

Road from the General's (George Washington's) mill by his mill dam and along his new road, to the Parish line 148

Road from the Ox road to the Parish line on Ellzey's Church road 148, 154, 167

Road from the crossroads at Cockerill's old place along Ellzey's Church road to the Parish line 163

Road from the old Court House to the Parish line 164

Extension of the Parish line from its present termination at Col. Broadwater's mill dam, along the Lawyer's road until it intersects the Ox road, and thence along the Ox road to Frying Pan Run and to the present County line 169

Road from the forks of the road below Patterson's to Dogue Run 117

Road from the fork of the old road near Benjamin Talbott's to the Church road near Lewis Sanders's (road from the forks of the old Court House road near Benjamin Talbott's to the Church road near Lewis Sanders's, for people to go to the new Church built by Mr. Edward Payne and to the mill, also a nearer and most convenient way to Colchester and Dumfries; road to go through the plantations of Benjamin Talbott, William Kitchen, Edward Washington and John Cotton); (also proposed was an alternate, and more difficult route through/near the plantations of Benoni Hally, William Kitchen, Mrs. Mary Ferguson, Edward Washington and John Cotton) 86, 87, 88

Road from the great road by Daniel Mills's to William Payne's old mill on Accotink Run 113(2)

Road from Ellzey's Church road opposite Richard Ratcliffe's land, passing by William Payne's mill, and from thence to the Alexandria road 168, 169

Road from the Back Lick Run to Falls Church (road from the Back Lick to the Church, going through Simon Pearson's plantation and round Edmund Butler's fence); (including intersection at Price's, leading from thence to the Church) 53, 57(2), 88, 99(2), 100

Road from Mark Canton's near Little River to Peyton's mill path 35

Road by Garrard Tramel's/Tramell's plantation (by Pimmets Run, and the orchard of Templeman's plantation, and to Mr. Broadwater's plantation) 25, 26

Road from Pimmets Run to Nicholas Garrett's plantation 65

Road from Mr. William Adams's to Pimmets Run 82, 84, 92

Road from Mr. William Adams's/William Adams's Gent. to Pimmets Run (including fork[?] road from the Alexandria road to the ferry at Georgetown) 96, 97, 100

Road from the Red house to Samuel Adams's mill on Pimmets Run, and from thence to the road leading from the Little Falls to Alexandria (road beginning near the Red house, thence to Pimmets Run and the mill, thence through Samuel Adams's plantation to the long branch of Pimmets Run, along John Butcher's fence to Mathew Earp's, and across the Church road to the Alexandria road) 118, 119

Road from the Great Falls of Potomac to Alexandria (road from the lower ford over Difficult Creek/Difficult Run via the Great road, near Edward Lanham's/Lannum's field, thence through the lands of William Swink, Joseph Simpson, Gustavus Scott Esqr. [rented by Daniel Jenkins], John Jackson, and Robert Carter, Esq. [rented by Charles Thrift], to the place on Scott's Run where William Shortridge's old mill stood, thence through the lands of Turberville, Scott, George Hunter, John Jackson, across Pimmets Run, through the lands of Richard Conway, Philip Darrell, William Gunnell, Mr. John West, Mr. Sampson Trammell, and Col. James Wren, to the Great road leading from Leesburg to Alexandria, where Col. Wren's store stands, thence by the Great road [Turnpike road] from the store to Alexandria) 121, 124(2), 128, 129, 130, 131(3), 139, 140, 141

NOTE: Entries for Piney Branch and Piney Run are combined

Road from the post in the Mountain road to Piney Branch 26

Road from the Piney Branch to Little River 27

Road from the ford of Piney Branch into the Ox road above the Pohick road 76, 77

Road from the Alexandria road to the Ox road near the Piney Branch 100

Road from Piney Run to the Main road 116

Road from Piney Branch into the Ox road near the Church 121, 122

Road from Pinny (Piney) Branch to the new Church 136

Road from John Pinkstone's to the Mountain road 88

Road from Herryford's (Hereford's) ferry to the Church of Pohick and into the main road by William Godfrey's 9

Road from Pohick Church to Hereford's ferry 46, 109

Road from Pohick/Pohick Church to Colchester 60, 62, 101, 104

Road from Pohick Church to the Ox road at Keen's hill 83, 87

Road from Court's ferry to Pohick Church 84

Road from the Wolf Pit hill to Pohick Church 106

Road from Pohick Church to the Ponds (road from the Ponds on the Pohick road, to Pohick Church) 114, 138, 147

Road from Pohick new Church to intersect the road from Robert Boggess's to Gen. (George) Washington's mill (road from Robert Boggess's to the General's [George Washington's] mill, and from that road to Pohick Church) 137, 148

The road from Pohick old Church (old Pohick Church) to Holland Point ferry 137, 148

Roads from Accotink to Pohick:
- Road from Pohick to Accotink and crossroad, thence the road to the warehouse from the crossroad (and parts thereof) 2, 14
- Road from Accotink to Pohick Run (including the fork of the road at Boggess's and the fork to Mr. Barnes's) 32
- Roads from Pohick/Pohick Run to Accotink 62, 74, 93, 117
- Road from Pohick to Accotink in the Back road 85
- River side road from Pohick to Accotink 105
- Sections of the Colchester road: from the Turnpike gate to Dogue Run; from Dogue Run to Accotink; from Accotink to Pohick Run; from Pohick Run to Colchester 165

Road from Pohick/Pohick Run to Occoquan ferry 2, 19, 84, 117

Road from Edward Violet's/Violett's to Pohick warehouse 15, 18, 43

Road near Pohick (near Col. Grayson's mill dam) 23

Road from the ford of Piney Branch into the Ox road above the Pohick road 76, 77

Road from Colchester to Boggess's mill on Pohick (road to leave the main road from Colchester to Alexandria above Giles Run and go through Mr. Bayly's land and George Mason's land above the plantation where Joseph Garberry lives then to cross the Church road at the Widow Atcheson's plantation and through her plantation to the mill) 102

Road from where John Hollis formerly lived to Pohick warehouse 106

Main road over the mill race at George Mason Jr.'s mill on Pohick 109

Road from Vernon's Mill on Pohick to the stand or meeting house 123

Road from the Turnpike road at Hallace's/Hollice's/Hollis's old field to where it intersects the Pohick road at John Fowler's plantation/fence 138, 147, 167

Road from the Ox road at Charles Beach's/Beache's/Beeche's (fence) to the Ponds on the Pohick road 138, 147, 167

Road from the head of Pohick Creek to the road leading from Alexandria to the Colchester road 159(2)

Road from the Ponds on the Pohick road, crossing the Stage road to the intersection of the road leading from Boggess's to Gen. Washington's mill 167

Road through Thomas Pollard's plantation (Road between Joseph Pollard's and Mr. Beckwith's plantations) 71, 73

Road from William Garner's/Gardner's to Ravensworth road (from the Ox road via the pond and Steptoe's quarter) 72, 73

Road from the Ponds to the Ox road 87, 115

Road from the Ponds (on the road from Cameron to the Falls Church) to Alexandria 112

Road from Pohick Church to the Ponds (road from the Ponds on the Pohick road, to Pohick Church) 114, 138, 147

Road from the Ox road at Charles Beach's/Beache's/Beeche's (fence) to the Ponds on the Pohick road 138, 147, 167

Road from the Ponds on the Pohick road, crossing the Stage road to the intersection of the road leading from Boggess's to Gen. Washington's mill 167

Road from Popeshead on the new road to the Mountain road 25

Road from Accotink to Popeshead 25, 87, 88

Road from the Mountain road to Sampson Turley's and a road out of that to the mouth of Popeshead 27

Road from Popeshead to the County line 58

Road from the Loudoun County line to Popeshead Run 92

Old Court House road from the mouth of Popeshead Run to the corner of John Gibson's fence at the road leading to the Newgate road 168, 169

Road from Posey's ferry to the main road 30

Road from Dogue Run to the Gum Spring, and from the spring to Posey's ferry 32, 33, 59, 117

Road from Dogue Run to (John) Posey's/Pozey's ferry landing and from thence to Little Hunting Creek 39, 44

Road from near Mrs. Manley's plantation to Posey's ferry 73, 74

Road from the lower part of the island on Potomac River belonging to John Trammel to the place where he is erecting his house 37, 40

Road from the Great Limestone Spring to Poultney's ferry 7

Road from the Back Lick Run to Falls Church (road from the Back Lick to the Church, going through Simon Pearson's plantation and round Edmund Butler's fence); (including intersection at Price's, leading from thence to the Church) 53, 57(2), 88, 99(2), 100

Road from Accotink to the school house near John Summers's (with intersection of the road at Price's leading from Accotink to Alexandria) 99

Road from Accotink above Price's to the Ox road 116

Road from Price's to Accotink 116

Road from the forks of the road above Summers's to Price's/Price's tavern 116, 149, 160

Road from the main road at or near Price's to Hepburn and Dundas's mill on Back Lick Run and thence into the main road leading to Alexandria (road beginning at the dividing line between the lands of William Fitzhugh and the heirs of Henry Fitzhugh on the road to Newgate, thence to the Colchester road and through/adjoining the dividing line of John and Barbara Ratcliff, the lands of Barbara Ratcliff and the land of the heirs of James Murray to Hepburn and Dundas's mill, and thence through Hepburn and Dundas's land, and the lands of David and Samuel Arell, Minor, Charles Lee Esq., to William Bird's fence on the Newgate road, and along the said road to Alexandria) 126, 127

Road from the White Oak Spring to Price's tavern 135

Road from Price's tavern to Holmes Run 135

Road from Accotink on the Turnpike road to Price's tavern 144

Road from the Alexandria road at John Hereford's lane, to Price's tavern on the old Court House road 147

Back Lick road from Price's ordinary to the fork of the road above Nicholas Fitzhugh's gate 151

Road from Price's ordinary to the forks of the road above Nicholas Fitzhugh's plantation 154

Sections of the Turnpike road:

- from the County line on the Turnpike road leading from Newgate to Alexandria as far as the Ox road 160
- from the Ox road, along said Turnpike road to the fork of the road at Hollice's old field near Price Skinner's 160
- from Hollice's old field in the fork of the road Price Skinner's to Price's tavern 160
- from Price's tavern to the forks of the road above Summers's 160

Colchester road (portion from Samuel Smith's lane, through Benoni Price's plantation) 155

Road from the Prince William road to the Ox Road 1

Prince William road between the Ox Road and where it comes into the road to the Court House 1

Road from the mountain to the road cleared by Prince William County 22

Road from the fork of the road that leads to Prince William into the Mountain road 61, 64

Road from Alexandria to Difficult bridge near Capt. W. H. Terrett's (road from Princess Street passing near Baldwin Dade's old tobacco house, and Ramsay's quarter, and to a white oak in the road above Capt. W. H. Terrett's old field) 119, 120

Road from the main road at or near Price's to Hepburn and Dundas's mill on Back Lick Run and thence into the main road leading to Alexandria (road beginning at the dividing line between the lands of William Fitzhugh and the heirs of Henry Fitzhugh on the road to Newgate, thence to the Colchester road and through/adjoining the dividing line of John and Barbara Ratcliff, the lands of Barbara Ratcliff and the land of the heirs of James Murray to Hepburn and Dundas's mill, and thence through Hepburn and Dundas's land, and the lands of David and Samuel Arell, Minor, Charles Lee Esq., to William Bird's fence on the Newgate road, and along the said road to Alexandria) 126, 127

Road from the old Court House to the fork of the road by John Ratcliff's plantation 55, 56

(Alteration of) Ox road from the corner between Simpson Halley and Richard Ratcliffe to the Alexandria road 157(2)

Road from Ellzey's Church road opposite Richard Ratcliffe's land, passing by William Payne's mill, and from thence to the Alexandria road 168, 169

Road between Ravensworth and the Ox Road 1

Road from Fitzhugh's quarter on Ravensworth to Robert Boggess's 12

Ravensworth road 30

Road from William Garner's/Gardner's to Ravensworth road (from the Ox road via the pond and Steptoe's quarter) 72, 73

Road out of the old Court House road near Benoni Halley's to the Ravensworth road near John Hollis's 72

Road cleared by William Reardon 3, 18

Road by William Reardon's plantation 19

Road from Alexander Henderson, Gent.'s plantation on Bull Run into the main road (road from Henderson's plantation through/adjoining the lands of Yelverton Reardon, William Moon, Marmaduke Beckwith, James Keen, Sampson Turley, William Simpson, and Peter Ryley to the main road) 104, 107, 108

Road from the Falls warehouse to the Red house 103

Road from the Red house to the Falls landing 114, 132

Road from the forks at Difficult to the Red house 115

Road from the Red house to Samuel Adams's mill on Pimmets Run, and from thence to the road leading from the Little Falls to Alexandria (road beginning near the Red house, thence to Pimmets Run and the mill, thence through Samuel Adams's plantation to the long branch of Pimmets Run, along John Butcher's fence to Mathew Earp's, and across the Church road to the Alexandria road) 118, 119

Road from the Red house to the Turnpike road 132

Road from the old Court House to the Red house 156

Road through James Richards's plantation to Grymes's mill (Grymes's mill road)153, 174

Road from the crossroad where Ricketts lives to the Maddams branch 95

Road from the forks of the road above Dickens's to Ricketts's 115

Road from Ricketts's to the Leesburg road 115, 136

Road from Ricketts's to Accotink 115

Road from Ricketts's to the Alexandria road 115

Road from the Falls Church by Ricketts's to the Ox road 134

Road from Ricketts's to Frazer's 136

Road from the Sycamore near the Little Falls warehouse to the river 158

Forks of the Back and River side roads on the road from Hunting Creek to the Gum Spring 97

Road dividing at Boggess's (into the River side and Back roads to Alexandria) 97

River side road from Pohick to Accotink 105

Road from the lower side of Accotink Run on the River side road to Dogue Run 117

Road from the new Church to the River side road 117

Road from Cameron Run or Creek by Clish to where General Washington's new road intersects the River side old road 137

Road from General (George) Washington's ferry to his mill, from thence to his tumbling dam, thence along his new road to intersect the riverside old road above the Gum Spring (and parts thereof: includes road from the Gum Spring to General Washington's mill) 137, 166

Road from Four Mile Run to Mason's ferry opposite to Rock Creek 18

Road from the main road to Rock Creek 49

Road from the Falls Church to Rock Creek 63

Road from Four Mile Run to Rock Creek/Rock Creek ferry 69, 115

Road from the forks of the road at the Falls Church to the old road from Rock Creek to Alexandria 100

Road from the new road to Rockey Run Church 8

Road from Alexandria to Rockey Run Chapel 17, 18

Road from Sumers's/John Summers's to the Mountain road near Rocky Run Chapel 22, 25

Road from Accotink to Rocky Run Chapel 25

Road from Williams's Gap to Rocky Run Chapel 26

Road from the new road between Col. Tayloe's quarter and James Lane's, and thence on the lower side of Lane's plantation to the Mountain road a little above Capt. Newton's quarter on Great Rockey Run 13

Road from the lower side of Little Rockey Run to Cub Run 13

Mountain road (portion between Richard Brown's and Little Rockey Run) 13, 37

Road between Little Rockey Run and Brown's ordinary 42

Road from Samuel Love's mill on Rocky Run, into the Turnpike road 169

Road from the Colchester road to the road to Alexandria (road from the Colchester road near John Ward's mill, thence through William Rogers's plantation to Richard Sanford's fence, and to the road leading to Alexandria opposite to Capt. Harper's fence) 124

Road from Alexander Henderson, Gent.'s plantation on Bull Run into the main road (road from Henderson's plantation through/adjoining the lands of Yelverton Reardon, William Moon, Marmaduke Beckwith, James Keen, Sampson Turley, William Simpson, and Peter Ryley to the main road) 104, 107, 108

NOTE: Entries for Sanders and Saunders are combined

Road by Lewis Sanders Jr.'s plantation 19

Road by Lewis Saunders's plantation 23

Road from the fork of the old road near Benjamin Talbott's to the Church road near Lewis Sanders's (road from the forks of the old Court House road near Benjamin Talbott's to the Church road near Lewis Sanders's, for people to go to the new Church built by Mr. Edward Payne and to the mill, also a nearer and most convenient way to Colchester and Dumfries; road to go through the plantations of Benjamin Talbott, William Kitchen, Edward Washington and John Cotton); (also proposed was an alternate, and more difficult route through/near the plantations of Benoni Hally, William Kitchen, Mrs. Mary Ferguson, Edward Washington and John Cotton) 86, 87, 88

Road from the fork of the old Court House road to the old Church road near Lewis Sanders's/Saunders's 94, 117

Road from the Colchester road to the road to Alexandria (road from the Colchester road near John Ward's mill, thence through William Rogers's plantation to Richard Sanford's fence, and to the road leading to Alexandria opposite to Capt. Harper's fence) 124

NOTE: Entries for Sangster and Songster are combined

Road that leads from Hollis's old field to Sangster's 84, 85

Road from Thomas Sangster's/Songster's to the Wolf Shoals/Wolf Run Shoals (on Bull Run/Occoquan) 85, 116, 139, 145, 164

Road from Thomas Sangster's to Occoquan (leading out from the Mountain road) 86, 98, 105, 107

Road from Thomas Sangster's to Thomas Throgmorton's mill on Bull Run 95

Ox road where it intersects at Thomas Sangster's (intersection leading from Colchester to the [Blue Ridge?] Mountains) 98

Road from Thomas Sangster's to Frying Pan (leading out from the Mountain road) 98

Road from the Loudoun County line to Thomas Sangster's 98, 100

Road from the Loudoun County line to Thomas Sangster's (and fork road at Thomas Sangster's leading to Colchester) 98

Road from the forks of the road below Sangster's to the Mountain road by Hally's 114

Road from Difficult to Sangster's shop/Songster's 114, 124

Road from Bull Run at Col. John Hooe's old mill to the Mountain road near Thomas Sangster's/Songster's (road from the Mountain road to Bull Run near Hooe's mill) 117, 138, 145, 164

Road from Difficult bridge/Difficult Run on the Ox road, to Songster's/Thomas Sangster's/Songster's shop 135, 145, 166

Road from the Ox road near Sangster's/Thomas Songster's to Hallace's/Hollies'/Hollis's/Hollice's old field (at the Turnpike road) 138, 147, 157, 167

Road from the Ox road above William Simpson's to where it intersects the road from the Wolf Run Shoals to Thomas Songster's by Sampson Turley's plantation 139

Road from the County line on the Mountain road, to Thomas Sangster's 145

Road from the Ox road above William Simpson's to where it intersects the road from Sangster's to Bull Run 145

Road from Sangster's/Thomas Songster's shop, on the Ox road, to the White Oak Harbor 146, 166

Road from William Deneale's mill to the road leading from Hollis's old field to Thomas Sangster's 151, 152

Road from the County line at the Turnpike road to Thomas Songster's 166

Road from Yates's ford on Bull Run, to Songster's (through the lands of Hector Kincheloe and Jesse Kincheloe the heirs of Daniel Kincheloe, and Robert Wickliffe) 173, 174

Road from the old Court House to the Falls warehouse (road coming from Bull Run through the Court House corn field, Mr. Scott's land and John Urton's land, into the old Falls road at the head of Sasse Branch, thence along the old roads by Thomas Ashbury's, through the Honourable Philip Ludwell Lee's plantation, and to the Falls) 48, 49

Road from the old school house to the road at Cameron 33

Road from Accotink to the old school house 34, 42, 60

Road from the Beaver Dam of Goose Creek to the school house above Anderson's and from thence to the top of the Blue Ridge 41

Road from William Donaldson's to Summers's old school house 54(2)

Road from the old school house near Francis Summers's to the old Court House (with intersection of the road leading to the old Court House road) 99

Road from the school house near John Summers's to Accotink Run 82

Road from Accotink to the school house near John Summers's (with intersection of the road at Price's leading from Accotink to Alexandria) 99

Road from Scott's Run to Allen Davis's lane 114

Road from the Great Falls of Potomac to Alexandria (road from the lower ford over Difficult Creek/Difficult Run via the Great road, near Edward Lanham's/Lannum's field, thence through the lands of William Swink, Joseph Simpson, Gustavus Scott Esqr. [rented by Daniel Jenkins], John Jackson, and Robert Carter, Esq. [rented by Charles Thrift], to the place on Scott's Run where William Shortridge's old mill stood, thence through the lands of Turberville, Scott, George Hunter, John Jackson, across Pimmets Run, through the lands of Richard Conway, Philip Darrell, William Gunnell, Mr. John West, Mr. Sampson Trammell, and Col. James Wren, to the Great road leading from Leesburg to Alexandria, where Col. Wren's store stands, thence by the Great road

[Turnpike road] from the store to Alexandria) 121, 124(2), 128, 129, 130, 131(3), 139, 140, 141

Road from the old Court House to the Falls warehouse (road coming from Bull Run through the Court House corn field, Mr. Scott's land and John Urton's land, into the old Falls road at the head of Sasse Branch, thence along the old roads by Thomas Ashbury's, through the Honourable Philip Ludwell Lee's plantation, and to the Falls) 48, 49

Road from the old Court House (via fence and line of John Mason's fence and Mr. Scott's line to the Falls road) to the Falls warehouse 72, 74

Road by John Seale's plantation 67, 68

Road from Mr. Sebastian's quarter to Maj. Charles Broadwater's quarter 71, 72, 73

Road from Samuel Adams's mill, to the place where Benjamin Sebastian formerly lived 125

Road from Thomas Kelley's road below William Ellzey's plantation to Segalon Run, across the creek at Dawson's ford and continuing to Nicholas Minor's 41

Road from Shanadore road to the Great Limestone 27

Road from the landing opposite Charles Shoemaker and Joseph Kirkbride's mill upon Occoquan River to the town of Alexandria 141

Road from the old Court House to William Shortridge's 86

Road from the Great Falls of Potomac to Alexandria (road from the lower ford over Difficult Creek/Difficult Run via the Great road, near Edward Lanham's/Lannum's field, thence through the lands of William Swink, Joseph Simpson, Gustavus Scott Esqr. [rented by Daniel Jenkins], John Jackson, and Robert Carter, Esq. [rented by Charles Thrift], to the place on Scott's Run where William Shortridge's old mill stood, thence through the lands of Turberville, Scott, George Hunter, John Jackson, across Pimmets Run, through the lands of Richard Conway, Philip Darrell, William Gunnell, Mr. John West, Mr. Sampson Trammell, and Col. James Wren, to the Great road leading from Leesburg to Alexandria, where Col. Wren's store stands, thence by the Great road [Turnpike road] from the store to Alexandria) 121, 124(2), 128, 129, 130, 131(3), 139, 140, 141

Road from Georgetown ferry beginning at Lubber Branch and extending to Dowdall's corner on the hill by Four Mile Run (running on the line between Thomas Darne and Samuel Shreve and then straight until it intersects the Georgetown road below the Alexandria road) 141, 142

Road from the Great Falls of Potomac to Alexandria (road from the lower ford over Difficult Creek/Difficult Run via the Great road, near Edward Lanham's/Lannum's field, thence through the lands of William Swink, Joseph Simpson, Gustavus Scott Esqr. [rented by Daniel Jenkins], John Jackson, and Robert Carter, Esq. [rented by Charles Thrift], to the place on Scott's Run where William Shortridge's old mill stood, thence through the lands of Turberville, Scott, George Hunter, John Jackson, across Pimmets Run, through the lands of Richard Conway, Philip Darrell, William Gunnell, Mr. John West, Mr. Sampson Trammell, and Col. James Wren, to the Great road leading from Leesburg to Alexandria, where Col. Wren's store stands, thence by the Great road [Turnpike road] from the store to Alexandria) 121, 124(2), 128, 129, 130, 131(3), 139, 140, 141

Road from Alexander Henderson, Gent.'s plantation on Bull Run into the main road (road from Henderson's plantation through/adjoining the lands of Yelverton Reardon, William Moon, Marmaduke Beckwith, James Keen, Sampson Turley, William Simpson, and Peter Ryley to the main road) 104, 107, 108

Road from Holmes Run, to the Turnpike gate, where William Simpson lives 136

Road from the Ox road above William Simpson's to where it intersects the road from the Wolf Run Shoals to Thomas Songster's by Sampson Turley's plantation 139

Road from the Ox road above William Simpson's to where it intersects the road from Sangster's to Bull Run 145

(Alteration of) the Ox road near William Simpson's house and running through his land 155(2)

Sections of the Turnpike road:
- from the County line on the Turnpike road leading from Newgate to Alexandria as far as the Ox road 160
- from the Ox road, along said Turnpike road to the fork of the road at Hollice's old field near Price Skinner's 160
- from Hollice's old field in the fork of the road Price Skinner's to Price's tavern 160
- from Price's tavern to the forks of the road above Summers's 160

Colchester road (portion from Samuel Smith's lane, through Benoni Price's plantation) 155

NOTE: Songster: see Sangster

Road from the Ponds on the Pohick road, crossing the Stage road to the intersection of the road leading from Boggess's to Gen. Washington's mill 167

Road from Vernon's mill on Pohick to the stand or meeting house 123

Road from the old Court House road near John Monroe's, to intersect with the road to Alexandria near Balty Stearns's 62, 63

Road from William Garner's/Gardner's to Ravensworth road (from the Ox road via the pond and Steptoe's quarter) 72, 73

Road through Mr. James Steptoe's land 15

Road from Holmes Run to the Stone bridge below the Turnpike 161

Road from the old storehouse to the forks of the road 110

Road from Hunter's old storehouse to the Little Falls 81

Road from the Sugar Lands to the new Chapell 18

Road from Sugarland Run to Goose Creek/Goose Creek ferry and from Goose Creek to Limestone Run (including digging down the banks at Goose Creek ferry) and parts thereof 3, 6, 8, 24

Road from Sugarland Run to the Chappel on the Beaver dam 14

Road from Difficult Run to Sugarland Run 19

Road from Sugarland Run to the new Church 20

Road from the old Court House to Difficult Run, from Difficult Run to Sugarland Run and from Sugarland Run to Broad Run 37, 39

Road (fork?) from the Falls road to the Sugar Land road 52

Road from William Donaldson's to Summers''s old school house 54(2)

Road from the forks of the road above Summers's to the road that leads to the old Court House 115

Road from the forks of the road above Summers's to Price's/Price's tavern 116, 149, 160

Road from the forks of the road above Summers's to Bride's (Bird's?) bridge 149

Sections of the Turnpike road:
- from the County line on the Turnpike road leading from Newgate to Alexandria as far as the Ox road 160
- from the Ox road, along said Turnpike road to the fork of the road at Hollice's old field near Price Skinner's 160

- from Hollice's old field in the fork of the road Price Skinner's to Price's tavern 160
- from Price's tavern to the forks of the road above Summers's 160

Road from the forks of the road above Summers's to Holmes Run 161

Road from the old school house near Francis Summers's to the old Court House (with intersection of the road leading to the old Court House road) 99

Road from the north run of Hunting Creek to the forks of the road above (Francis?) Summers's 116

Road from the fork of the road near Francis Summers Jr.'s to Accotink Run 106

Road from the forks of the road near Francis Summers Jr.'s to the foot of Cameron hill 106

Road from Sumers's/John Summers's to the Mountain road near Rocky Run Chapel 22, 25

Road from the bridge at Cameron to John Summers's 46

Road from the fork of the road leading to John Summers's to the old Court House road 82

Road from the school house near John Summers's to Accotink Run 82

(Fork?) road from Cameron to John Summers's 97

Road from Accotink to the school house near John Summers's (with intersection of the road at Price's leading from Accotink to Alexandria) 99

Road from John Summers's house to the Colchester road 155

Road from Mathew Boswell's shop on Elzey's Church road to the mouth of the Widow Summers's lane 143

Road from the Great Falls of Potomac to Alexandria (road from the lower ford over Difficult Creek/Difficult Run via the Great road, near Edward Lanham's/Lannum's field, thence through the lands of William Swink, Joseph Simpson, Gustavus Scott Esqr. [rented by Daniel Jenkins], John Jackson, and Robert Carter, Esq. [rented by Charles Thrift], to the place on Scott's Run where William Shortridge's old mill stood, thence through the lands of Turberville, Scott, George Hunter, John Jackson, across Pimmets Run, through the lands of Richard Conway, Philip Darrell, William Gunnell, Mr. John West, Mr. Sampson Trammell, and Col. James Wren, to the Great road leading from Leesburg to Alexandria, where Col. Wren's store stands, thence by the Great road

[Turnpike road] from the store to Alexandria) 121, 124(2), 128, 129, 130, 131(3), 139, 140, 141

Road from the Sycamore near the Little Falls warehouse to the river 158

NOTE: Entries for Talbot and Talbott are combined

Road from the mouth of John Hurst's lane to the upper part of Talbot's fence 113

Road by/through Benjamin Talbot's/Talbott's plantation 60, 90

Road that goes through Benjamin Talbott's plantation from the old Court House to Bull Run 67

Road from the fork of the old road near Benjamin Talbott's to the Church road near Lewis Sanders's (road from the forks of the old Court House road near Benjamin Talbott's to the Church road near Lewis Sanders's, for people to go to the new Church built by Mr. Edward Payne and to the mill, also a nearer and most convenient way to Colchester and Dumfries; road to go through the plantations of Benjamin Talbott, William Kitchen, Edward Washington and John Cotton); (also proposed was an alternate, and more difficult route through/near the plantations of Benoni Hally, William Kitchen, Mrs. Mary Ferguson, Edward Washington and John Cotton) 86, 87, 88

Road from Benjamin Talbott's to Mr. Thomas Lewis's mill on Difficult Run (road from Benjamin Talbott's plantation thence along the Main road to the end of Samuel Talbott's fence thence through Mr. John Waller's plantation, and along the old mill path to Mr. Thomas Lewis's mill on Difficult Run) 94, 95

Road from Benjamin Talbott's plantation to the Loudoun County line at Amos Fox's mill on Difficult Run 112(2)

Road from Benjamin Talbott's to Mr. Thomas Lewis's mill on Difficult Run (road from Benjamin Talbott's plantation thence along the Main road to the end of Samuel Talbott's fence thence through Mr. John Waller's plantation, and along the old mill path to Mr. Thomas Lewis's mill on Difficult Run) 94, 95

Road through the upper end of Samuel Talbott's plantation to the lower end 104

Road from Vincent Lewis's road to Col. Tayloe's quarter 3

Road from the new road between Col. Tayloe's quarter and James Lane's, and thence on the lower side of Lane's plantation to the Mountain road a little above Capt. Newton's quarter on Great Rockey Run 13

Road by Garrard Tramel's/Tramell's plantation (by Pimmets Run, and the orchard of Templeman's plantation, and to Mr. Broadwater's plantation) 25, 26

Road turned by Mr. William Henry Terrett (and the old road in the vicinity) 21, 23(2)

Road from Alexandria to Difficult bridge near Capt. W. H. Terrett's (road from Princess Street passing near Baldwin Dade's old tobacco house, and Ramsay's quarter, and to a white oak in the road above Capt. W. H. Terrett's old field) 119, 120

Road from Edward Thompson's to Thomas Davis's 10

Road from Edward Thompson's to the top of the Blue Ridge 16

Road from Joseph Thompson's plantation into the county road 104

Road from the Great Falls of Potomac to Alexandria (road from the lower ford over Difficult Creek/Difficult Run via the Great road, near Edward Lanham's/Lannum's field, thence through the lands of William Swink, Joseph Simpson, Gustavus Scott Esqr. [rented by Daniel Jenkins], John Jackson, and Robert Carter, Esq. [rented by Charles Thrift], to the place on Scott's Run where William Shortridge's old mill stood, thence through the lands of Turberville, Scott, George Hunter, John Jackson, across Pimmets Run, through the lands of Richard Conway, Philip Darrell, William Gunnell, Mr. John West, Mr. Sampson Trammell, and Col. James Wren, to the Great road leading from Leesburg to Alexandria, where Col. Wren's store stands, thence by the Great road [Turnpike road] from the store to Alexandria) 121, 124(2), 128, 129, 130, 131(3), 139, 140, 141

Road from Charles Thrift's near the Little Falls road to the new road leading from the Great Falls to Alexandria (road from the Little Falls road to the Great Falls road, running through Charles Thrift's land/Robert Carter's land rented by Thrift to the Great Falls road at John Jackson's land; alteration of the road running through John Turberville's land) 131, 140

Road from Thomas Sangster's to Thomas Throgmorton's mill on Bull Run 95

Road from John Hurst's plantation above the Falls Church to Tolston mill upon Difficult 81

Road from young Gerrard Trammell's to Tolston mill upon Difficult 81

Road from the Traff hill to Amos Fox's mill on Difficult Run 156

NOTE: Entries for Tramel, Tramell, Trammel and Trammell are combined

Road from Trammell's old field to the Falls road 54

Road from the Falls Church to Trammell's 116

Road by Garrard Tramel's/Tramell's plantation (by Pimmets Run, and the orchard of Templeman's plantation, and to Mr. Broadwater's plantation) 25, 26

Way by Mr. Wren's plantation and through Gerrard Trammell's old field (to be established a bridle way for James Donaldson, Gent.) 57

Road from William Adams's Gent. to Gerrard Trammell's 66

Road from young Gerrard Trammell's to Tolston mill upon Difficult 81

Road from Gerrard Trammell's bridge towards William Adams's 105

Road from Gerrard Tramell's to Gullatt's fence 115

Road from the lower part of the island on Potomac River belonging to John Trammel to the place where he is erecting his house 37, 40

Road crossing the bridge made by John Trammel Jr./Broad Run bridge 22, 24

Road from the Great Falls of Potomac to Alexandria (road from the lower ford over Difficult Creek/Difficult Run via the Great road, near Edward Lanham's/Lannum's field, thence through the lands of William Swink, Joseph Simpson, Gustavus Scott Esqr. [rented by Daniel Jenkins], John Jackson, and Robert Carter, Esq. [rented by Charles Thrift], to the place on Scott's Run where William Shortridge's old mill stood, thence through the lands of Turberville, Scott, George Hunter, John Jackson, across Pimmets Run, through the lands of Richard Conway, Philip Darrell, William Gunnell, Mr. John West, Mr. Sampson Trammell, and Col. James Wren, to the Great road leading from Leesburg to Alexandria, where Col. Wren's store stands, thence by the Great road [Turnpike road] from the store to Alexandria) 121, 124(2), 128, 129, 130, 131(3), 139, 140, 141

Road from John Gest's house to the road by William Tramell's 35

Road from the Turnpike gate at the intersection of the Georgetown road to the white oak at the Widow Tucker's 133

Road from the Falls Church to (the white oak at) Widow Tucker's 133, 162

Road from the Widow Tucker's to the Turnpike [to?] Alexandria 162

Road through Mr. Turbeville's plantation/old field 56, 57

Road from the Falls warehouse to the fork of the road by Turberville's quarter 106

Road from the Great Falls of Potomac to Alexandria (road from the lower ford over Difficult Creek/Difficult Run via the Great road, near Edward Lanham's/Lannum's field, thence through the lands of William Swink, Joseph Simpson, Gustavus Scott Esqr. [rented by Daniel Jenkins], John Jackson, and Robert Carter, Esq. [rented by Charles Thrift], to the place on Scott's Run where William Shortridge's old mill stood, thence through the lands of Turberville, Scott, George Hunter, John Jackson, across Pimmets Run, through the lands of Richard Conway, Philip Darrell, William Gunnell, Mr. John West, Mr. Sampson Trammell, and Col. James Wren, to the Great road leading from Leesburg to Alexandria, where Col. Wren's store stands, thence by the Great road [Turnpike road] from the store to Alexandria) 121, 124(2), 128, 129, 130, 131(3), 139, 140, 141

Road from Charles Thrift's near the Little Falls road to the new road leading from the Great Falls to Alexandria (road from the Little Falls road to the Great Falls road, running through Charles Thrift's land/Robert Carter's land rented by Thrift to the Great Falls road at John Jackson's land; alteration of the road running through John Turberville's land) 131, 140

Road from the old Court House to Turley's mill 47

Road from Sampson Turley's plantation to the Mountain road (road from Sampson Turley's plantation on Johnimore into the Mountain road) 27, 89

Road from the Mountain road to Sampson Turley's and a road out of that to the mouth of Popeshead 27

Road from Alexander Henderson, Gent.'s plantation on Bull Run into the main road (road from Henderson's plantation through/adjoining the lands of Yelverton Reardon, William Moon, Marmaduke Beckwith, James Keen, Sampson Turley, William Simpson, and Peter Ryley to the main road) 104, 107, 108

Road from Alexander Henderson's plantation on Bull Run by Sampson Turley's to the Ox road 117

Road from the Ox road above William Simpson's to where it intersects the road from the Wolf Run Shoals to Thomas Songster's by Sampson Turley's plantation 139

Road from the old Court House to William Turner's tavern 134

Road from William Turner's tavern to the Parish line on the old Court House road 147, 158, 167

The road from Four Mile Run to the Turnpike gate 132

Road from the Turnpike gate at the intersection of the Georgetown road to the white oak at the Widow Tucker's 133

Road from Holmes Run, to the Turnpike gate, where William Simpson lives 136

Road from the Turnpike gate down the Colchester road to Dogue Run at Mrs. Moon's 136

Sections of the Colchester road: from the Turnpike gate to Dogue Run; from Dogue Run to Accotink; from Accotink to Pohick Run; from Pohick Run to Colchester 165

Road from the Great Falls of Potomac to Alexandria (road from the lower ford over Difficult Creek/Difficult Run via the Great road, near Edward Lanham's/Lannum's field, thence through the lands of William Swink, Joseph Simpson, Gustavus Scott Esqr. [rented by Daniel Jenkins], John Jackson, and Robert Carter, Esq. [rented by Charles Thrift], to the place on Scott's Run where William Shortridge's old mill stood, thence through the lands of Turberville, Scott, George Hunter, John Jackson, across Pimmets Run, through the lands of Richard Conway, Philip Darrell, William Gunnell, Mr. John West, Mr. Sampson Trammell, and Col. James Wren, to the Great road leading from Leesburg to Alexandria, where Col. Wren's store stands, thence by the Great road [Turnpike road] from the store to Alexandria) 121, 124(2), 128, 129, 130, 131(3), 139, 140, 141

Turnpike road from Alexandria to Georgetown ferry 126

Turnpike road from Alexandria to Difficult 126

Turnpike road from Alexandria to Newgate 126

Road from Cameron Run into the Turnpike road 127

Turnpike road/roads 130, 133, 152, 156(3), 158(2), 159(2)

Road from the Red house to the Turnpike road 132

Road from the entrance of Avory's road near Marle's house to the Turnpike road leading to the Falls Church 132

Road from the County line to the White Oak Spring (part of the Turnpike road leading to Alexandria) 135

Road from the Ox road near Sangster's/Thomas Songster's to Hallace's/Hollies'/Hollis's/Hollice's old field (at the Turnpike road) 138, 147, 157, 167

Road from the Turnpike road at Hallace's/Hollice's/Hollis's old field to where it intersects the Pohick road at John Fowler's plantation/fence 138, 147, 167

Road from Accotink on the Turnpike road to Price's tavern 144

Road from Hollis's old field to Accotink, on the Turnpike road 144

Road from the County line on the Turnpike road leading to Alexandria and to the Ox road 144

Road from the Ox road on the Turnpike road to Hollis's old fields, opposite to where another road intersects the Turnpike road leading from the Ox road 144

Road from the Turnpike road to the Ox road, near the Brick Church 145

Road from Difficult bridge (on the Turnpike road) to the old Court House Run/Court House Run 148, 161

The road from Bird's (Bride's?) bridge to the Turnpike 149

Road from the Turnpike (near Alexandria) to Four Mile Run 150, 162

Road from Marle's to the Turnpike road leading to Leesburg (road from Marl's to the Leesburg road) 150, 152, 162

Road from William Hartshornes mill to the Turnpike road 151

Sections of the Turnpike road:
- from the County line on the Turnpike road leading from Newgate to Alexandria as far as the Ox road 160
- from the Ox road, along said Turnpike road to the fork of the road at Hollice's old field near Price Skinner's 160
- from Hollice's old field in the fork of the road Price Skinner's to Price's tavern 160
- from Price's tavern to the forks of the road above Summers's 160

Road from Holmes Run to the Stone bridge below the Turnpike 161

Road from the Widow Tucker's to the Turnpike [to?] Alexandria 162

Road from the forks of the road below Fitzhugh's, to the intersection of the Turnpike road at Frazier's old place 163

Road from the County line at the Turnpike road to Thomas Songster's 166

Road from Samuel Love's mill on Rocky Run, into the Turnpike road 169

Road from Daniel Lewis's mill to the Turnpike road, at the Mine Branch 171

Road from the Mountain road by Col. Colvill's Tuskarora quarter to the ford of Goose Creek by Samuel King's 11

Road (from Goose Creek?) near Tuscarora Run 42

Road from the old Court House to the Falls warehouse (road coming from Bull Run through the Court House corn field, Mr. Scott's land and John Urton's land, into the old Falls road at the head of Sasse Branch, thence along the old roads by Thomas Ashbury's, through the Honourable Philip Ludwell Lee's plantation, and to the Falls) 48, 49

Road by John Urton's plantation 62

Road from Bond Veale's to the upper side of Difficult Run 9

Road from Vernon's mill on Pohick to the stand or meeting house 123

Road to Broad Run Chapel and from thence to Vestal's Gap 30

Main roads from Vestal's and Williams's Gaps leading to Alexandria and Colchester 102(2), 111

Road from Edward Violet's/Violett's to Pohick warehouse 15, 18, 43

Path through the plantation of Verlinda Wade 33, 35

Road from Benjamin Talbott's to Mr. Thomas Lewis's mill on Difficult Run (road from Benjamin Talbott's plantation thence along the Main road to the end of Samuel Talbott's fence thence through Mr. John Waller's plantation, and along the old mill path to Mr. Thomas Lewis's mill on Difficult Run) 94, 95

Road from Samuel Bayly's land to Ward's mill 153

Road from Ward's mill to the road leading from Fitzhugh's quarters near Mrs. French's 154

Road from John Ward's mill to the road that leads to Alexandria 123

Road from the Colchester road to the road to Alexandria (road from the Colchester road near John Ward's mill, thence through William Rogers's plantation to Richard Sanford's fence, and to the road leading to Alexandria opposite to Capt. Harper's fence) 124

Road from the neighborhood of Occoquan Forrest crossing Accotink at John Ward's mill and down to Alexandria 143

Road from Pohick to Accotink and cross road, thence the road to the warehouse from the crossroad (and parts thereof) 2, 14

Roads from William Gardner's to the place where the road to Colchester leaves the Ox road and from that place to Colchester warehouse 74

Road from Colchester/Colchester warehouse to Wolf Pit hill 117, 139

Road from the Falls warehouse up to the Falls road leading to Difficult 44

Road from the Falls warehouse to the Falls landing 44, 107, 110(2)

Road from Alexandria to the Falls warehouse 48, 76

Old road from the Falls warehouse to Alexandria 49

Road from the old Court House to the Falls warehouse (road coming from Bull Run through the Court House corn field, Mr. Scott's land and John Urton's land, into the old Falls road at the head of Sasse Branch, thence along the old roads by Thomas Ashbury's, through the Honourable Philip Ludwell Lee's plantation, and to the Falls) 48, 49

Road from the old Court House to the road leading from the Chestnut tree to the Falls warehouse 67

Road from the old Court House (via fence and line of John Mason's fence and Mr. Scott's line to the Falls road) to the Falls warehouse 72, 74

Road from the Falls Church to the road leading from Alexandria to the Falls warehouse 79, 114, 132

Road from the Alexandria road to the Four Mile Run (including fork leading to Georgetown and the Falls warehouse) 99

Road from the Falls warehouse to the Red house 103

Road from the Falls warehouse to the fork of the road by Turberville's quarter 106

Road from the Sycamore near the Little Falls warehouse to the river 158

Road from Madden's shop to the Little Falls warehouse 164

Bridle way from Colchester to Occoquan warehouse 48, 49

Road from Giles Run to Occoquan warehouse 59

Road from Alexandria to John Hooe's ferry on Occoquan, near the old warehouse landing 152(2), 153(2)

Road from Edward Violet's/Violett's to Pohick warehouse 15, 18, 43

Road from where John Hollis formerly lived to Pohick warehouse 106

308

Road from the Ellicot and Co. bridge on Occoquan into the road leading from Washington's shop to Robert Boggess's mill and from thence to the intersection with the Alexandria Post road 169

Road from the fork of the old road near Benjamin Talbott's to the Church road near Lewis Sanders's (road from the forks of the old Court House road near Benjamin Talbott's to the Church road near Lewis Sanders's, for people to go to the new Church built by Mr. Edward Payne and to the mill, also a nearer and most convenient way to Colchester and Dumfries; road to go through the plantations of Benjamin Talbott, William Kitchen, Edward Washington and John Cotton); (also proposed was an alternate, and more difficult route through/near the plantations of Benoni Hally, William Kitchen, Mrs. Mary Ferguson, Edward Washington and John Cotton) 86, 87, 88

Roads on George Washington's own lands (altered and opened by him per his survey and plat) 127

Road from Robert Boggess's to General (George) Washington's mill 137, 166

Road from Pohick new Church to intersect the road from Robert Boggess's to Gen. (George) Washington's mill (road from Robert Boggess's to the General's [George Washington's] mill, and from that road to Pohick Church) 137, 148

Road from General (George) Washington's ferry to his mill, from thence to his tumbling dam, thence along his new road to intersect the riverside old road above the Gum Spring (and parts thereof: includes road from the Gum Spring to General Washington's mill) 137, 166

Road from Cameron Run or Creek by Clish to where General (George) Washington's new road intersects the River side old road 137

Road from the General's (George Washington's) mill by his mill dam and along his new road, to the Parish line 148

Road from General (George) Washington's new road into the old road 156

Road from the Ponds on the Pohick road, crossing the Stage road to the intersection of the road leading from Boggess's to Gen. Washington's mill 167

Roads near Lund Washington's quarter 75

Road from the County line to the fork of the road by Waters's ordinary 58

Road from Alexandria to Lewis Ellzey's meadow branch and the ford over Holmes Run (road from Duke Street crossing the gut below John West's meadow, passing Dutch Caty's house and Captain T. West's field to the road at the mouth of Col. Gilpin's lane, thence with the road to a branch near the hop-yard, and along the side of the hill to Holmes Run opposite the mouth of Messers Bird and Hawkins's lane) 119, 120

Road from the Great Falls of Potomac to Alexandria (road from the lower ford over Difficult Creek/Difficult Run via the Great road, near Edward Lanham's/Lannum's field, thence through the lands of William Swink, Joseph Simpson, Gustavus Scott Esqr. [rented by Daniel Jenkins], John Jackson, and Robert Carter, Esq. [rented by Charles Thrift], to the place on Scott's Run where William Shortridge's old mill stood, thence through the lands of Turberville, Scott, George Hunter, John Jackson, across Pimmets Run, through the lands of Richard Conway, Philip Darrell, William Gunnell, Mr. John West, Mr. Sampson Trammell, and Col. James Wren, to the Great road leading from Leesburg to Alexandria, where Col. Wren's store stands, thence by the Great road [Turnpike road] from the store to Alexandria) 121, 124(2), 128, 129, 130, 131(3), 139, 140, 141

Road from Roger West's house into the main road leading to Alexandria, at the ford of Cameron Run 154, 155

Road from Alexandria to Lewis Ellzey's meadow branch and the ford over Holmes Run (road from Duke Street crossing the gut below John West's meadow, passing Dutch Caty's house and Captain T. West's field to the road at the mouth of Col. Gilpin's lane, thence with the road to a branch near the hop-yard, and along the side of the hill to Holmes Run opposite the mouth of Messers Bird and Hawkins's lane) 119, 120

Road from Vincent Lewis's road to William West's house 3

Road from the road near William West's to the road that leads to Dumfries 23

Road from Alexandria to Difficult bridge near Capt. W. H. Terrett's (road from Princess Street passing near Baldwin Dade's old tobacco house, and Ramsay's quarter, and to a white oak in the road above Capt. W. H. Terrett's old field) 119, 120

Road from the Turnpike gate at the intersection of the Georgetown road to the white oak at the Widow Tucker's 133

Road from the Falls Church to (the white oak at) Widow Tucker's 133, 162

Road from Sangster's/Thomas Songster's shop, on the Ox road, to the White Oak Harbor 146, 166

Road from the White Oak Harbor/old White Oak Harbor, on the Ox road to Colchester (at the old sign post) 146, 166

Road from the County line to the White Oak Spring (part of the Turnpike road leading to Alexandria) 135

Road from the White Oak Spring to Price's tavern 135

Road from Yates's ford on Bull Run, to Songster's (through the lands of Hector Kincheloe and Jesse Kincheloe the heirs of Daniel Kincheloe, and Robert Wickliffe) 173, 174

Road from Williams's Gap to the upper fork of Goose Creek 6

Road from the Court House to Williams's Gap 8, 14

Road from Williams's Gap to Rocky Run Chapel 26

Main roads from Vestal's and Williams's Gaps leading to Alexandria and Colchester 102(2), 111

Road that leads from the Court House to Mrs. Willis's 6

Road from the Ox road to Colchester (alteration made by going through John Ford's plantation, along an old mill path, and into the Ox road at the Wolf Pit hill) 50

Road from Colchester/Colchester warehouse to Wolf Pit hill 117, 139

Road from the Wolf Pit hill to Pohick Church 106

Road from the (upper part of) Wolf Pit hill to Loofborough's/David Loofbourrow's 116, 139

Road from Thomas Sangster's/Songster's to the Wolf Shoals/Wolf Run Shoals (on Bull Run/Occoquan) 85, 116, 139, 145, 164

Road from Occoquan above the Wolf Run Shoals to Alexander Henderson's mill 114

Road from the Ox road above William Simpson's to where it intersects the road from the Wolf Run Shoals to Thomas Songster's by Sampson Turley's plantation 139

Road from the Wolf Run Shoals to the shallow ford on Bull Run 157(2)

Road from the Wolf Trap Run to the main road 2

Road from Wolf Trap Run to the old Court House 122

Road from (John?) Hunter's fence to Collins's fence (road to begin at William Deneale's corner on Wolf Trap (Run) and thence with Lewis's line and thence along the line of Deneale and Fairfax to their corner near Collins's fence) 118(2)

Road from the Ox road at the Wood Cutting to Occoquan and the County line leading to the mill 167

Way by Mr. Wren's plantation and through Gerrard Trammell's old field (to be established a bridle way for James Donaldson, Gent.) 57

Road from the Great Falls of Potomac to Alexandria (road from the lower ford over Difficult Creek/Difficult Run via the Great road, near Edward Lanham's/Lannum's field, thence through the lands of William Swink, Joseph Simpson, Gustavus Scott Esqr. [rented by Daniel Jenkins], John Jackson, and Robert Carter, Esq. [rented by Charles Thrift], to the place on Scott's Run where William Shortridge's old mill stood, thence through the lands of Turberville, Scott, George Hunter, John Jackson, across Pimmets Run, through the lands of Richard Conway, Philip Darrell, William Gunnell, Mr. John West, Mr. Sampson Trammell, and Col. James Wren, to the Great road leading from Leesburg to Alexandria, where Col. Wren's store stands, thence by the Great road [Turnpike road] from the store to Alexandria) 121, 124(2), 128, 129, 130, 131(3), 139, 140, 141

Divisions of the road from the Great Falls of Potomac to Alexandria
- from Col. Wren's store to the Falls road 131, 155
- from the Falls road to Difficult 131

Road from Col. Wren's tavern up the Great Falls road to Madden's shop 163

Road through James Wren's plantation 46

Road from Yates's ford on Bull Run, to Songster's (through the lands of Hector Kincheloe and Jesse Kincheloe the heirs of Daniel Kincheloe, and Robert Wickliffe) 173, 174

www.ingramcontent.com/pod-product-compliance
Lightning Source LLC
Chambersburg PA
CBHW080534300426
44111CB00017B/2726